The Economy of Green Cities

Local Sustainability

Volume 3

Series Editor: Monika Zimmermann

For other ICLEI publication, go to
http://www.springer.com/series/10034

Richard Simpson • Monika Zimmermann
Editors

The Economy of Green Cities

A World Compendium on the Green Urban Economy

Springer

Editors
Richard Simpson
ICLEI - Local Governments for Sustainability
World Secretariat
Kaiser-Friedrich-Str.7
53113 Bonn, Germany

Monika Zimmermann
ICLEI - Local Governments for Sustainability
World Secretariat
Kaiser-Friedrich-Str.7
53113 Bonn, Germany

Chapter 7 "Working in Green Cities: Improving the Urban Environment While Creating Jobs and Enhancing Working Conditions," by Edmundo Werner, is published with the kind permission of the International Labour Office (ILO). The copyright remains with ILO.

Chapter 9 is based on the Cities Chapter of UNEP's Green Economy Report, Towards a Green Economy: Pathways to Sustainable Development and Poverty Eradication (2011). The copyright remains with UNEP.

ISSN 2211-2782 ISSN 2211-2790 (electronic)
ISBN 978-94-007-9889-2 ISBN 978-94-007-1969-9 (eBook)
DOI 10.1007/978-94-007-1969-9
Springer Dordrecht Heidelberg New York London

Printed on acid-free paper

Springer is part of Springer Science+Business Media (www.springer.com)

Foreword

We now live in an urban age where cities and their economies dominate the global economy. Throughout their history cities have been centers of commerce and trade, a function that has enabled their growth and prosperity. Urban areas and metro regions are the hubs of their respective country's economy. Economic activities in cities are a key growth engine to a nation's economy. At the same time, due to the concentration of human population, consumption and economic activity, urban areas are the main contributors to greenhouse gas emissions as well as placing enormous demands on the world's natural resource base. Just as cities are the source of many environmental problems, their characteristics also offer many of the solutions, and should be recognized as such.

Following the recent economic crisis that has impacted most of the developed world economies, there has been a renewed interest in new forms of growth. Of these Green Growth and a transition to Green Economy are the most widely discussed. At the same time, new challenges are emerging in developing countries where local and national governments are struggling to keep pace with rapid urbanization, especially in providing housing, infrastructure and social services. Combined with the urgent need to reduce carbon emissions in cities to combat global environmental challenges, the urban economy needs to transition to a more sustainable and green growth trajectory. Yet, there has been little conceptualization in the literature regarding the Green 'Urban' Economy.

A transition to a Green Urban Economy needs to manifest itself in cities worldwide. It needs to realize the opportunities that exist in an urban system in terms of synergies, concentration of people and economic activities, and the strong interactions and symbiotic relationships between sectors. A Green Urban Economy needs to oversee a recasting of the economic structure of cities in favor of less carbon intensive activities, mainly by fostering 'clean tech' and green services and products as key growth sectors. By shifting behavior to more sustainable consumption patterns, which minimize environmental impact, urban economies can transition towards green economic development.

'The Economy of Green Cities: A world compendium on the Green Urban Economy' is a rich and overdue contribution to the discussion of what the Green

Economy can mean for urban areas. A key theme in the 2012 Rio+20 conference, the Green Economy can must resonate strongly in cities worldwide for it to be a vital tool for sustainable development and poverty eradication. Whether energy efficiency in buildings or new and innovative waste management systems, this publication is an excellent addition to the conceptualization of the Green Urban Economy – an imperative for cities in achieving economic viability, social equity and environmental harmony. It is a useful companion for local decision makers at the urban level to inform local policymaking.

By addressing our economic, social and environmental challenges, a Green Urban Economy is a holistic approach to analyze the urban system through an economic lens. Energy efficiency in the physical infrastructure including transport, waste management and water infrastructure as well as the protection of ecosystems highlight a fundamental principle of the Green Urban Economy: valuing environmental goods and services while contributing to climate protection. At the same time, new business models that promote green market development and a new economic structure can in turn create new jobs and migrate labor to less carbon intensive economic activities. This is essential for maintaining and enhancing competitiveness – a prerequisite for sustainable urban areas, economic growth and social development.

It gives me great pleasure to contribute to the publication of this book, which bridges the gap between the promotion of the Green Economy globally and the manifestation of this new growth strategy at the urban level. It is an essential recognition of the role of cities and urban economies in pioneering the Green Economy and how green cities are an imperative solution in not only addressing global environmental challenges but also in ensuring local socio-economic prosperity.

ICLEI – Local Governments for Sustainability President
David Cadman

Author Bios

Dr. Essam Hassan Mohamed Ahmed joined the Climate Change Unit at the Egyptian Environmental Affairs Agency (EEAA) in 2007. Since then he has been involved in CDM projects, Mitigation and Adaptation studies, Green ICTs, Green Cities and Transportation sectors in particular. Dr. Ahmed is also lead author of WGIII; 5th IPCC Assessment Report. Before joining the EEAA, he worked in the Egyptian Military Armed Forces as Engineering Officer for 20 years.

Vito Albino has a Laurea degree in Mechanical Engineering, and was visiting scholar at the University of Cincinnati, USA (1986), visiting professor at the University of South Florida, USA (1994), and from 1988 to 2000 he taught Engineering Management at the University of Basilicata in Italy. He is now the Professor of Innovation and Project Management at Politecnico di Bari in Italy. He is a member of several national and international associations and has authored papers published for national and international journals and conference proceedings in the fields of innovation, energy-materials flow analysis, and industrial clusters.

Stefan Anderberg is Associate Professor of Human Geography and Co-director at the Lund University Centre for Sustainability Studies as well as Director of the Research School of Sustainability Science. His research is related to global sustainability issues such as the development of international trade and land use in the context of climate change, sustainable local and urban development, and the environmental history of cities and regions. It focuses on the development of resource flows in different scales, and related governance issues, on different methods for assessing resource flows, and on how cities, communities, regions and companies respond to the challenges of sustainable development and climate change.

Steven Bland graduated from Cambridge University (UK) in 2009 with a Starred first B.A. (Hons) in Geography. He then completed a Professional Masters (MProf) in Leadership for Sustainable Development, run by the leading sustainable development charity Forum for the Future. Steven joined the ICLEI World Secretariat in 2010. His responsibilities include preparing and delivering training programs on Cities and Climate Change, project proposal writing and being in charge of program development for the ICLEI World Congress 2012. Previous work experience includes short term posts at the Department for Energy and Climate Change (UK), Birmingham City Council (UK), and as a research assistant at the Cambridge University Geography Department. Steven is an Associate at the Local Government Information Unit think tank (UK).

David Cadman is the President of ICLEI – Local Governments for Sustainability. Elected in 2002, he had served as Vancouver City Councillor for 9 years, and 6 years as a Councillor with COPE, the Coalition of Progressive Electors prior to his election. He represented Vancouver on the Greater Vancouver Regional District and Metro Vancouver board of directors, serving on both the Land Use and Transportation and Environment and Energy committees.

A social and environmental activist for over 30 years, David has served as the President of the Society Promoting Environmental Conservation (SPEC). He has also served at the national and international level of the United Nations Association, and was awarded the United Nations Peace medal and United Nations 50th Anniversary medal.

David was born in Montreal and grew up in Toronto. He attended the University of the South in Tennessee and then moved to Geneva where he earned a master of arts degree in Interdisciplinary and International Development. He studied adult education at the Sorbonne in Paris and is fluent in French.

After spending several years in Tanzania and Kenya developing literacy programs, David returned to Canada in 1976 and settled in British Columbia. He worked for the Social Planning and Research Council in establishing the Community Development Institute.

Rosalie Callway has worked in the field of sustainable development for over 14 years. She worked with 'Stakeholder Forum' an international sustainability NGO where she edited and published the book 'Governance for Sustainable Development: A foundation for the future'. Moving on to the UK Local Government Association's 'European and International Unit', Rosalie supported a range of policy and practice exchanges between local government internationally. She now works as a consultant promoting sustainable institutional development and strategic environmental management. Rosalie is currently an associate advisor on 'Public Engagement and Sustainability' with 'Involve', a charity focusing on public engagement, participation and dialogue. She is also an associate on 'low carbon communities' for the '2DegreesNetwork' – a sustainability portal.

Paul Chambers leads Auckland Council's organization-wide corporate, community and business sustainability efforts, which encompasses environmental consideration, financial efficiency, social responsibility and community involvement. Activities include program development and implementation, analysis, reporting and compliance, as well as other related responsibilities to foster and develop sustainability and social responsibility throughout the council's operations. Paul has recently returned from the U.S where he was selected to be a Climate Change Fellow by the Fulbright Program funded by the US State Department.

Dr. Amelia Clarke has been working on environment and sustainability issues since 1989, and was recognized as 1 of 50 Canadian environmental leaders by the Green Living magazine. She is now a faculty member in the School of Environment, Enterprise and Development (SEED) at the University of Waterloo and is Director of their Master of Environment and Business (MEB) program. She holds a Ph.D. in Management from McGill University. Amelia Clarke was also the President of Sierra Club Canada from 2003 to 2006. She has served on numerous advisory committees, including for The Canadian Earth Summit Coalition (2011–2012), the Atlantic Canada Sustainability Initiative (2007–2010); the International Forum on Applied Sustainable Development (2007); the World Peace Forum (2006); the federal government's National Advisory Committee to the U.N. Habitat's World Urban Forum III (2006); and the Canadian delegation during the United Nations World Summit on Sustainable Development (Johannesburg) negotiations (2002). In 1996, she founded the Sierra Youth Coalition, and is proud of having launched their sustainable campus program which has worked with more than 80 university campuses.

Dr. Sean Connelly is an Assistant Professor with the Geography Department at the University of Regina and an Associate with the SFU Centre for Sustainable Community Development. Sean's research interests are in urban and environmental geography, planning and sustainable community development. He is co-editor of *Seeds of Transition: The Convergence of the Social Economy and Sustainable Community Development (Athabasca University Press, forthcoming).* Sean previously worked with ICLEI-Local Governments for Sustainability in their Johannesburg office in preparation for the Local Government Session of the World Summit on Sustainable Development.

Rosa Maria Dangelico has a Laurea degree in Management Engineering and holds a Ph.D. in Innovation Management and Product Development. She has been post-doc Research Fellow and contract Professor at Politecnico di Bari in Italy, Research Fellow 2009 of the Aspen Institute, visiting scholar at De Groote School of Business – McMaster University, Canada, and at IESE Business School – University of Navarra in Spain. She is currently an Assistant Professor in Management Engineering at Sapienza University of Rome. Her main research interests concern the integration of environmental sustainability issues into firms' strategies and new product development.

Golda Arpudhalin Edwin received her M.Sc. degree from the Department of Zoology at the Pondicherry University is currently pursuing her Ph.D. Her area of research interest is sustainable water management practices with special reference to greywater management using constructed wetlands. She has authored several articles and presented in national/international conferences and workshops. She has undertaken community development projects such as rain water harvesting, health and hygiene workshop with guidance from the center for sustainable development, Guatemala. Her honors include general proficiency for two consecutive years and she is also a gold medalist in the Masters Degree program. She is a member of the International Water Association.

Dr. Matthew A. French is a development practitioner specialising in affordable housing, slum upgrading and community planning in developing countries. His professional qualifications include a Bachelor of Architecture (BArch) and Masters of Architecture (MArch, research) from Victoria University of Wellington, New Zealand, and a Ph.D. from the University of Cambridge, United Kingdom. His research and practice explores the interrelationship between environmental and socio-cultural dimensions of sustainable urban development. His funded research projects include the technical development of low-cost seismic reinforcing for adobe housing in Chile, Peru, and India (2005); an ethnographic investigation of the thermal performance and comfort of urban poor housing in Argentina (2007); and the environmental sustainability of slums in Kenya with respect to bio-climatic housing and settlement design (2010). He has coordinated capacity development workshops on participatory design of slum upgrading in Kenya and Brazil ('Change by Design' 2011). His current portfolio at the United Nations Human Settlements Programme (UN-Habitat) focuses on the development of normative guides, tools, and knowledge products in three thematic areas: housing policy, slum upgrading, and environmentally sustainable housing.

Professor Andy Gouldson leads the Centre for Low Carbon Future's work on the economics of low carbon cities. He is also Director of the ESRC Centre for Climate Change Economics and Policy at the University of Leeds in the UK. He has worked on issues of environmental policy and management for over 20 years, specializing in the design and delivery of policies that improve the relationship between economic development and the environment, particularly at the business and local levels.

Bernd Hendriksen, is Practice Leader in The Netherlands, Climate Change & Sustainability, KPMG Advisory N.V. For over 12 years Bernd has been advising KPMG's Netherlands based clients on sustainable development. In 2011, Bernd was appointed leader of KPMG's climate change and sustainability practice in the Netherlands, where he manages a team of over 40 professionals that serve both local and international clients. His experience covers a vast range of sectors, including Energy and Natural Resources, Building and Construction, Transport, Government, Urban Planning and NGOs. He regularly shares his latest research findings and best practices through public speaking engagements across the Netherlands and Europe. Bernd was the lead author on the CDP Cities 2011 Global Report – after KPMG was commissioned by the Carbon Disclosure Project (CDP) to gather and analyze data from a group of the largest cities in the world (C40) in order to publish their inaugural report. Prior to working for KPMG, Bernd was responsible for the development of large-scale energy/environmental projects for local municipalities such as Arnhem.

Asif Kabani has 15 years of experience of working with donor governments and has previously worked with UN DESA, UN office Geneva at IGF, UNICEF, UNHCR, UNODC, and UNDP. Kabani also worked as Consultant, Specialist, Technical expert and Program Manager with the Asian Development Bank (ADB) and Aga

Khan Development Network (AKDN). He has an M.Sc. in Development Studies, MBA in Information Technology, MBA in Management and post graduate diploma in public policy and internet governance. He is a fellow of UN and Diplo Foundation, Geneva, and a member of the DevEx Group, Washington D.C. He plays a key role in representing civil society and donor governments at international forums and debates. His key focus is public policy development with particular focus on sustainable development and with fostering relationships between Governments, businesses, civil society and academics.

Maliha A. Kabani is currently working as Education Specialist with IKEA-Save the Children International, for children working in cotton fields in Pakistan. Kabani has more than 10 years' experience working with the grassroots communities, as well as working as consultant and technical expert on education projects for various donor governments. She has done flood assessment and program development for 2010 floods in Pakistan. She worked as ECO and Program Coordinator for various NGOs and is a member of UNESOC and WSIS, Geneva. She is regular speaker at national and international events and a regular contributor on the subject of Green Economy.

Shay Kelleher is a qualified Regional and Urban Planner. Educated in University College Cork, University College Dublin and Trinity College Dublin he holds a degree in Geography, a master's degree in Regional and Urban Planning and a M.Sc. in Environment and Development. He has worked with Ireland's Department of the Environment and as Town Planner. He has undertaken research in Ireland, Britain, Europe and Africa and has strong interests in urban planning, regional development, development economics, urban and regional economics, and environmental sustainability.

Andrew Kemp is a Project Officer for the contracts team at the Federation of Canadian Municipality's Green Municipal Fund. He is responsible for negotiating funding agreements and evaluating municipal sustainability plans, feasibility studies and capital projects. He has previously worked for a leading Canadian economic development consulting firm promoting local economic sustainability in Canada and has several years of experience in business evaluation in the financial services sector. Academically, Andrew Kemp is in the final stages of completing a master's degree in local economic development from the University of Waterloo.

Caroline King is a Researcher at the University of Oxford working on the geography of environmental change in the Northern Sahara. Her research interests environmental analysis and assessment in dryland ecosystems.

Santhosh Kodukula (Sunny) has a Master's Degree in Industrial Ecology and Environmental Management from Mahidol University, Thailand. During his Master's Sunny specialized in Urban and Regional Planning with focus on Urban Transport Planning and Management.

Currently, Sunny works as the EcoMobility Officer at the ICLEI – Local Governments for Sustainability World Secretariat office in Bonn, Germany. Prior to joining ICLEI EcoMobility, he worked with the GIZ Sustainable Urban Transport

Project (SUTP) providing technical assistance and implementing project activities in Asia and the Pacific.

At the GIZ-SUTP, Sunny was involved in the design and delivery of over 70 training courses on sustainable urban mobility and transport targeted at policy-makers and urban and transport planners in various countries. He has also worked extensively in the areas of non-motorised transport and pedestrianisation.

His research interests include public space modeling, planning for pedestrians and cyclists, and transport and livability. Sunny has contributed to various published papers, reports and newspapers, and was also involved in transport research at the university level.

Christophe Lalande has been working at UN-HABITAT over the past 7 years on promoting sustainable urban development. His professional expertise is in housing policy and development, in particular the design and implementation of slum upgrading and prevention programmes in developing countries, and in producing policy guidelines, methodologies and tools to address the specific challenges faced by the urban poor, indigenous peoples and the most vulnerable groups in cities in the realisation of their right to adequate housing. Christophe Lalande is coordinating UN-HABITAT's new initiative for green buildings and sustainable housing to respond to the global challenges of housing, slum upgrading and climate change, from the perspective of the environmental and socio-cultural sustainability of affordable housing solutions. Christophe is a national of France. He holds a Bachelor degree in Political Sciences (2001) and a Masters in Public Policy and Political Sociology (2003) from the Institut d'Etudes Politiques de Paris.

Art von Lehe is an attorney and a LEED Accredited Professional. His background in environmental science, law, and policy combined with knowledge of both the federal and local landscapes informs his current work with local governments. Mr. von Lehe serves as Senior Program Officer for ICLEI-Local Governments for Sustainability, USA's STAR Community Index, where he develops content for a strategic local government sustainability planning and performance management system. Before working on the STAR Community Index, Mr. von Lehe worked to provide ICLEI's over 600 city and county member network with information, analysis, and support regarding key policies relevant to sustainable local action. Before joining ICLEI, Mr. von Lehe analyzed emerging climate and clean air policy at the U.S. Environmental Protection Agency in Washington, DC. He also served as project manager with the Coastal Conservation League in Charleston, South Carolina, where he advised a large portfolio of local governments on climate and sustainability measures covering a range of relevant sectors. Mr. von Lehe received a bachelor's degree in environmental studies from the University of North Carolina at Asheville, a master's degree in environmental law from Vermont Law School, and a J.D. degree from the University of South Carolina School of Law.

Ulrich Mans is a Ph.D. candidate at the Amsterdam Institute for Social Science Research (AISSR). His research focus rests on sustainability transitions and the importance of renewable energy in emerging market cities. He has a professional background in international development and conflict resolution. In the past, he has been employed by various think tanks on sustainability, governance, and societal transitions in post-conflict countries. With several years of practical experience as a consultant, he has also worked for government agencies, NGOs and the United Nations Environment Program.

Dr. Sean Markey is an Associate Professor with the Resource and Environmental Management program and an Associate with the Centre for Sustainable Community Development at Simon Fraser University. Sean's research concerns issues of local and regional economic development, rural and small-town development, and community sustainability. He has published widely in academic journals and is the principle author of *Investing in Place: Economic Renewal in Northern British Columbia (UBC Press, forthcoming); Second Growth: Community Economic Development in Rural British Columbia* (UBC Press, 2005); and co-editor of *The Next Rural Economies: Constructing Rural Place in Global Economies* (Cabi Press, 2010). Sean continues to work with municipalities, non-profit organizations, Aboriginal communities and the business community to promote and develop sustainable forms of community economic development. He also serves on the Board of Directors with the Vancity Community Foundation.

Kes McCormick is an Assistant Professor at the International Institute for Industrial Environmental Economics at Lund University, Sweden. With a background in political science and environmental science, he engages in a combination of research, education, communication and innovation activities. Broadly speaking he works in the fields of sustainability and governance. Specifically, he concentrates on exploring sustainable urban transformation and accelerating the implementation of renewable energy (and energy efficiency). Within sustainable urban transformation, his research activities are focused on models and tools for collaboration between a diversity of stakeholders, particularly local governments, universities, urban citizens and business.

Corina McKendry is an assistant professor at Colorado College, a liberal arts school in Colorado Springs, Colorado, USA. She teaches courses in environmental politics and political economy in the Political Science Department and for the interdisciplinary Environmental Program. Her primary research interest is the relationship between urban environmentalism and the changing role of cities in the global economy. She is also interested in the social equity implications of urban greening initiatives. She received her Ph.D. from the Department of Politics at the University of California Santa Cruz in 2011.

Sara Meerow works as a junior researcher in the Department of Human Geography, Planning and International Development Studies at the University of Amsterdam, where she completed a post-graduate degree in International Development Studies. Her master's research in Bangkok, Thailand focused on power sector resilience and policies for distributed, renewable electricity generation. She is also currently working on a research project on Caribbean urban innovation management.

Axel Michaelowa is head of the group on international climate policy at the chair "Political Economy and Development", University of Zurich and Center for Comparative and International Studies (CIS). Besides his research work, he is senior founding partner of the consultancy "Perspectives". Axel is a lead author of the chapter on international agreements in the 5th Assessment Report of the Intergovernmental Panel on Climate Change due in 2014 and wrote on mitigation policies in the 4th Assessment Report of the Intergovernmental Panel on Climate Change published in 2007. He is member of the editorial boards of the journals "Climate Policy", "Climate and Development" and "Greenhouse Gas Measurement and Management" as well as of the board of the research network "Climate Strategies". For the last two decades, Axel has been working on climate policy issues and published over 100 articles, studies and book contributions, mainly on the Kyoto Mechanisms and especially the Clean Development Mechanism (CDM). Axel is a member of the CDM Executive Board's Registration and Issuance Team and has contributed to the development of eight approved CDM baseline methodologies. As one of the leading European experts on CDM institution building in developing countries, Axel supported several governments in Asia and North Africa/Middle East.

Philip Monaghan is Founder and CEO of Infrangilis. He is a recognized strategist in the fields of the economic development and sustainability. He is acclaimed author of 'How Local Resilience Creates Sustainable Societies' (2012) and 'Sustainability in Austerity' (2010).

Hans Mönninghoff has 12 years experience working as consulting engineer in the fields of water and energy. He has been Director of Services for Environmental Affairs for the City of Hannover for 23 years, and Director of Economic Affairs for the Green Party for 7 years.

Dr. Raquel Moreno-Peñaranda is a Research Fellow at UNU-IAS OUIK. Her work focuses on sustainable natural resource management, looking at the linkages between nature and society. One of her core projects studies the role of urban ecosystems in sustainability and local well-being, paying special attention to local culture for enhancing local ecosystem services and biodiversity, and thus contributing to improve quality of life in cities and reducing urban ecological footprints. She has carried out extensive consultancy, advisory, and research co-ordination for several local and sub-national governments in different countries, international environmental NGOs, civil society organizations, and multilateral development institutions, including the Global Environmental Facility in Washington, DC. Originally from Spain, where she graduated in Biology, she earned a Ph.D. in Energy and Resources from the University of California Berkeley.

Jan De Mulder works as Public Governance counsel at the Representation of the Government of the Region of Flanders (Belgium) to the European Union. Previously he worked for more than 10 years with the Flemish Environment Administration as a legal advisor. He is also engaged as a voluntary researcher with the Department of Public International Law, University of Ghent, Belgium.

Nandhivarman Muthu had obtained his M.Sc. degree in Zoology and has over 10 years of work experience as an environmentalist, researcher and an activist for sustainable development. He is currently pursuing his doctorate from Pondicherry University, India. His research interest includes an extensive study to evolve and implement Green Campus Initiative in Pondichery University. The main objective of this research is to assess and provide factual solutions to the educational institutions like schools and colleges to make it sustainable with special reference to water, energy and waste.

Lena Neij is Professor and Director at the International Institute for Industrial Environmental Economics at Lund University, Sweden. Her research is focused on the dynamics of energy systems and the analysis of governance and policies for more sustainable development. The core issue in her research is the development and assessment of methods for analyzing the dynamics of energy systems in view of technical change (i.e. development, introduction, and diffusion of new technology) and policy measures for effecting and accelerating technical change. Her research, which is system-based and interdisciplinary, covers energy for sustainable development in general, with a focus on renewable energy technologies, energy efficiency in buildings and sustainable urban development.

Dr. Kenneth K. Odero is a Spatial Scientist. He holds a Ph.D. in Urban and Regional Planning from the University of Nairobi. Between 1999 and 2006, Dr. Odero taught Regional Economics, Regional Planning, and Institutional Development at the University of Zimbabwe. Previously Dr. Odero worked full-time as a consultant for ProPerArt Associates (Pvt.) Ltd. In 1997, Dr. Odero worked on a short-term contract as a Technical Advisor in the International Labor Organization's Rural Travel and Transport Program. Shortly after graduating (1988–1990) he worked for the Government of Kenya as a Land Use Planner. In 1990, Dr. Odero joined the Institute for Development Studies at the University of Nairobi as a Research Fellow where he researched and published on, inter alia, the agrarian question, urban management, sustainable development and rural transportation. Dr. Odero has an interdisciplinary orientation having been exposed to a wide range of development areas both as a scholar and professional. He is currently founder and President of Climate XL Inc and Executive Director of Climate Excellence Africa, a not-for-profit international climate change organization. He also lectures in regional planning at the University of Zimbabwe.

Eric Copius Peereboom, is Senior Consultant, Climate Change & Sustainability, KPMG Advisory N.V. Eric has been active in the public sector (local, national, EU) and private sector for over 12 years working as a consultant, and has performed many complex policy and strategy oriented advisory projects in the field of sustainability. Within KPMG, Eric has played a vital contribution in the development of KPMG's proposition Sustainable Cities. Projects he leads generally focus on vision and policy development, scenario development, benchmarking, policy monitoring and evaluation, and cooperation between the public and private sector. Eric worked alongside Bernd Hendriksen in the development of the CDP Cities 2011 Global Report, and has also been engaged in other KPMG studies that have focused on energy & climate strategy in the public sector.

Dr. Gopalsamy Poyyamoli (M.Sc. Marine Biology and Ph.D. Ecology) has been working as an Associate Professor in the Department of Ecology and Environmental Sciences, Pondicherry University, India, for more than two decades. He is focusing on Social Ecology and Sustainability with special emphasis on rural, urban and Industrial Ecosystems.

Dickella Gamaralalage Jagath Premakumara is a development planner, specializing in participatory environmental planning and sustainable urban development in Asia. He has over 15 years experience in working with academic, government, NGOs and bilateral and international agencies. At present, he is a policy researcher at the Kitakyushu Urban Centre of the Institute for Global Environmental Strategies (IGES) in Japan, where he is actively involved in strategic policy research in urban environmental planning and management. He focuses on drawing lessons from the practices and experiences of local actions taken by the local stakeholders at city-level, especially in the area of municipal solid waste management and sustainable material-cycle society, and makes policy recommendations to promote effective policy strategies that could be of great use for other cities in the Asia-Pacific Region. He is an active member of various regional networks and publishes regularly for academic and policy journals.

Rula Taher Qalyoubi, Ph.D. is an environmental economist for the Policy and Strategy Sector of the Environment Agency, Abu Dhabi, United Arab Emirates. She was formally Associate Professor of Economics at the University of Wisconsin – Stout, Wisconsin, USA. As a policy analyst and adviser, Qalyoubi wrote policy briefs regarding environmental health and climate change. She provided expert advice to government, NGOs and businesses regarding energy alternatives and indices, drafted gap assessments in environmental standing policies and procedures and lead efforts on waste management initiatives to achieve sustainability goals, and directed efforts in renewable technology implementation assessments and economic transformation using benefit cost studies.

Jonathan Dagadu Quartey is an Economist with special interest in environmental and natural resource issues. He is a certified trainee of the United Nations University World Institute for Development Economics Research (UNU-WIDER) in the Economics of Climate Change and also Climate Change Science and the modeling of Biophysical Impacts. He obtained his Master's degree in Economics from the Kwame Nkrumah University of Science and Technology (KNUST), Ghana in 2000. Since 2001, Quartey has taught Environmental and Resource Economics, Macroeconomics and Development of Economic Thought at the Kwame Nkrumah University of Science and Technology. Prior to 2001, he had worked for the Ghana Education Service. His research areas are Forestry Economics, Economics of Public utilities: Energy and Water, Economics of Climate Change and Economics of Natural Resources. Quartey has made a number of presentations on the Environment and Natural Resources in local and international conferences. He is currently pursuing a part-time Ph.D. program in Development Studies.

Dr. Golam Rasul, Head, Economic Analysis Division, is a development economist. He is a national of Bangladesh. Prior to his appointment as Head of the Economic Analysis Division in 2009, he served as a policy development specialist at ICIMOD for 5 years. Dr. Rasul holds a Ph.D. in regional and rural development planning from the Asian Institute of Technology (AIT), Thailand. He worked for more than a decade in the Bangladesh Civil Service in different ministries and in field administration in different capacities where he was involved in the formulation and implementation of development planning and programming. Along with development work, he has been actively involved in research in areas that include agriculture, natural resource management, poverty alleviation, and sustainable development in Bangladesh and the South Asian region. His research findings have been published in many international journals including World Development, Environmental Management, Journal of Environmental Management, Environmental Conservation, Sustainability: Science, Practice and Policy, Society and Natural Resources, Regional Development Studies, Development in Practice. Five of his papers have appeared as 'most read papers' in their respective journals. He is currently working on green economy, valuation of ecosystem services, sustainable mountain development, food security and regional cooperation. He is member of the OECD DAC Task Team on Green Growth and Poverty Reduction as one of the developing country experts.

Philipp Rode is Executive Director of LSE Cities and Senior Research Fellow at the London School of Economics and Political Science. He is Ove Arup Fellow with the LSE Cities Programme and co-convenes the LSE Sociology Course on 'City Making: The Politics of Urban Form'. As researcher and consultant he manages interdisciplinary projects comprising urban governance, transport, city planning and urban design. Rode organised Urban Age conferences in partnership with Deutsche Bank's Alfred Herrhausen Society in ten cities bringing together political leaders, city mayors, urban practitioners, private sector representatives and academic experts. The focus of his current work is on cities and climate change which includes his role as coordinating author of the cities and buildings chapters for UNEP's Green Economy Report. He manages the Urban Age research efforts and recently co-authored 'Transforming Urban Economies' (2011) and 'The Global MetroMonitor' (2010); and published the reports 'Cities and Social Equity' (2009) and 'Integrated City Making' (2008). He has previously worked on several multidisciplinary research and consultancy projects in New York and Berlin and was awarded the Schinkel Urban Design Prize 2000.

Dr. Mark Roseland Director of the Centre for Sustainable Community Development and Professor of Resource and Environmental Management at Simon Fraser University. Dr. Roseland lectures internationally and advises communities and governments on sustainable development policy and planning. The 4th edition of Dr. Roseland's Toward Sustainable Communities was published in Spring 2012 and released at the ICLEI – Local Governments for Sustainability 2012 World Congress and International Urban Researchers Symposium, associated with the Rio+20 Earth Summit. Dr. Roseland is leading development of the Pando | Sustainable Communities researcher-practitioner network. At SFU, he is a founding member of the SFU Community Trust's Board of Directors, responsible for the award-winning UniverCity sustainable community development project.

Boshra Salem is a Professor at the University of Alexandria. Her research interests include ecosystem management, environmental impact assessment and biodiversity conservation in the North Western Desert of Egypt. She is a member of the UNESCO-MAB Committee in Egypt and was the National Team Leader for the UNU-UNESCO-IHP GWAHS-CS project in Egypt.

Maike Sippel has a background as an architect and city planner and did her Ph.D. in Economics on linkages between local and global climate policy. She has been working on greenhouse gas mitigation and energy strategies for cities and climate policy issues over the last decade, at the Hamburg Institute of International Economics and then at Stuttgart University. Her work has covered cities and the CDM since 2003. Maike is also consulting on this issue with cities and international organizations. She has published some 20 articles, studies, book contributions and newspaper articles. Besides her research work, she has co-founded the NGO network "Global Marshall Plan Initiative". She is a Coordinator for Sustainable Development at the University of Applied Science, Konstanz, and an assistant professor on sustainability at the University of Applied Science Ravensburg-Weingarten.

Phoebe Stirling graduated from the Architectural Association (London, UK) in 2011 with a masters' degree in Histories and Theories of Architecture and Urbanism. She is currently part of the Urban Research Team with ICLEI at the organization's Word Secretariat in Bonn, Germany.

Makoto Ueda was born in Kyoto in 1961 and joined the City of Kyoto in 1986. Since then he has promoted commerce, human-rights, and tourism before becoming the Director of Kyoto's Climate Change Prevention Office in the Environment Policy Bureau. He is motivated and deeply committed to addressing global challenges. He earned a doctorate in Policy Science from Doshisha University in 2009.

Andrew Walters leads a range of sustainability projects across the organization to identify opportunities to embed resource efficiency and world's best environmental and sustainability practice across all of Auckland Council's operations, services, programs and facilities. Activities include travel planning, environmental management systems, sustainable procurement, Greenhouse Gas accounting. Andrew has significant expertise in waste management, environmental management systems, policy and strategy planning, and implementation and sustainable buildings in New Zealand and the UK. Andrew has held senior policy and advisory roles in Auckland City Council, Ministry for the Environment in the Government Urban and Economic Development Office.

Dr. Edmundo Werna has worked for over 25 years on different aspects of urban development, with particular attention to labor. He joined the International Labor Organization in 2004. He has a Ph.D. in urban development from the University of London (UK), an M.Phil. in development studies from the Institute of Development

Studies in Sussex (UK), and a Bachelor's degree in urbanism and architecture from the Federal University of Minas Gerais (Brazil). Prior to joining the ILO, Werna coordinated the urban development agenda of UNDP's UN Volunteers Programme. He also undertook consultancies for several organizations, including local governments, the World Health Organization, European Commission, the World Bank, UN-Habitat, UNCDF/UNDP and also the ILO. Werna also has extensive academic experience, published eight books in the UK and USA, plus many chapters in edited books and several articles in scientific journals. He is now a member of the editorial board of the journals *Habitat International* and *International Journal of Sustainable Urban Development*. Werna received a number of prizes from his practical and academic work, including the Earth Summit 2002 Sustainable Development Award, one project selected as Best Practice for the UN-Habitat/Dubai Awards (2004), one project chosen as best practice for the Hanover 2000 Exhibition as well as awards for three articles and chapter for one book.

Dr. Jonathan Woetzel is a Director with McKinsey & Company and is based in Shanghai. He currently leads McKinsey's Cities Special Initiative globally, convening with urban authorities in over 40 countries worldwide. In his 27 years with the Firm, 26 of which in China, Jonathan has advised clients in a range of industries including energy, materials, technology and industrial, helping transform local companies into global leaders and developing policy recommendations for government. Dr. Woetzel co-chairs McKinsey's Urban China Initiative with Columbia and Tsinghua Universities, a not-for-profit think tank on urban issues. He leads McKinsey's service to the Government of China and has led Asia Energy and Materials practice, the McKinsey Global Institute in Asia, and its Corporate Finance practice in China. He established McKinsey's China practice in the 1990s including managing its Shanghai location. Dr. Woetzel has written several books including *Capitalist China: Strategies for a Revolutionized Economy* (Wiley, 2004) and *Operation China: From Strategy to Execution* (Harvard Business Press, 2008), co-authored with Jimmy Hexter. He holds a Ph.D. in political science from the University of Southern California, where his work focused on China's political economy. He is a lecturer at Beijing University's Guanghua School of Management and at the China Europe International Business School.

Le-Yin Zhang obtained her doctorate from the University College London (UCL). She directs M.Sc. in Urban Economic Development in UCL and has many years' of professional experiences in a number of countries. Her climate change research focuses on the linkages between climate change and sustainable economic development in developing countries and cities.

Acknowledgments

Compiling a book on a newly created term, such as Green Urban Economy, and within a very short time is a challenge. The Urban Research team, led by Richard Simpson, of the ICLEI World Secretariat took up this challenge, and is now presenting a rich compendium on the Green Urban Economy.

To achieve this, the book is based upon an open call for contributions to draw out existing work from across the world from practitioners and researchers. Our thanks go to the contributors of this book that took the time to share their knowledge and expertise, their concepts and new thoughts, findings and facts, and good practice examples. We are grateful for the time they spent and the insights they share with us.

ICLEI's Media Partner, Springer, has supported us with the production of this book which follows two publications on the Resilient Cities 2010 and 2011 congresses within the series *Local Sustainability*.

We express our particular gratitude to the members of the Urban Research team at the World Secretariat. Phoebe Stirling for editing most of the contributions, and Shay Kelleher and Mitja Maletin for their comments and suggestions. In addition, a special thank you goes to Kristina Yuzva for her editorial support.

We are thankful to ICLEI for providing the supporting environment for this endeavor and to colleagues for providing valuable ideas and thoughts, in particular Konrad Otto-Zimmermann and Susanne Salz.

All of us look forward to continuing the dialogue and creative exchange with even more experts in the future.

Director, Capacity Center — Monika Zimmermann
Research Analyst, Capacity Center — Richard Simpson
ICLEI – Local Governments for Sustainability,
World Secretariat

Contents

Contributors

Essam Hassan Mohamed Ahmed Egyptian Environmental Affairs Agency (EEAA), Cairo, Egypt

Vito Albino Department of Mechanics, Mathematics and Management, Politecnico di Bari, Bari, Italy

Stefan Anderberg Lund University Centre for Sustainability Studies, Lund, Sweden

Jen Baker Office of Sustainability, Raleigh, NC, USA

Steven Bland ICLEI – Local Governments for Sustainability, Bonn, Germany

Shannon N. Bouton McKinsey & Company, Shanghai, China

Rosalie Callway Sustainability Consultant, London, UK

Paul Chambers Auckland Council, Auckland, New Zealand

Amelia Clarke Master of Environment and Business (MEB) Program, School of Environment, Enterprise and Development (SEED), University of Waterloo, Waterloo, ON, Canada

Sean Connelly Centre for Sustainable Community Development, Simon Fraser University, Burnaby, BC, Canada

Rosa Maria Dangelico Management Engineering, Sapienza – Università di Roma, Rome, Italy

Golda Arpudhalin Edwin Pondicherry University, Puducherry, India

Matthew A. French UN-Habitat, Nairobi, Kenya

Andy Gouldson School of Earth and Environment, University of Leeds, Leeds, UK

Lynn Graham Office of Sustainability, Raleigh, NC, USA

Bernd Hendriksen KPMG Advisory N.V., Rotterdam, The Netherlands

Asif Kabani Founder and Hon. VP, International Sustainable Development Resource Centre, Geneva

Head of Research and Project Management, Ministry of Information and Broadcast, Govt of Pakistan

Maliha A. Kabani Founder and Hon. President, International Sustainable Development Resource Centre, Geneva

Manager, Education - IKEA, Save the Children International (SCI), Country office, Pakistan

Shay Kelleher ICLEI – Local Governments for Sustainability, Bonn, Germany

Andrew Kemp Green Municipal Fund, Federation of Canadian Municipalities, Ottawa, ON, CanadaUniversity of Waterloo, Waterloo, ON, Canada

Niall Kerr School of Earth and Environment, Center for Climate Change Economics and Center for Low Carbon Futures, University of Leeds, Leeds, UK

Caroline King Oxford University Centre for the Environment, Oxford, UK

Santhosh Kodukula ICLEI – Local Governments for Sustainability, Bonn, Germany

Johan Kuylenstierna Stockholm Environment Institute, University of York, York, UK

Christophe Lalande UN-Habitat, Nairobi, Kenya

Art von Lehe ICLEI Oakland, California, USA

Anna Leonard Office of Sustainability, Raleigh, NC, USA

Molly Lindsay McKinsey & Company, Shanghai, China

Ulrich Mans Department for Geography, Planning and International Development Studies, University of Amsterdam, Amsterdam, The Netherlands

Sean Markey School of Resource and Environmental Management, Simon Fraser University, Surrey, BC, Canada

Kes McCormick International Institute for Industrial Environmental Economics, Lund University, Lund, Sweden

Corina McKendry Political Science Department, Colorado College, Colorado Springs, CO, USA

Sara Meerow Department for Geography, Planning and International Development Studies, University of Amsterdam, Amsterdam, The Netherlands

Axel Michaelowa University of Zurich, Zurich, Switzerland

Philip Monaghan Infrangilis, Manchester, M45BD, UK

Hans Mönninghoff Economic and Environmental Affairs, Hannover, Germany

Raquel Moreno-Peñaranda United Nations University-Institute of Advanced Studies-Operating Unit Ishikawa/Kanazawa (UNU-IAS-OUIK), 6F International Organizations Center, Yokohama, Japan

JanDeMulder Permanent Representation of Belgium to the EU/Flemish Representation, Brussels, Belgium

Nandhivarman Muthu Pondicherry University, Puducherry, India

Lena Neij International Institute for Industrial Environmental Economics, Lund University, Lund, Sweden

Kenneth K. Odero Climate XL Africa, Nairobi, Kenya

Richard Pearce Quantum Strategy and Technology Ltd., Warrington, UK

Eric Copius Peereboom KPMG Advisory N.V., Rotterdam, The Netherlands

Gopalsamy Poyyamoli Pondicherry University, Puducherry, India

Dickella Gamaralalage Jagath Premakumara Institute for Global Environmental Strategies (IGES), Kitakyushu Urban Centre, Kitakyushu, Japan

Julian Prosser Office of the City Manager, Raleigh, NC, USA

Rula Taher Qalyoubi, Ph.D. 3rd Group Consultancy Engineering (3G), United Arab Emirates

Jonathan Dagadu Quartey Department of Economics, Kwame Nkrumah University of Science and Technology, Kumasi, Ghana

Golam Rasul Economic Analysis Division, International Centre for Integrated Mountain Development (ICIMOD), Kathmandu, Nepal

Philipp Rode London School of Economics and Political Science, London, UK

Mark Roseland School of Resource and Environmental Management, Centre for Sustainable Community Development, Simon Fraser University, Surrey, BC, Canada

Boshra Salem University of Alexandria, Alexandria, Egypt

Richard Simpson ICLEI – Local Governments for Sustainability, Bonn, Germany

Maike Sippel University of Applied Science, Konstanz, Germany

Emily Steele Office of Sustainability, Raleigh, NC, USA

Phoebe Stirling ICLEI – Local Governments for Sustainability, Bonn, Germany

Corrado Topi Stockholm Environment Institute, University of York, York, UK

Makoto Ueda Climate Change Prevention Office, Environment Policy Bureau, Kyoto, Japan

Andrew Walters Auckland Council, Auckland, New Zealand

Edmundo Werna Natural Resources and Industries, Department of Sectoral Activities, International Labour Organization, Geneva, Switzerland

Jonathan Woetzel McKinsey & Company, Shanghai, China

Le-Yin Zhang Development Planning Unit, University College London, London, UK

Monika Zimmermann ICLEI – Local Governments for Sustainability, Bonn, Germany

List of Boxes

List of Figures

List of Tables

Part I
Agenda Setting for the Urban Future with a Green Urban Economy

Green Urban Economy: Agenda Setting for the Urban Future

Monika Zimmermann and Richard Simpson

Abstract Irrespective of a lack in any clear outcome from Rio+20 on Green Economy, or rather in the absence of any inter-national agreement thereon, the Green Urban Economy approach still embodies a call to local actors to investigate in particular economic benefits from environmental and social actions. It is a political and practice oriented concept, which connects the global significance of economic activities in urban centers and metropolitan regions with the challenges of the future. The proximity to economically active citizens and companies provide local governments a particular opportunity to realize their responsibility for city development and planning, which can influence private actions and lifestyles, as well as framework conditions for economic activities.

Keywords Agenda setting • Rio+20 • Introduction • Green urban economy

1 Green Urban Economy: Translating Green Economy to the City Level

Green Economy was one of the two key themes of the Rio+20 UN Conference on Sustainable Development in June 2012. ICLEI set out to translate this internationally formulated theme to the local government and city level. The "Rio Earth Summit" conference of the United Nations on Environment and Development in 1992 resulted in, amongst other, the "Agenda 21" as a call for action. ICLEI at the time translated "Agenda 21" to the local level with the concept "Local Agenda 21", which was anchored in Chapter 28 of Agenda 21.

M. Zimmermann • R. Simpson (✉)
ICLEI – Local Governments for Sustainability, Bonn, Germany
e-mail: capacity.center@iclei.org; urban.research@iclei.org

R. Simpson and M. Zimmermann (eds.), *The Economy of Green Cities: A World Compendium on the Green Urban Economy*, Local Sustainability 3,
DOI 10.1007/978-94-007-1969-9_1, © Springer Science+Business Media Dordrecht 2013

If the objective is to not only goal of Green Economy but rather to produce a its concrete realization, this cannot be achieved without the significant contribution of local governments. Thus a Green Economy must be accompanied and supported by a "Green Urban Economy". With this goal ICLEI introduced in 2011 the term "Green Urban Economy" to the international discussion, keeping its local government membership informed, substantiating the concept in various studies, and assembling the first case studies on local government action. With the Green Urban Economy the debate on the Green Economy was localized to towns, cities and urban mega-regions.

Is this "old wine in new skins"? Is the international discussion only out for a new terminology to disguise lack of action? Should "sustainable development" be completely dissolved and give way to a new primacy of the economy, as some fear? Should values give way to pure market forces? While some doubt a "Green Economy" for fear that social and ecological goals could become lost, others, especially developing countries, fear a new economic colonization of technologies from industrialized economies. Critical scientists have asked: can the economy be viewed from a urban perspective? Is it a step back to a resource economy, in which small cycles are closed and economies are reduced to the immediate exchange of goods within small spaces?

The answer is No. "Green Urban Economy" is a political and practice oriented concept, which connects the global significance of economic activity in urban centers and metropolitan regions with the challenges of the future (in which an increasing urban population follows an urban lifestyle) and the specific scope of action of their local governments in collaboration with other actors (Box 1). Their proximity to economically active citizens and companies provides local governments a particular opportunity to realize their responsibility for city development and planning, which strongly

Box 1 Green Urban Economy Agenda is Simultaneously Goal and Journey

- **Greening** the Urban Economy consists of a bundle of measures, laws, instruments, mechanisms, which reward ecologically and socially sensible action economically and therewith accelerate the transformation of the economy towards a sustainable urban economy. It relates to the preservation of environmental quality and supporting environmentalism.
- Greening the **Urban** Economy sets the emphasis on local government as innovator and driver, with the multiplicity of city actors at their side. It is based on the fact that today already more than the majority of economic goods and services are produced in an urban economy. It refers to towns, cities and mega-urban regions as a place of living, activity and administration unit in a period of rapid worldwide urbanization.
- Greening the Urban **Economy** includes many dimensions: decent jobs for the many, equitable access to resources and protection from risks, such as the consequences of climate change. It refers to the management of household or private affairs and especially expenses. It also relates to the management of resources of a community, country, etc., especially with a view to its productivity.

influence private action. Local governments can set framework conditions and instruments, which have an effect immediately and locally.

Irrespective of a lack in a clear outcome from Rio+20 on Green Economy, or rather in the absence of any inter-national agreement, the approach embodies a challenge to local governments and city actors to develop entirely new instruments (e.g. Carbon trading in Tokyo) and to apply existing instruments in a new way (e.g. providing city properties to buyers who guarantee high energy standards), thereby engaging economic actors and considerations. Environmentalists need to understand the language and dynamics of the economy, and combine immanent economic interests (such as cost savings) with ecological interests (such as saving of resources) and vice versa.

From 20 years of experience with "Local Agenda 21" the strengths and weaknesses of approaches and actions are known. In many places across the world, processes have started, goals have been set, integrated thinking promoted, and action encouraged. Governments have become attentive that only a combination of economic, social and ecological approaches can address the growing challenges of the world (ICLEI 2012a, b).

From 20 years of "Local Agenda 21" it is knows that only in a few places it has succeeded to bring the economy on board and to open the dialogue with those who produce and distribute – and thus influence consumption – that they might turn their practices around in such a way that environmentally and socially damaging actions are no longer economically rewarded. We also know that the power and financial resources of the market are becoming stronger in proportion to the financial influence of the public sector. This is another reason to make clear to local, regional and national governments their scope of action for targeted influence on the market and on the economy as a whole (Box 2).

Box 2 Green Urban Economy Core Messages

The Green Urban Economy is a process of transforming and transitioning existing and emerging urban economies to be more environmentally benign and socially desirable. The way how services and goods are produced, procured, distributed and consumed informs environmental and social objectives. What is being done well needs to be economically rewarded. Socially and environmentally benign economic activities need to become the norm, not the exception.

This calls for:

- Technology innovation: technological innovation to produce more efficiently and use abundant renewable resources sustainably;
- Green business and governance models: an investigation into current business and governance models to optimize these for economic returns while achieving environmental and social benefits (for example green procurement, rather than business as usual);
- New green business opportunities: identifying new business opportunities that value the environment and people.

Shrinking resources, rising prices and growing unequal access to resources, be these food, energy, water or similar, will affect the *urban* poor. Poverty eradication must be part of a green and inclusive economy, and not excluded from a greening of the economy.

2 Local Governments as Drivers of the Green Urban Economy

Urban areas are growing in importance. Already today the 150 smallest UN-member countries each have fewer inhabitants than a megacity the size of Mumbai or Tokyo. At the same time there is a backlog in providing decent housing and infrastructure to one billion urban dwellers, while the global urban population will increase by another three billion over the next 40 years. The logical conclusion is that within the next 40 years the same urban infrastructural capacity will have to be built as that over the last 4,000 years. This means an unimaginable investment into residential buildings, roads, provision and waste utilities, and expected public services. Food production and distribution also pose a new dimension of challenge. It is clear that this development cannot be realized with current levels of resource and energy wastage.

Local governments themselves will only be able to provide a small amount of the investment required. They can however provide framework conditions, incentivize, moderate, stimulate ideas, innovate and provide and set market incentives. Towards this cities need to be prepared: communal targets must be developed and new models built, governance styles improved, technical expertise acquired and new instruments tested for a multiplicity of different interests and actors. Local governments must continuously refine their roles and approaches.

Local governments do not yet have the structural capacity, competency, nor aptitude – so at least it could appear on first view. We know that the journey is still long. And yet across the world there are already many good examples for innovative local government action, as this book shows. Cities and towns take a lead, even if nations still hesitate (Box 3).

Box 3 Example of Local Action for Greening the (Urban) Economy: Municipalities…

- **… use their economic and financial instruments to advance low carbon development and resource efficiency;**
  Examples: Tokyo Carbon Trade, congestion charge.
- **… use their planning authority for designing more eco-efficient, compact cities;**
  Examples: Strategic development plans, EcoMobility and public transport services, by-laws such as buildings codes for renewables and energy efficiency.

(continued)

Box 3 (continued)

- **… use their purchasing power to get more and cheaper green products and services on the market;**
  Examples: Green purchasing, joint procurement.
- **… use municipal operations and utilities as drivers for eco-efficiency and (green) job creation;**
  Examples: Increasing the amount of local renewables in the energy provision for citizens, facilitating job training on installation technologies.
- **… measure and value ecosystems as a source for service provision;**
  Examples: Eco-tourism, urban agriculture, nature stewardship schemes.
- **… support local economic activities and businesses to attract, retain and increase green jobs;**
  Examples: Portland's Economic Development Strategy, local inventories on green jobs, encouraging the development of corporate social responsibilities.
- **… support and advise (local) business on training and capacity building;**
  Examples: Local agendas for employment, facilitating relevant training courses to support implementation.
- **… value jobs in the informal sector and transfer them with caution and sensitivity into formal employment if applicable;**
  Examples: Health and social protection of workers on landfill sites or in waste recycling.
- **… encourage research and development for designing new products, technologies, services and operational models to transform or innovate the local economy;**
  Examples: Smart bike and car sharing technology systems, smart meters, smart grids.
- **… work with (local) business to demonstrate that resource-efficient production, dissemination and consumption can result in economic benefits;**
  Example: EcoProfit® programs.
- **… inform their citizens to become green consumers;**
  Examples: Information and awareness campaigns, local awards, strengthening local NGOs and partners, promoting eco or fair trade produce.
- **… develop indicators and measurement systems to demonstrate that resource efficient, environmentally and socially sound actions serve all.**

3 Green Urban Economy: New and Then Again Not So New

Green Urban Economy is an approach towards direct action, which complements and strengthens the current efforts of local government and city actors moving towards sustainability. There will be as many questions as answers: is the compact city really more resource efficient and bring economic benefits? Does the urban lifestyle compensate the efficiency advantage of urban density and where are the economic incentives? Is energy saving enough, or do we need a much more radical view on proportional access to the finite amount of resources on this planet? What pricing mechanisms can be used for resources, energy, waste management, water etc to prevent social inequalities?

4 Situating This Book in a Global Context

Green Urban Economy requires even more creativity, new thinking, arguments, numbers and success indicators, as well as a critical examination of the reality, in which there will not only be win-win-win situations. Therefore varying perspectives and practical experiences from the emerging fields of environmental economics, urban development, green cities, and concrete examples from city governments are brought here together in order to document the different layers of knowledge and to support both researchers and practitioners in exploring and further detailing such an agenda. *The Economy of Green Cities: A World Compendium on the Green Urban Economy* subsequently brings a diverse set of contributors together to start a discussion on the implications of a Green *Urban* Economy, irrespective of no clear international outcome from Rio+20.

Nonetheless, *The Economy of Green Cities: A World Compendium on the Green Urban Economy* was inspired by the international discussions on the Green Economy. The Green Economy was popularized by key international organizations' publications and initiatives such as the United Nations Environment Programme's (Oct, 2008) *Global Green New Deal* publication and the launch of the *Green Economy Initiative*, and the UNEP/ILO/IOE/ITUC (2008) *Green Jobs: Towards decent work in a sustainable, low-carbon world.*

The Green Economy notion gained greater momentum with the first Preparatory Committee meeting of the United Nations Convention on Sustainable Development in May 2010 based upon the United Nations General Assembly December 2009 resolution to make "Green Economy within the context of sustainable development and poverty eradication" and "institutional framework for sustainable development" key priority themes and thematic challenges for discussion at Rio+20.

Within the international negotiation context of Rio+20 the term "Green Economy" had already been critically questioned on grounds of lacking a clear definition, being insufficient as a tool, viewed as another form of protectionism and donor conditionality. There was also an initial hesitancy regarding the extent to which it would address the needs of countries for job growth, increasing incomes and poverty eradication. These discussions have played out very differently among regions in the run up to Rio+20.

While "Green Economy" only gained international traction since 2008–2009, the term can be traced back two decades earlier to two prominent publications *Blueprint for a Green Economy* (Pearce et al. 1989), and *The Green Economy: Environment, Sustainable Development and the Politics of the Future* (Jacobs 1991).

5 Moving the Debate to Urban Practice

The "Green Economy" in its original conception could be seen as rooted in the scholarly thought of environmental economics. Environmental economics acknowledges that the environment and the economy necessarily interact i.e. environmental degradation is not an incidental consequence of economic activity, but a central consequence of the way consumption and production are organized (Jacobs 1991). Thus environmental protection requires that the environmental perspective is integrated into economic policy i.e. clarifying the way that economic processes cause environmental degradation and how which economic and public policy can be designed to achieve environmental objectives (Jacobs 1991). It implies that economic policy needs to be changed to avert environmental damages, pollution, and the exhaustion of natural resources caused by economic activities.

This book seeks to open up and instigate discussions on how a Green Economy can be relevant to urban areas and cities, and how a Green Urban Economy can be situated within existing and emerging knowledge and experiences.

The result is new intellectual alliances across and among disciplines with a specific practice oriented focus. This publication is in so far not academic in nature, but rather seeks to draw out experiences, ideas and visions, as well as examples to accelerate and to inform the debate of a Green *Urban* Economy. This publication seeks to create a momentum and new amalgamation of intellectual thought, practice and knowledge, and most importantly arguments for changing the way the economy and the environment in cities is viewed and correspondingly treated without neglecting their essential contribution to social objectives. As a result it is provocatively positioned within a changing socio-political landscape to encourage and to enable deeper, richer and outcome oriented debate.

6 The Structure of This Compendium

Part II correspondingly draws together different perspectives on some guiding principles, opportunities and ideas of what greening the urban economy can look like, in order to instigate and to further the debate amongst researchers, policy makers and practitioners.

Part III highlights the essential requirement that a range of different actors have to be involved with and work together towards greening the urban economy. Such a spectrum must encompass public, private and civic institutions and groups; each has a range of roles to play, but most importantly each has the ability to maximize their potential through collaboration, cooperation and coordination.

Part IV assesses how Green Economy principles can be applied to various urban sectors which make up the urban economic system, enabling an economy that is sustainable, low carbon and resource efficient. In addition, it analyzes the various strategies and approaches that can be utilized to implement a Green Economy in urban centers in terms of policy and technical innovation for example using economic valuation of ecosystem services to inform local decision making.

In Part V six city case studies of action are presented. They demonstrate that local governments are already implementing measures to promote green economic development, but also the potential that still exists in urban areas for innovative, sustained and transformational action. The case studies also demonstrate that in practice, the implementation of green urban planning and economic policies will be spatially diverse, with different approaches, measures and roles being taken.

References

ICLEI (2012a) Local Sustainability 2012 – taking stock and moving forward, ICLEI Global Report. ICLEI, Freiberg

ICLEI (2012b) Local Sustainability 2012 Case study series: showcasing progress in local sustainability, ICLEI Global Report. ICLEI, Freiberg

Jacobs M (1991) The green economy: environment, sustainable development and the politics of the future. UBC Press, Vancouver

Pearce D, Markandya A, Barbier EB (1989) Blueprint for a green economy. Earthscan Publications Ltd, London

Part II
Setting a Green Economy Agenda for Cities

Introduction: A Green Economy for Green Cities

Richard Simpson

Keywords Green cities • Sustainable development • Introduction • Green urban economy

With the international sustainable development discussions exploring the concept and new driver "Green Economy within the context of poverty eradication and sustainable development", it is important and relevant to reflect upon its meaning for cities and urban areas. Cities and urban areas are a key dimension as by 2050 two-thirds of all humans will be living there. They represent 2% of the Earth's land surface, 50% of the global population, and 70–80% of the global energy consumption and CO_2 emissions. At the same time there is a backlog in providing decent housing and infrastructure to one billion urban dwellers, while the global urban population will increase by another three billion over the next 40 years: a near doubling of the urban population to 6.2 billion.

Alone the richest 100 urban economic areas are estimated to account for US$19,874 billion (gross domestic product at purchasing power parity) (PwC 2009), almost 30% of global GDP in 2008. An even greater amount circulates when including urban areas made up of small and medium sized towns and cities, not only seemingly endless mega-cities.

Urban dynamics play out in regions, countries and cities very differently according to local and regional trends and conditions. In the process of managing the rapid growth of existing cities, designing and building of new cities, and the rehabilitation and revitalization of old, often even shrinking, cities, presents an opportunity and urgency to "get it right". For an economy to thrive, it requires functioning cities. It implies that the urban economy across sectors needs to realize opportunities, have

R. Simpson (✉)
ICLEI – Local Governments for Sustainability, Bonn, Germany
e-mail: urban.research@iclei.org

R. Simpson and M. Zimmermann (eds.), *The Economy of Green Cities: A World Compendium on the Green Urban Economy*, Local Sustainability 3,
DOI 10.1007/978-94-007-1969-9_2, © Springer Science+Business Media Dordrecht 2013

enabling framework conditions, and actors to work collaboratively together, so that all contribute to stepping up a level in their urban sustainability performance.

It is important to recognize that economic production and consumption patterns strongly interact with social and ecological systems. Societies are dependent upon the performance of their urban economies, while the urban economy is dependent on ecosystem services and resources. Urban economies rest and draw upon both local resources and cities' vast hinterlands. The resulting tensions between the human and Earth system are expressed in numerous ways for example by rising energy costs, energy or water security concerns, environmental risks like climate change, shrinking natural resource base, declining ecosystem services etc.

Such tensions between economic forces, economic benefits and socially undesirable outcomes are deeply interwoven with the choices made on infrastructure investments, production, distribution and consumption cycles, and the inclusion of environmental and social considerations therein. Choices are being made in cities on what to invest where, what to produce and distribute how, and what to consume why, as well as what to throw away where. For example, India alone is likely to invest US$300 billion in urban infrastructure over the next 20 years (Asia Economics Analyst 2007). As such, capital spent on urban development needs to be spent in a way that it takes long-term economic, environmental and social viability into account. Choices of investments, asset management, business management and governance can be utilized for more optimal and more socially desirable outcomes.

For example in the wake of the global financial crisis 2008–2010 and subsequent economic slowdown, the concept of a Green Economy was provided with fresh impetus following wide-spread discussions on a 'Green' New Deal, to enable a "green recovery". Large investments were seen as necessary to support the rccovery of the world economy. These financial investments offered an opportunity to invest in green economic sectors.

Due to the concentration of people, knowledge, infrastructures, resources and economic activities, cities offer unique opportunities to address and respond to many environmental and social challenges. Cities and urban areas, large and small, are not only the source of resource consumption and pollution, they also offer many innovative responses, solutions, and provisions.

The political and economic foundations of cities will determine for example whether four billion new urban houses in the next 40 years can be built, whether global warming can be limited to 2 °C compared to 1990 levels, and whether the state and quality of our ecosystems and resources can be sustainably maintained. Urban economic activities and their local and global environmental implications need to be rethought, reshaped and remodeled.

Part II draws together different perspectives on guiding principles, ideas and opportunities of what a Green Urban Economy can look like to further and instigate the debate amongst researchers, policy makers and practitioners.

Part II opens with Kenneth Odero tracing the notion of "green" to the Garden City movement associated with Sir Ebenezer Howard (1850–1928) and exploring important linkages and central issues of a green urban economy for an equitable and sustainable future. Following, Corina McKendry illustrates on hand of two selected post-industrial cities how the environmental and economic discourse has moved away from an antagonistic relationship towards one of mutual compatibility. An attractive and healthy environment emerges as an economic value to invest in.

Kes McCormick, Stefan Anderberg and Lena Neij explore conceptual connections between sustainable urban transformations and development for a green urban economy, and their key drivers and urban structures. It concludes with a call for further discussion and action on visioning, collaboration, sharing, learning, reconnecting and evaluation. It is deeply connected to the notion of green cities which is subsequently explored by Essam Ahmed, where achieving sustainable development can and needs to become a strategic economic policy agenda.

According to Edmundo Werna the resulting benefits to cities can transfer to its workforce especially when a net positive balance of decent jobs is created in the process. Benefits to a city's workforce can be indirect in form of safer working conditions and a healthy environment, as well as directly through the additional business opportunities that create jobs for example in the construction, waste or urban forestry and agricultural industry. It is pivotal that any green urban economy agenda is strongly situated within the context of poverty eradication and is inclusive of low-income groups and marginalized groups. Asif Kabani and Maliha Kabani highlight how for example green buildings and retrofitting need to include low-income households, who lack resources to upgrade their homes to achieve energy cost-savings.

The implications of these principle goals are detailed by Philipp Rode on hand of the resulting opportunities and challenges that exist in urban sectors and across the economic, social and environmental spheres. His contribution is a condensed version of the Cities Chapter of UNEP's seminal publication *Towards a Green Economy: Pathways to Sustainable Development and Poverty Eradication*. It underscores the need for enabling strategies and a regulatory framework for national, regional and local governments for greener cities. This chapter already provides an insight into possible strategies and instruments.

The final contributions highlight how through measurement systems the advances and performance can be measured and tracked. Vito Albino and Rosa Maria Dangelico illustrate this on hand of their Green Cities Practices Matrix. Jonathan Woetzel, Shannon Bouton and Molly Lindsay with the Urban Sustainability Index, which suggests that performance in Chinese cities are rapidly implementing a balanced approach between sustainability and growth. They suggest no negative association between economic growth and the index's sustainability measures; rather they go hand in hand.

The chapter overview and the contributions make clear that a strong focus is evolving around the notion of green cities, and the multiple benefits a greening of the urban economy and its sectors can realize. The contributions set an ambitious but also an encouraging agenda to accelerate the urban transformation towards greater sustainability. The Green Economy and therewith a Green Urban Economy is only at the beginning of an important debate geared towards practice and action. The varying perspectives make clear that the economy and its urban sectors need to reduce greenhouse gas emissions, become more resource efficient, grow or transform in a way that is inclusive, equitable and creates jobs.

In the process of such an urban economic transformation or growth it can and has to realize a number of social and environmental objectives, directly and indirectly. The outlook is positive not least as a changing economic make up also interacts with the way how the environment is treated and seen. An attractive and healthy urban environment and economic sectors emerge as an economic asset to invest in. The discourse and operationalization of a green urban economy can further facilitate and drive this process. Further research, documentation and exchange on principles, goals and benefits can provide the necessary stimulus to enrich the debate.

References

Asia Economics Analyst (2007) Issue no: 07/13, 6 July 2007, Goldman Sachs Economic Research

PricewaterhouseCoopers UK Economic Outlook November (2009) Available via: https://www.ukmediacentre.pwc.com/imagelibrary/downloadMedia.ashx?MediaDetailsID=1562. Cited Jan 2012

New Urban Spaces: The Emergence of Green Economies

Kenneth K. Odero

Abstract 'Green' in the context of cities has long been associated with open recreational spaces, urban forestry, ambient air quality, and similar attributes that portray a 'healthy' and arguably, sustainable urban environment. Seen as a design concept, green (as opposed to 'brown') cities epitomized a conscious and sustainable approach to urban planning. With the emergence of climate change as the defining issue at the turn of the twenty-first century, the notion of green cities has progressively taken on a new meaning underlined by an emphasis on "green growth", "low carbon development", "climate compatible development", etc. The new focus of the *green cities* concept – energy efficiency, reduction of greenhouse gas emissions, and green jobs – has transformed how we think about urbanization, especially how we (re)create and use urban spaces. Therefore, this chapter will trace the origins of this shift, explore the multifaceted dimensions of a green urban economy, and critically analyze the significance of *greening of cities* in terms of policy and practice.

Keywords Urban spaces • Green cities • Climate change • Sustainable production and consumption

1 Background

The idea of *aesthetics* and *efficiency* has been central to urban planning thought and practice. The impetus marking the foundations of urban planning was the creation of order and a healthy living environment. This was particularly true for the early settlements that sprung up around factories during Europe's *Industrial Revolution* in

K.K. Odero (✉)
Climate XL Africa, Nairobi, Kenya
e-mail: odero@climatexl.org

R. Simpson and M. Zimmermann (eds.), *The Economy of Green Cities: A World Compendium on the Green Urban Economy*, Local Sustainability 3,
DOI 10.1007/978-94-007-1969-9_3,

the eighteenth century that were largely unplanned and spontaneous. Being without municipal services such as reticulated water and sewer systems, these early urban settlements quickly turned squalid, festered with filth and diseases. To mitigate the squalor and accompanying disease outbreaks, measures were taken to plan, regulate and improve living conditions around settlements marking the birth of modern town planning. Improving the aesthetic value of such factory towns was just as important as was the need to address their unhygienic conditions. Thus, different designs were proposed and implemented, some of which turned out to be major schemes with perhaps the most prominent ones being the *Garden City of Movement* associated with Sir Ebenezer Howard (1850–1928) and *Contemporary City* by Charles-Édouard Jeanneret (1887–1965) known by his more popular name Le Corbusier see Howard (1902).

Another thread of no less significance to urban planning was theoretical analysis relating to the question of the efficiency of central places. Anchored around notions of *location* and *distance*, the relative location of places and (production) activities, as well as their *interaction* was a central focus of the works of Johann Heinrich von Thünen (1783–1850), Alfred Weber (1868–1958), Walter Christaller (1893–1969) and August Lösch (1906–1945), among others. The central focus of their analyses was to answer the question 'where' and 'why' specific economic activities are taking place in a given spatial system by taking into account spatial demand and supply, spatial pricing and output, choice of location, land use and spatial equilibrium of production probably the other references go here. In spite of their predictive capability, in general models resulting from this rich discourse did not necessarily lead to the development of efficient cities other than in the narrow transport cost saving sense. From a purely ecological perspective, the footprints of most cities built after the Second World War are unsustainable giving rise to the contemporary concerns about city greening. The latter should however not be mistaken for the aesthetic characterization of garden cities contemplated by Howard and Le Corbusier even though there are important overlaps worth mentioning. For example, the compactness of their designs and necessity of reducing greenhouse gas emissions in an ecologically constrained world.

This relationship is by no means trivial. Although rooted within the positivist conception, locational analysis might still be relevant today from a normative perspective when the greatest challenge facing the world is how to plan and build green cities that are efficient. For this to happen, many of the assumptions made in locational analysis may have to be relaxed sufficiently taking into account the risks associated with a variable and changing climate. This way we have a better chance of changing individual and collective production and consumption behavior patterns sufficiently to realize green urban economies. This is the core agenda of developing new urban spaces in the twenty-first century.

2 The Central Urban Question

There is no doubt that the threat posed by climate change will continue to bear heavily on decisions made by firms and households going forward. Given that the rest of the world that is not already urban is rapidly urbanizing, by 2025 almost two-thirds of the world's population will be living in urban settlements. According to figures from the Population Reference Bureau (PRB 2011), the world is adding the largest numbers to its population than in any time in history. Despite the fact that the annual population growth rate has declined to 1.2% per year, today world population grows by about 83 million annually.

Current production and consumption patterns that are highly skewed and dependent on fossil fuel are simply not sustainable. Although cities are the engines of growth and are likely to be so at least for the foreseeable future, such a role by definition is not infinite due to the unsustainability embedded in our production and consumption practices. The implication is straightforward: current practices must change in order to preserve the environment. In other words, we must literally evolve to a *green culture* that is aligned with the limits of our planet's natural resource base. While there are encouraging signs that this is happening in some cities, the pace is nowhere close to what might be considered adequate to save the Earth. Similarly, there is a wide gap between what major industrialized countries have pledged in terms of emission reduction and their current levels of emission. Paradoxically, it is this state of affairs that has given rise to green economics as an area of focus. But what is the connection between cities, the economy and environmental sustainability that make the green urban economy such a central issue?

3 The World's Ecological Footprint

According to the Global Footprint Network (2011), presently humanity uses the equivalent of 1.5 planets to provide the resources we use and absorb our waste. This means it now takes the Earth 1 year and 6 months to regenerate what we use in a year. The result is collapsing fisheries, diminishing forest cover and depletion of fresh water systems, the vital ecosystem services that life and our very own existence depends on. Accordingly, the world is running an *ecological deficit* and this is not only unsustainable, but it also clearly shows that there are limits to growth and that we need another growth pattern that is smatter, eco-friendly and that utilizes green technology.

4 From Rhetoric to Action

In as much as there is plenty of evidence of unsustainable practices, equally, there is no shortage of policy recommendations and inaction. From the 1972 *Stockholm Declaration* of the UN Conference on the Human Environment, through to the 1992 UN Conference on Environment and Development (Rio Earth Summit that came up with the Rio Declaration and *Agenda 21*) and on to the World Summit on Sustainable Development in 2002 that proposed the *Johannesburg Plan of Implementation,* there remains a substantial gap between intentions and accomplishments suggesting things must be done differently to reverse the trend. There is nothing inevitable about the circumstances we are in today. It is by and large the result of bad choices and bad instincts. Hence, in designing cities for the future we must shift from asking *where* (to borrow from locational analysis) economic activities are taking place, to *what* kinds of activities and towards what end? Urban research, policy and design must address itself to the fundamental questions about *what* (product) and *how* (technology) to produce.

5 Information Not Enough

The creation of new urban spaces (i.e. green urban economies) should focus on institutional design by which is meant rules and regulations to structure production and consumption of goods and services. To achieve both social and economic development as well as environmental conservation, major shifts in the way in which economic institutions allocate (natural) resources, incentives *and* information are required. Without meaningful institutional innovation, green urban economies risk failing to address the central problem of ecological debt. Payment for environmental services by individuals and corporations, for example, must now become part of the menu for guiding investors and investments in *natural* capital and infrastructure. Very often the natural infrastructure upon which all humans and civilizations rely, be it for provisioning such as food, fuel, and fresh water; for regulating such as climate stabilization, water purification and air quality; or supporting (e.g. soil formation and primary production), are undervalued if not entirely forgotten. This neglect led to 'brown' as opposed to *green* cities.

6 Beyond Green Economics

Can cities that currently spend circa 10% of their gross domestic product on servicing their annual energy bill cut down their energy consumption sufficiently to a scale that it makes a difference on atmospheric carbon concentration? By posing the question thus we are forced to think beyond dollars and cents. Instead we start to ask harder questions about not just investments and jobs, but how we can create a future that is both *equitable* and *sustainable*. Concerns about squalor and the quality of life inspired early generations of philanthropists and designers (e.g. E. Howard and Le Corbusier). In modern times, in addition to ecologically unsustainable urbanization we are also faced with the contagious phenomenon of the 'Occupy Movement' directed against economic and social inequality particularly in cities. It would be an absurdity to try to build new urban spaces without addressing the grievances that fuel the now commonplace protests. We must consciously formulate a broader agenda on sustainability that is inclusive and progressive.

7 No Quick Fixes

There is a real danger that the world might be getting preoccupied with 'techno-fixes'. Technologies, some which have not been tried and tested, are quickly and rather easily finding their way into markets. While there may not be adequate time for these 'technologies' to go through their life cycle to maturity, given the reality of climate change and its unavoidable impacts, there is even a greater risk of maladaptation if diligence is not exercised. Considering that the world is faced with a double crisis of urbanization and climate change like never before, there might be no second chance to correct 'mistakes' once made.

8 Conclusion

Greening the urban economy is a modern imperative. The *Garden City Movement* associated with Ebenezer Howard and other historical antecedents provide useful lessons for both policy and practice. For example, by asking the right questions and modifying some of the underlying assumptions about the relationship between cities, the economy and environmental sustainability, locational models and analyses can be judiciously applied taking into account the ecological footprint of cities. Institutional analysis and innovation can also play a role in integrating payment of ecosystem services as one way of trying to settle our ecological debt. But, there are no quick fixes to the globalization and climate crises. Social and economic development must go hand-in-hand with environmental conservation.

References

GFN (2011) Global Footprint Network website. Available via: http://www.footprintnetwork.org/en/index.php/GFN/page/world_footprint/. Cited 9 Jan 2011

Howard E (1902) Garden cities of tomorrow. Sonnenschein & Co., Ltd., London

PRB (2011) 2011 World Population Data Sheet. Population Reference Bureau, Washington, DC

Environmental Discourse and Economic Growth in the Greening of Postindustrial Cities

Corina McKendry

Abstract The main theme of this chapter is how the physical greening of postindustrial cities in the form of urban parks, the reclamation of water features for recreation and wildlife, and enhancement of pedestrian spaces has been critical to changing how people understand the relationship between cities, nature, and economic growth. For over a century cities were seen as antithetical to nature, and environmental protection was assumed to be a barrier to economic success. This assumption has changed significantly in the last several years, and the reintroduction of nature into city centers has played an important role in this change. This chapter argues that the physical greening of formerly industrial cities helped to transform the common sense understanding of the relationship between cities, nature and economic growth from contradictory to mutually beneficial. In turn, this has enabled broader and less visible sustainability initiatives to be undertaken based on the argument that they will help to promote competitiveness and growth in formerly industrial spaces. The recent history of Chicago, Illinois and Birmingham, England are used to illustrate this argument.

Keywords Ecological modernization • Environmental discourse • Natural amenities • Urban entrepreneurialism

1 Introduction: Changing Perceptions

Since the early days of the industrial revolution, the pollution, disease, and grime that defined industrialized cities established them in the public imagination as the antithesis of nature, and much of urban planning was focused on separating

C. McKendry (✉)
Political Science Department, Colorado College, Colorado Springs, CO, USA
e-mail: Corina.McKendry@ColoradoCollege.edu

R. Simpson and M. Zimmermann (eds.), *The Economy of Green Cities: A World Compendium on the Green Urban Economy*, Local Sustainability 3,
DOI 10.1007/978-94-007-1969-9_4,

people from the environmental ills of the industrial city (i.e. Howard 1965 [1902]; see also Ward 1998). The public health movement at the turn of the last century, the environmental movement, and improved regulation on pollution in the middle of the twentieth century alleviated many of the most egregious urban environmental problems. However, the belief that cities were antithetical to nature maintained its stronghold. Exacerbating this perceived duality between cities and nature was how city leaders prioritized the use of urban space. Economic growth was their primary focus, and economic growth meant having a substantial industrial base. The infrastructure to service industry therefore received top priority in the construction of the built environment of the city. Furthermore, as there was a general assumption that environmental protection and economic growth were incompatible, local governments were seen as too weak and too much at the mercy of entrenched (polluting) industrial interests to undertake significant environmental protection initiatives (Andrews 2006; Dryzek et al. 2002; Portney 2003; Vogel 2003).

This assumption about the incompatibility of cities, the environment, and economic growth has changed significantly in the last several years, of course, and many of the world's most economically vibrant cities are also taking on leadership roles in addressing climate change and other global environmental issues. Examining the recent history of Chicago, Illinois and Birmingham, England, this chapter argues that the reintroduction of nature into city centers played an important role in this change. In the 1990s deindustrializing cities such as Chicago and Birmingham began using physical greening of formerly industrial spaces as a way to spur economic revitalization. In so doing, they have helped to transform the common sense understanding of the relationship between cities, nature and economic growth from contradictory to mutually beneficial. This, then, has enabled broader, less visible sustainability initiatives to be undertaken based on the argument that they will help to promote competitiveness and growth in formerly industrial cities.

2 Emergence of an Urban Understanding of Economic and Environmental Compatibilities

In Birmingham and Chicago, as in many deindustrializing cities of the global North, the impetus for the physical greening of the city was the changing global economy. As globalization led industries to abandon their former strongholds, local governments worked to transform their cities to be competitive in the new knowledge-based economy (Hall and Hubbard 1998; Harvey 1989). Greening was undertaken as part of this entrepreneurial development agenda (Jonas and While 2007; McKendry 2011). As such, economic growth was the focus of city greening, not contributing to environmental progress per se. However, as stated above, this established the groundwork for the broader urban sustainability movement, still founded on the assumption that such efforts would have economic as well as environmental benefits.

The continuity of this progression from physical greening to sustainability can be seen most clearly in city leaders' use of ecological modernization discourse.

Ecological modernization assumes that environmental improvements can be compatible with, and even beneficial for, economic growth (Eckersley 2004; Hajer 1995; Mol and Spaargaren 2002). As will be seen below in the cases of Birmingham and Chicago, this discursive framing first emerged as justification for extensive public investment in the physical greening of the city. It continued to be used to explain and build support for less visible sustainability initiatives such as efforts to address climate change.

Birmingham and Chicago offer excellent illustrations of this process. Both cities were former industrial powerhouses whose economies suffered significantly in the last decades of the twentieth century. Both cities undertook extensive physical greening programs with the goal of spurring economic growth, and in each of these cities the use of ecological modernization discourse has spread from physical greening to efforts to establish the cities as leaders in urban sustainability. Tracing the continuity of this discourse in Birmingham and Chicago will help illustrate this important shift in how the relationship between the economy and the environment is understood in cities.

2.1 *Birmingham, England, United Kingdom*

In the decades after World War II Birmingham's economy boomed, its success based largely on the auto industry. During this time the environment of the city reflected the primacy of the car to the local economy. The city center, which was severely damaged by bombing during the war, was rebuilt in service of the automobile (Hendriks 1999; Kennedy 2004). Huge roads were driven through the city, destroying many neighborhoods and small industries and cutting the city center off from surrounding areas. The most notable of these was the Inner Ring Road that circled the city center. As these roads were being constructed, pedestrians were seen as an impediment to vehicles, and dozens of pedestrian underpasses (called "subways") were built beneath the new roads to minimize pedestrian disruption to the flow of traffic (Hendriks 1999). The new freeways were complemented by enclosed, concrete shopping malls and the destruction of much downtown housing. Between the malls, Ring Road, and subways, the latter of which were unappealing if not downright dangerous, downtown became virtually inaccessible to pedestrians, turning "central Birmingham [into] a diesel-choked, economic wasteland" (Flatman 2008, p. 28).

The emphasis on building Birmingham for cars was a reflection of the auto-based economy that dominated the city at the time. By the 1980s, however, Birmingham's economy had gone into a tailspin with the collapse of the British auto industry. As the decade progressed, it became clear to city leaders that manufacturing was no longer the economic driver it once had been and that Birmingham would need to be transformed yet again. The 1985 *Economic Strategy for Birmingham* outlined two principal tactics for job creation and economic growth: to promote business tourism through becoming a major conference and exhibition center and to attract knowledge workers and investment by "the creation of a new, international, city image" through the construction of a "high-quality city-centre environment" (Hajer 1995, p. 246).

As intended, the creation of this new city center significantly transformed the physical environment of Birmingham. The Inner Ring Road, built with such gusto a few decades earlier, was torn down. The city center was pedestrianized with numerous public plazas filled with trees, fountains, public art, and green space. These areas connected the main tourist and shopping attractions of the city center while making downtown an enjoyable "place in which people are happy to walk and explore," (Flatman 2008, p. 38; see also Holyoak 2004), a striking contrast to the city center of only a few decades earlier. The plazas and green spaces were accompanied by the creation of Birmingham's most striking environmental amenity: the transformation of the abandoned industrial canals that cut through the city into sites of consumption, leisure, and nature (see Birmingham City Council and British Waterways 2002; Webster 2003).

Since this time, Birmingham has succeeded in becoming a major European convention and tourist destination, and, as city leaders had hoped, Birmingham's environmental amenities have been a crucial component to the success of these new industries. In the early 1990s, Birmingham became the destination for more day visitors than any other UK city (Martin and Pearce 1992) and some 32 million visitors come to Birmingham each year, a number that has been steadily growing (Marketing Birmingham 2009a). A recent marketing survey found that the canals are visitors' favorite feature of the city center (Card Group 2007). A comment made by Bill Clinton after visiting Birmingham illustrates the successful transformation from "economic wasteland" to vibrant city center. After attending a G8 meeting at the new downtown convention center in 1998, Clinton said of the city, "I was astonished when I saw how beautiful Birmingham was. The buildings, the art, the use of water: it is an extraordinary jewel of a city… I was bowled over when I was there. It is quite wonderful" (quoted in University of Birmingham n.d.). For city leaders, the goal of the reintroduction of nature into Birmingham was, quite explicitly, economic growth (Hubbard 1995).

2.2 *Chicago, Illinois, United States*

Throughout most of its history Chicago was by no means known for its beauty or its green spaces. Rather, it was famous for the putrid stench of the Chicago River, contaminated with human excrement and waste from slaughterhouses, steel mills and other heavy industry, for Al Capone, and for patronage politics. Being known as a rough, industrial city is fine as long as industry is providing a steady source of income and employment. But Chicago's image problem, exacerbated by racial tensions and white flight, came to be a serious issue when the industries that had been supporting much of the Chicago economy collapsed in the last decades of the twentieth century. The loss of industry escalated the pressure on Chicago leaders to find new ways to attract investment and rebuild the city to succeed in the postindustrial global economy. Greening was soon to become a part of this transformation, contributing significantly to the city's postindustrial rebirth.

The greening of Chicago as an economic development strategy can be traced back to 1993 when the City of Chicago, in partnership with the Park District and Cook County Forest Preserve District, developed a plan for the expansion of parks and green space. The major document issued by this collaboration, entitled *CitySpace*, was released in 1998. One of the principle claims of the report was that the provision of parks, trails, and other such natural amenities would be necessary for attracting the business and skilled labor necessary to reinvigorate the city's ailing economy (City of Chicago et al. 1998). Within 10 years most of the particular projects *CitySpace* recommended had been implemented. The most important of these was the construction of the 24.5 acre, $450 million Millennium Park. Technically the largest green roof in the world, the park connected downtown to the lakefront, covering land that had long been held by railroads, once the drivers of Chicago's economic centrality (Gilfoyle 2006; see also Cronon 1991).

Other projects inspired by the report included the development of inland waterways (particularly the Chicago River) for recreation and the reclamation of vacant lots for community parks. As the report predicted, this greening of Chicago made an important contribution to the city's success as a postindustrial, global metropolis (Economist 2002; Sassen 2004). Property values around Millennium Park skyrocketed, major industries such as Boeing moved their headquarters to Chicago, and Chicago was heralded by the media as "the green star by which aspiring cities sail" (Gordon 2007).

3 Nature, Cities, and Economic Growth

The physical greening of these two cities certainly had environmental merit in terms of increased vegetation and its associated benefits, reduced auto use, etc. But throughout the greening of these cities, the discourse of ecological modernization was the dominant theme in city leaders' justifications of these investments in a greener city. It was the important *economic* benefits of these changes, including attracting tourists, businesses, conferences, and middle class residents that were highlighted and prioritized. To reiterate, ecological modernization's claim that economic growth and environmental protection are mutually beneficial represented a significant change from earlier assumptions about the relationship between cities and nature. Particularly at the height of their industrial eras, both cities were enthusiastically dedicated to heavy industry and concrete. Nature and the environment were, at best, irrelevant to the imperative of economic growth. Shortly after the physical greening of Birmingham and Chicago came to be justified by its economic merits, however, the language and assumptions of ecological modernization emerged to promote city sustainability efforts that moved beyond physical greening.

Birmingham, for example, has declared itself the birthplace of the "low carbon revolution" and has proclaimed its goal of becoming "the first sustainable Global

City" (Birmingham City Council 2010, p. 23) by becoming a leader in efforts to address climate change. Climate change mitigation as a development strategy emerged in 2006 when a coalition of local political and business leaders developed a climate change action strategy for the city. Its goal was to "bring to Birmingham businesses that [would] contribute both to the fight against global climate change and help develop the city as a carbon-neutral centre of excellence" (Birmingham Post 2006). To do so, the city council has invested significant resources into developing an extensive network of low-emission combined heat and energy generation plants (CHP), encouraging the production and use of electric vehicles, and installing solar panels on council owned housing.

CHP has been one of the most notable successes of Birmingham's sustainability efforts (Birmingham Evening Mail 2008). Though also touted for its greenhouse gas reduction benefits, CHP seems to have been initially established as a way to save money on the council's substantial energy bill and to help reduce the cost of energy to city center businesses. City leaders saw the potential competitive edge that the provision of inexpensive power could give to Birmingham. As one city official working on sustainability in Birmingham explained, "If we can provide a site with guaranteed cheap energy, then [we'll] attract businesses into Birmingham. And if [we] do that before Manchester, or before London, then [we] attract inward investment and that's creating jobs, that's putting [us] at the center of the low carbon" economy (Be Birmingham staff member, personal communication, July 21, 2008). The compatibility of economic competitiveness and environmentally beneficial development had become clearly established in the minds of Birmingham leaders.

Similarly, throughout the latter years of his long tenure, Chicago's mayor Richard M. Daley frequently proclaimed his goal of making Chicago the greenest city in the country. As so clearly articulated in the *CitySpace* report, Chicago's goal of becoming the greenest city in America began unabashedly as an economic goal. This continued as the city moved towards addressing climate change and other broader environmental issues, and the discourse of ecological modernization came to saturate city initiatives. One place this could be seen was in the mission statement of the Chicago Department of Environment (DOE). Under Daley, the DOE's mission was "to protect human health and the environment, improve the urban quality of life, and *promote economic development*" (Chicago Department of Environment n.d., emphasis added). Excerpts from press releases from May 2008 through December 2010 also illustrate the extent to which the economic benefits of environmental programs were emphasized by the Chicago government:

> 'Strategically investing in recycling is both an environmentally and economically sound decision. Programs like Chicago's Blue Cart initiative are helping to turn what could have become our trash into a reusable commodity that is spurring private investment and putting more and more people to work.' (City of Chicago 2008)

> The 'Green Office Challenge' and 'Earth Hour' [are] things that can be done in difficult economic times to help save companies and residents money, enhance the quality of life of residents and position the city for future economic growth and prosperity. (City of Chicago 2009)

> 'In the last year, we've taken many steps that show it is possible to implement responsible environmental policies that also create economic growth and protect taxpayers,' [Daley]

> said. He said one of those steps has been the implementation of the Chicago Climate Action Plan …'The benefits of this plan go beyond the important goal of improving the environment. Implementing the plan will save companies and residents money, enhance our quality of life and position Chicago for future economic growth,' Daley said. (City of Chicago 2010)

Undoubtedly, for Chicago officials, like their Birmingham counterparts, the success of their environmental programs is "defined in economic terms – whether it's reducing the cost of city operations or increasing the tax base by attracting new businesses and the residents who will work there" (Davies 2007; see also Saulny 2007). Yet the wide embrace of ecological modernization at the local level has enabled the implementation of a number of valuable programs that may reduce the environmental impact of these urban areas.

4 Conclusion

It is relatively easy to see the economic benefits of a greener city, particularly as environmental amenities such as parks, trees, and green space increase property values, generate tourist revenue, and benefit local businesses. Indeed, it is notable that the provision of such amenities was justified by city leaders in both Chicago and Birmingham not principally for their general benefits to quality of life, let alone to the environment, but as strategies for economic revitalization. This continued as these cities began to address climate change. Despite growing public concern about the environment, economic growth remains a necessity for the maintenance of local state legitimacy.

It is therefore not surprising that city leaders have justified their sustainability efforts largely in terms of their economic benefits. But there was nothing inevitable about this discursive shift to ecological modernization on the city level. Though the Obama administration has asserted that a greening of the economy will pull the country out of recession, the political resistance the administration has encountered has made it clear that ecological modernization has not become national policy in the United States. In contrast, in the UK the Stern Review and subsequent government policies have argued for the economic necessity and long-term benefits of addressing environmental concerns, particularly climate change.

Ecological modernization is clearly more widely accepted in the UK. The ability of UK cities such as Birmingham to go farther than most US cities in addressing climate change can partly be attributed to the national acceptance of the economic necessity of doing so. But despite these national differences, the use of ecological modernization discourse by city governments initially emerged from the need to physically reinvent the city by incorporating nature into the built environment. The benefits of this greening were clear to many. The benefits of climate adaptation and mitigation are harder to see, yet they offer a continuation of the discourse of green economic growth that have marked the environmental transformation of postindustrial cities.

References

Andrews RNL (2006) Managing the environment, managing ourselves: a history of American environmental policy, 2nd edn. Yale University Press, New Haven, CT
Birmingham City Council (2010) Making Birmingham Green: Birmingham Climate Change Action Plan 2010+, Birmingham
Birmingham City Council, British Waterways (2002) Holliday Wharf to Warwick Bar City Centre Canal Corridor Development Framework, Birmingham
Birmingham Evening Mail (2008) How Brum's fuelling a greener future, 20 May. Available via: http://www.lexisnexis.com/hottopics/lnacademic/? Cited 17 Jan 2011
Birmingham Post (2006) Vision for city reputation as a global leader, 25 November, p 3. Available via: http://www.lexisnexis.com/hottopics/lnacademic/? Cited 17 Jan 2011
Card Group (2007) Birmingham Visitor Survey: A Review of Visitors to the City. Birmingham, UK.
Chicago Department of Environment (n.d.) Welcome to DOE. Available via: http://egov.cityofchicago.org/city/webportal/portalEntityHomeAction.do?BV_SessionID=@@@@0140960301.1205358508@@@@&BV_EngineID=cccfadedighjjlecefecelldffhdfgk.0&entityName=Environment&entityNameEnumValue=13. Cited 12 Mar 2008
City of Chicago (2008) Press release: city shifting to Blue Cart Recycling program by end of 2011
City of Chicago (2009) Press release: Daley Repeats Call for Federal, State Action to Rebuild Infrastructure
City of Chicago (2010) Press release: Mayor Daley Marks 40th Anniversary of Earth Day
City of Chicago, Chicago Park District, Forest Preserve District of Cook County (1998) CitySpace: On Open Space Plan for Chicago. Available via: http://www.cityofchicago.org/content/dam/city/depts/zlup/Sustainable_Development/Publications/CitySpace/CitySpace1a.pdf
Cronon W (1991) Nature's metropolis: Chicago and the Great West, 1st edn. W. W. Norton, New York
Davies J (2007) A tale of four cities. AMR Research. Available via: http://www.amrresearch.com/Content/View.asp?pmillid=20932. Accessed 20 Dec 2001
Dryzek JS, Hunold C, Schlosberg D, Downes D, Hernes H (2002) Environmental transformation of the state: the USA, Norway, Germany and the UK. Polit Stud 50:659–682
Eckersley R (2004) The green state: rethinking democracy and sovereignty. MIT Press, Cambridge, MA
Economist (2002) Green machine. The Economist, 15 August
Flatman B (2008) Birmingham: shaping the city. RIBA Publishing, Birmingham
Gilfoyle TJ (2006) Millennium park: creating a Chicago landmark. University of Chicago Press, Chicago
Gordon J (2007) Is your city green? MSN City Guides. Available via: http://cityguides.msn.com/citylife/greenarticle.aspx?cp-documentid=4848590. Cited 3 Oct 2007
Hajer MA (1995) The politics of environmental discourse: ecological modernization and the policy process. Oxford University Press, New York
Hall T, Hubbard P (eds) (1998) The entrepreneurial city: geographies of politics, regime, and representation. Wiley, Chichester
Harvey D (1989) From managerialism to entrepreneurialism: the transformation in urban governance in late capitalism Geografiska Annaler, Series B. Hum Geog 71(1):3–17
Hendriks F (1999) Public policy and political institutions: the role of culture in traffic policy. Edward Elgar, Cheltenham
Holyoak J (2004) Street, subway, and mall: spatial politics in the Bull Ring. In: Kennedy L (ed) Remaking Birmingham: the visual culture of urban regeneration. Routledge, New York, pp 13–24
Howard E (1965 [1902]) Garden cities of to-morrow. MIT Press, Cambridge, MA
Hubbard P (1995) Urban design and local economic development: a case study in Birmingham. Cities 12(4):243–251
Jonas AEG, While A (2007) Greening the entrepreneurial city? Looking for spaces of sustainability politics in the competitive city. In: Krueger R, Gibbs D (eds) The sustainable development

paradox: urban political economy in the United States and Europe. Guilford Press, New York, pp 123–159

Kennedy L (2004) The creative destruction of Birmingham. In: Kennedy L (ed) Remaking Birmingham: the visual culture of urban regeneration. Routledge, New York, pp 1–10

Marketing Birmingham (2009a) Birmingham Visitor Economy Overview. Birmingham, UK

Martin S, Pearce G (1992) Policy Review: The internationalization of local authority economic development strategies: Birmingham in the 1980s. Regional Studies, 26(5):499–503

McKendry C (2011) Smokestacks to green roofs: city environmentalism, green urban entrepreneurialism, and the regulation of the postindustrial city. PhD dissertation, University of California, Santa Cruz

Mol APJ, Spaargaren G (2002) Ecological modernization and the environmental state. In: Mol APJ, Buttel FH (eds) The environmental state under pressure. JAI, Amsterdam, pp 33–52

Portney KE (2003) Taking sustainable cities seriously: economic development, the environment, and quality of life in American cities. MIT Press, Cambridge, MA

Sassen S (2004) A global city. In: Madigan C (ed) Global Chicago. University of Illinois Press, Urbana, pp xii, 223

Saulny S (2007) In miles of alleys, Chicago finds its next environmental Frontier New York Times, 26 November. Available via: http://www.nytimes.com/2007/11/26/us/26chicago.html?scp=1&sq=Chicago+miles+of+alleys&st=nyt. Cited 23 Apr 2008

University of Birmingham (n.d.) Birmingham's global profile. Available via: http://www.birmingham.ac.uk/university/about/global.aspx. Cited 10 Feb 2011

Vogel D (2003 [1986]) National styles of business regulation: a case study of environmental protection. Beard Books, Washington, DC

Ward SV (1998) Selling places: the marketing and promotion of towns and cities, 1850–2000. E & FN Spon, London

Webster F (2003) Reinventing Birmingham, England, in a globalized information economy. In: Breton R, Reitz JG (eds) Globalization and society: processes of differentiation examined. Praeger, Westport, CT, pp 185–199

Sustainable Urban Transformation and the Green Urban Economy

Kes McCormick, Stefan Anderberg, and Lena Neij

Abstract This chapter explores the connections between the concepts of sustainable urban transformation and the green urban economy, proposes a framework for understanding how these concepts 'fit' together, and makes some practical suggestions for local governments and for national and international policy.

Keywords Sustainable urban transformation • Green urban economy • Sustainable development • Green economy

1 Introduction

The concepts of sustainable urban transformation and the green urban economy have emerged as key themes on the international political and economic agenda in recognition of the increasingly important role of cities in global development (UNEP 2011). In fact, the twenty-first century has been called the 'urban century' (UN-Habitat 2008). The purpose of this chapter is to explore the relationship between sustainable urban transformation and the green urban economy. What do these concepts actually mean and how do they relate to each other? An initial observation is that the green urban economy and sustainable urban transformation are overlapping concepts. However, this chapter suggests that the green urban economy

K. McCormick (✉) • L. Neij
International Institute for Industrial Environmental Economics,
Lund University, Lund, Sweden
e-mail: kes.mccormick@iiiee.lu.se; lena.neij@iiiee.lu.se

S. Anderberg
Lund University Centre for Sustainability Studies, Lund, Sweden
e-mail: stefan.anderberg@lucsus.lu.se

R. Simpson and M. Zimmermann (eds.), *The Economy of Green Cities: A World Compendium on the Green Urban Economy*, Local Sustainability 3,
DOI 10.1007/978-94-007-1969-9_5,

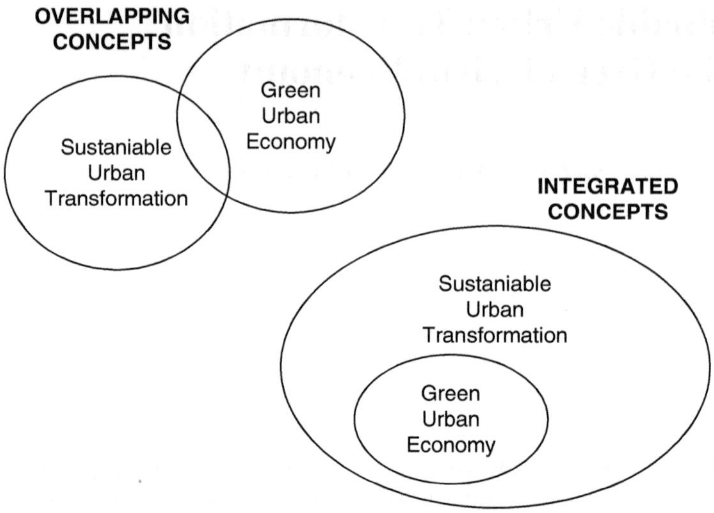

Fig. 1 This figure shows different ways to understand the relationship between sustainable urban transformation and the green urban economy – depicted as either overlapping concepts or integrated concepts

can be considered as an integrated part of the broader transformation of cities and urban areas towards sustainability (see Fig. 1). Furthermore, this chapter presents a basic framework for understanding sustainable urban transformation, which involves two interconnected dimensions – drivers of change and sustainable urban structures. This chapter concludes with reflections and 'food for thought' on advancing sustainable urban transformation and the green urban economy.

2 Background and Discussion

The strategic importance of cities in relation to sustainable development and the green economy has been increasingly recognized (UN-Habitat 2008; UNEP 2011). After two centuries of urbanization spreading around the world, the majority of the global population currently lives in urban areas, and urban centers will continue to grow. Cities play a dominant role in global consumption, production and pollution (Sukhdev 2009). The importance of cities is generally expected to increase due to the role of metropolitan areas as growth centers of the emerging globalizing service economy. For this reason, policies formulated by international bodies and national governments need to be implemented at the community and city level (Roseland 1997). The local level has therefore been identified as a key for sustainable development and there is a general agreement that effective and integrated solutions can only be found and efficiently implemented at the local level (UN-Habitat 2010; ICLEI 2011; Wheeler and Beatley 2010).

Cities are often associated with social and economic problems such as poverty and segregation, tensions between different groups, and economic vulnerability, as well as ecological problems related to pollution, resource use, congestion and spatial competition (Legner and Lilja 2010). They are also associated with economic and cultural wealth, and a dynamic development that can provide opportunities for technological and social innovation (Sukhdev 2009). The concentration of population, activities and resource use in cities bring potentials for important efficiency increases as well as for multi-purpose solutions combining different sustainability goals. New urban technology and infrastructure may also be replicable or useful in urban areas in different regions, as has historically been the case with district heating, wastewater treatment, and public transport systems (Wheeler and Beatley 2010). Larger cities also often have particular weight through their consumption, head office functions, or cultural influence. In particular, the populations in major cities can play an important role for developing new consumer cultures and attitudes.

Around the world cities have very different starting points and conditions for sustainable development or the green economy (Yang 2010; Tuts and Altinger 2011). Widespread poverty, over-population, unhealthy housing conditions, inadequate infrastructure, hygienic problems, poor water quality and uncontrolled pollution are examples of problems that still dominate cities in the developing world (UN-Habitat 2010). Many of these problems however have decreased in cities in industrialized countries in Europe and North America during the twentieth century. This has been primarily due to stable and more equally distributed economic growth, improved organization, town planning, and investment in infrastructure, construction and urban renewal. A similar development has taken place in parts of Latin America and Asia. In Europe today, urban sustainability problems primarily consist of segregation and growing social tensions, local traffic problems, continuous growth of solid waste generation, and the large and often inefficient consumption of energy and resources with linkages to climate change and global environmental problems.

Importantly, urban sustainability problems are not necessary characteristics of urbanization but can rather be considered as results of poor governance and planning (Rode and Burdett 2011). There are many possibilities to improve the situation. The design of cities plays a significant role in relation to the (positive and negative) impacts of urban development as well as how urban citizens interact and live together. Development opportunities for individual cities can also vary over time. Cities are constantly influenced by diverse processes of transformation – a changing structure of population, economy, culture, lifestyles and national policies that may lead to altered urban functions and new local needs and opportunities (UN-Habitat 2010). Interpreting these different development processes, responding to related demands, and identifying and realizing opportunities are constant challenges for urban governance and planning. Cities around the world are also influenced in different ways by large scale transformation processes, such as global economic development and downturns, but the vulnerability of and opportunities for specific cities may also differ due to internal factors, such as the local economic structure as well as external relations and geographic location.

Economic transformation provides opportunities that can be innovatively used for furthering sustainable development. After a long stagnation due to an often dramatic industrial restructuring, many major cities in Europe and North America have experienced renewed growth related to development of a profitable service economy, and related revitalization of inner city and harbor areas. Revitalization of older housing or former harbor and industrial areas can be used for creating attractive city areas, realizing the enormous potential for energy savings within cities, and developing distributed energy systems. The revitalization of buildings and districts also opens up opportunities for the development of public space and social inclusion. In recent years, the approach of economic development in the urban setting has gained attention, and cities are being highlighted as successful growth engines. Furthermore, the political leadership in many cities are now actively working with strategies to increase their attractiveness and competitiveness in the context of sustainable development and the green economy (ICLEI 2011).

Nevertheless, the increased awareness on global challenges and ambitions on sustainability have not translated into powerful initiatives that are decisively shifting urban development in a sustainable direction. A complex array of reasons are responsible for this situation, including: a lack of urgency for undertaking the 'radical' changes that are needed and therefore inadequate political, business or broader social support; fragmentation in research activities as well as in urban practice and planning; limited coordination between international, national and local levels; and a significant separation between science and practice. So, while there are experiences with sustainable city initiatives and urban transformation, there are few examples where transformative change has been adequately connected to sustainability goals to realize strategic potentials. A key contribution of the concept of sustainable urban transformation is to provide a structural focus on urban sustainability efforts.

3 Definitions and Frameworks

Sustainable development is an elusive concept with diverse definitions (Koglin 2008). But it is also a concept that has captured the attention and 'imagination' of the world. Straightforwardly, it means that current development should not harm the interests of future generations. It has been commonly depicted as an integration of economic, social and environmental spheres. Furthermore, institutional and time dimensions have also become more prominent in sustainability discussions, which highlight the importance of governance and democracy as well as processes and actions over time (Waas et al. 2011). The green economy can be understood as a new way of looking at the contribution or role of economic activities to sustainable development. A critical aspect of the green economy is to recognize and measure the value of ecosystems and natural resources as well as the economic benefits of a wider perspective on human, ecological and economic capital (UNEP 2011).

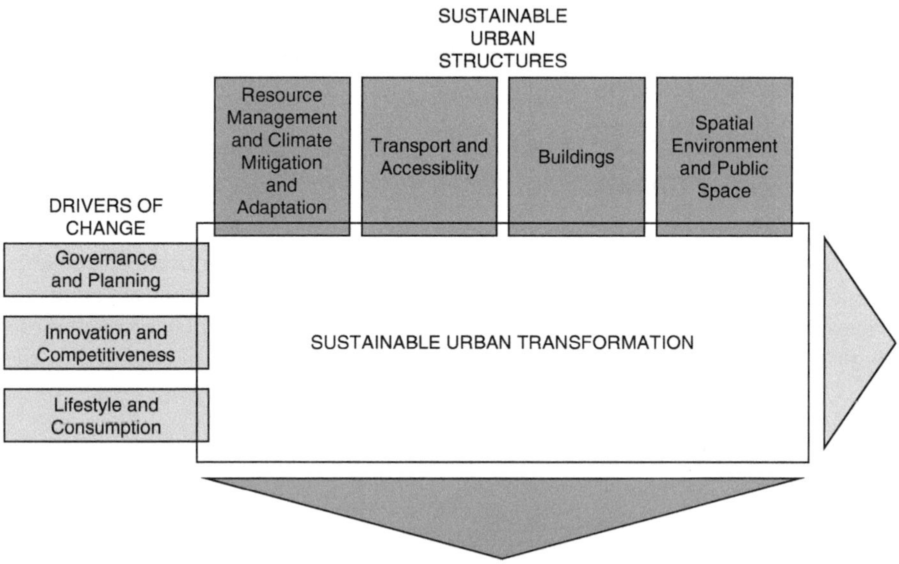

Fig. 2 This figure shows two dimensions of sustainable urban transformation – drivers of change and sustainable urban structures

The concept of sustainable urban transformation places the emphasis on understanding cities as a source of possibilities for sustainability, promoting active collaboration among diverse stakeholders, and integrating different perspectives and bodies of knowledge and expertise. A green urban economy can be defined as realizing "opportunities to enhance human well-being and local natural resources, while reducing future costs, ecological scarcities and environmental risks" (ICLEI 2011). As discussed in this chapter, the green urban economy and sustainable urban transformation are intimately connected and can be considered as integrated concepts. Furthermore, sustainable urban transformation can be defined in two dimensions – drivers of change and sustainable urban structures (see Fig. 2). There are two immediate points to make about this basic framework. First, it is the interactions between the different elements of the framework that is particularly important but this is difficult to represent in a diagram. Second, the green urban economy permeates through this framework.

It is important to differentiate between sustainable urban development and sustainable urban transformation. This is not simply a matter of semantics. Camagni (1998) provides a constructive definition of sustainable urban development as "a process of synergistic integration and co-evolution among great subsystems making up a city (economic, social, physical and environmental), which guarantees the local population a non-decreasing level of well-being in the long-term, without compromising the possibilities of development of surrounding areas and contributing by this towards reducing the harmful effects of development on the biosphere." The emerging concept of sustainable urban transformation places a stronger emphasis on structural transformation processes – broad, multi-dimensional and radical change – that can effectively direct urban development towards sustainability. Sustainable urban transformation can therefore be considered an evolution of sustainable urban development.

3.1 Drivers of Change

The drivers of change (or processes) depicted within the framework for sustainable urban transformation encompass governance and planning, innovation and competitiveness, and lifestyle and consumption. Below are some short explanations of these key drivers and their relationships with each other.

3.1.1 Governance and Planning

For achieving ambitious targets for sustainable cities, there is a need to analyze and practice different strategies including effective strategic planning and integration of policy instruments. Such efforts should be interconnected across sectors and be adapted for specific urban and national policy conditions to ensure empowerment, engagement and collaboration of relevant stakeholders. Bugliarello (2010) identifies three key policy challenges: policies must be ambitious but politically and economically realistic in deciding on appropriate balances; policies must be developed quickly and with flexibility for rapidly changing urban conditions; and it is imperative to eliminate contradictory policies.

3.1.2 Innovation and Competitiveness

There are significant challenges facing cities and local municipalities with regard to reconciling economic growth and maintaining or restoring the local and global environment (Wheeler and Beatley 2010). Innovation and clean technology are considered as necessary for not only developing a green economy but also as keys to fostering urban competitiveness in a globalizing economy. Therefore sustainable urban economic development must encourage symbiotic relationships among industries, governments, universities and citizens to ensure sustainable management of human, ecological and economic capital, and turn density and urban systems into eco-efficiency (Simpson 2010).

3.1.3 Lifestyle and Consumption

Research related to socio-economic and cultural development in the urban setting is important and needs to be further developed to effectively support the planning and implementation of sustainable urban governance strategies. The negative implications of over-consumption are particularly evident in cities (Rode 2009). UN-Habitat

(2008) suggests that "harmony within cities hinges not only on prosperity and its attendant benefits, but on two pillars that make harmony possible: equity and sustainability". By defining an improved quality of life and creating visions of sustainable lifestyles it will be possible to outline how to design, support and govern more sustainable cities.

3.2 Sustainable Urban Structures

The sustainable urban structures highlighted within the framework for sustainable urban transformation include resource management and climate mitigation and adaptation, transport and accessibility, buildings, and spatial environment and public space. Below are some short explanations of these key structures and their relationships with each other.

3.2.1 Resource Management and Climate Mitigation and Adaptation

Sound resource management and design of urban structures that mitigate and adapt to climate change are major challenges for cities. Urban systems must be multifunctional and be able to integrate ecological, economic, recreational and aesthetic values (WWF 2010). Key areas include: shifting urban energy systems towards renewable sources; increasing energy and material efficiency; ensuring sustainable management of the quality and sufficiency of water supply; and transforming waste management into sustainable material and energy usage.

3.2.2 Transport and Accessibility

The transportation sector accounts for significant environmental and social impacts. Sustainable urban transport research and practice has focused on specific problems such as pollution, road safety and on various measures and their effects. However, in order to create sustainable mobility in the urban context, a more integrated approach is needed, which simultaneously addresses energy security, environmental and social impacts, accessibility issues, urban conditions, and equitable economic development (Sukhdev 2009).

3.2.3 Buildings

The challenge for the building and construction sector is to create affordable, attractive, comfortable and sustainable buildings, which help their occupants to mitigate contributions to climate change, utilize renewable energy, reduce excessive material consumption as well as incorporate principles of reuse, whilst adapting to changing environmental realities (Rode et al. 2011). The efficiency of the proposed strategies also requires an understanding of human behavior and consumption in the context of the built environment.

3.2.4 Spatial Environment and Public Space

Urban development planning increasingly focuses upon the spatial environment in terms of the revitalization of districts and city centers, urban public spaces and the interconnection of fragmented urban landscapes, and to develop a continuous and welcoming web of humane livability within the urban experience (UN-Habitat 2008; Roseland 1997). This encompasses preserving existing 'green' spaces (such as parks and gardens) and 'blue' features (such as ponds and canals) and integrating new 'green' and 'blue' structures into cities in innovative ways that stimulate social interactions.

4 Reflections

Sustainable urban transformation can be thought of as a 'design' problem on a grand scale. In other words, intelligently designed cities can respond to the major environmental, social and economic challenges of the twenty-first century (Rode 2009). Ultimately, achieving these goals demands a structural transformation of urban 'systems'. As stated, the transformation of cities towards sustainability should be understood as being broad, multi-dimensional and radical change that equates to a significant shift in development paths. There is clearly an increasing emphasis on the role of cities in regards to sustainable development and the green economy. This chapter concludes with six main themes to further the discussion and action on sustainable urban transformation and the green urban economy – visioning, collaborating, sharing, learning, reconnecting and evaluating.

4.1 Visioning

Visions and ideas about the future, and how to change direction and move towards sustainability are vital for mobilizing individuals and organizations (Wheeler and Beatley 2010). Importantly, visions need to focus on creating a better world that is desirable and optimistic, and cities that are livable and exciting, rather than on the threat of climate change or economic disaster. In particular, Roseland (1997) argues that local communities must be deeply engaged in defining sustainability from their local perspective. Developing long-term visions and implementing short-term actions that are consistent with achieving such visions is therefore an underlying foundation for advancing sustainable urban transformation.

4.2 Collaborating

The processes of sustainable urban transformation and a green urban economy are to a large extent still divided and fragmented, power is unevenly distributed, and public engagement is more ad-hoc than strategic. There is a strong need for new innovative approaches and methods to traditional planning that focus on interaction and engagement between academia and practitioners in a fruitful way, and involve urban citizens and the business sphere. Achieving the transformative changes that are required in cities will depend on participation of local communities in planning and decision-making (Sukhdev 2009; Roseland 1997). Shifting from such rhetoric to reality will demand significant efforts by multiple stakeholders.

4.3 Sharing

While the context of sustainable urban transformation varies considerably from country to country, and city to city, many strategies are similar (Wheeler and Beatley 2010). It is therefore imperative that knowledge of and experiences with sustainability in cities is shared and utilized effectively to further develop visions and strategies, and most importantly, implementation and action. Additionally, showcases of sustainability in cities of different sizes, types, contexts and importance deserve attention. A comprehensive and dynamic database of all such cases is needed to show lessons from positive and negative experiences.

4.4 Learning

A necessary condition for sustainable urban transformation is to develop functional knowledge exchange, communication and learning processes around key aspects of ecological, social and economic sustainability. Knowledge resides in two main forms in cities – the first is hard data stored in documents or computers, while the second is soft data stored in professional and social networks that connect to a range of stakeholders in the community, not just within universities or local governments (Campbell 2009). Improving knowledge and learning around sustainable urban transformation therefore demands both individual learning processes and learning in social and organizational contexts.

4.5 Reconnecting

A major challenge underlying any move towards sustainable urban transformation is that urban life is totally disconnected with the environmental and socio-economic systems on which it depends (Camagni 1998; Legner and Lilja 2010). There is a 'desperate' need to reconnect people and communities with the design and management of sustainable urban structures, particularly through greater participation and engagement, but also through planning and designing cities in ways that connect people and their constructed 'environments' to natural ecosystems.

4.6 Evaluating

There remains a shortage of comprehensive and comparable evaluations on activities and initiatives that aim to promote sustainability in cities and urban areas. Evaluations are essential for the verification of results and impacts, and for learning about processes of change (Walton et al. 2005). Improved understanding and learning are vital for the modification and improvement of measures and, not least, for future decisions about how to realize sustainable urban transformation. While there is a great diversity of tools applicable to the assessment of sustainability in the urban context, there is a need for frameworks that can integrate these existing tools.

5 Conclusion

In conclusion, creating sustainable urban environments, integrating various goals and initiatives, and identifying and achieving synergy effects poses important challenges in terms of developing new approaches to city development. These must be strongly based on active collaboration between stakeholders, and the integration of different perspectives and bodies of knowledge and expertise (Rode and Burdett 2011). Methods for public participation, collaboration between practitioners and researchers, and the involvement of the business sphere are all needed as well as linking strategies for competitiveness with sustainability, and strengthening processes for more systematic learning. WWF (2010) concludes that "depending on how we develop and manage our urban infrastructures during the next three decades, they could become either a force for environmental destruction or a primary source of ecological rejuvenation". Cities therefore represent both a challenge and an opportunity.

References

Bugliarello G (2010) The future of sustainability: some urgent sociotechnological challenges. Sustainability 3:351–358

Camagni R (1998) Sustainable urban development: definition and reasons for a research programme. Int J Environ Pollut 10:6–26

Campbell T (2009) Learning cities: knowledge, capacity and competitiveness. Habitat Int 33:195–201

ICLEI (2011) Green urban economy. Available via: http://www.iclei.org/
Koglin T (2008) Sustainable development in general and urban context: literature review. Lund University, Lund
Legner M, Lilja S (2010) Living cities: an anthology in urban environmental history. FORMAS, Stockholm
Rode P (2009) City making as climate policy. In: Proceedings of the urban age conference, Istanbul, Turkey, 4–6 November 2009
Rode P, Burdett R (2011) Cities: investing in energy and resource efficiency. In: UNEP. Towards a green economy: pathways to sustainable development and poverty eradication. Available via: http://unep.org/greeneconomy/
Rode P, Burdett R, Goncalves J (2011) Buildings: investing in energy and resource efficiency. In: UNEP. Towards a green economy: pathways to sustainable development and poverty eradication. Available via: http://unep.org/greeneconomy/
Roseland M (1997) Dimensions of the eco-city. Cities 14:197–202
Simpson R (2010) A green economy for cities. Available via: http://www.stakeholderforum.org/sf/outreach/
Sukhdev P (2009) Green economy for an urban age. In: Proceedings of the urban age conference, Istanbul, Turkey, 4–6 November 2009
Tuts R Altinger L (2011) Towards a green economy: promoting sustainable urban development and green infrastructure investment. In: Proceedings of the UN conference for sustainable development, New York, 7–8 March 2011
UNEP (2011) Towards a green economy: pathways to sustainable development and poverty eradication. Available via: http://unep.org/greeneconomy/
UN-Habitat (2008) State of the world's cities 2008/2009: harmonious cities. Earthscan, London
UN-Habitat (2010) State of the world's cities 2008/2009: bridging the urban divide. Earthscan, London
Waas T, Huge J, Verbruggen A, Wright T (2011) Sustainable development: a bird's eye view. Sustainability 3:1637–1661
Walton JS, El-Haram M, Castillo NH, Horner RMW, Price ADF, Hardcastle C (2005) Integrated assessment of urban sustainability. Eng Sustain 158(2):57–65
Wheeler S, Beatley T (2010) Introduction. In: Wheeler S, Beatley T (eds) Sustainable urban development reader. Routledge, New York
WWF (2010) Reinventing the city: three prerequisites for green urban infrastructures. Available via: http://www.panda.org/
Yang Y (2010) Sustainable urban transformation: driving forces, indicators and processes. ETH, Zurich

Green Cities: Benefits of Urban Sustainability

Essam Hassan Mohamed Ahmed

Abstract In the wake of the global financial crisis of 2008, the concept of a green economy was provided with fresh impetus following wide-spread discussions on a "Green" New Deal to enable a "Green Recovery". Large investments were seen as necessary to support the recovery of the world economy offering an opportunity to invest in green economy sectors. Nowadays many developed countries have adopted the Green Cities concept as strategic tool to face environmental challenges in general and climate change impacts in particular. With climate change concerns gaining mainstream attention, pressure is mounting for greenhouse gas (GHG) reduction to be taken into account during the course of urban development. From an environmental perspective, we must protect natural resources and include sustainability in development strategies.

Key strategies for achieving green cities will be to reduce consumption of resources such as water and electricity, waste management, and green buildings which use natural materials in their structure. City governments need to coordinate policies and decisions with other levels of government but more importantly need to be equipped with strategic and integrated planning capacities. Finally, governments can set a leadership example by using public procurement in the construction and management of facilities to drive the greening of the building sector.

Keywords Green cities • Green building • Urban sustainability • Green economy

E.H.M. Ahmed (✉)
Egyptian Environmental Affairs Agency (EEAA), Cairo, Egypt
e-mail: esamhasan62@yahoo.com

R. Simpson and M. Zimmermann (eds.), *The Economy of Green Cities: A World Compendium on the Green Urban Economy*, Local Sustainability 3,
DOI 10.1007/978-94-007-1969-9_6, © Springer Science+Business Media Dordrecht 2013

1 Introduction

Around the globe, countries are looking to improve local economies and the comfort and quality of life of their citizens. As a result of the global financial crisis of 2008, the concept of a green economy was provided with fresh impetus following wide-spread discussions on a "Green" New Deal, to enable a "Green Recovery". Large investments were seen as necessary to support the recovery of the world economy, offering an opportunity to invest in green economy sectors.

The Stern Review (2006) shows that "the benefits of strong, early action on climate change outweigh the costs". A scenario of 2–3°C warming by the end of the century could result in a permanent loss of up to 3% in global economic output. With warming of 5–6°C this could reach an average 5–10% loss in global GDP. Poorer countries will suffer higher losses: according to the Economics of Ecosystems and Biodiversity study "economic analysis indicates that maintaining healthy ecosystems is often the less expensive option" (TEEB 2011). At the national level the impact of environmental policies within the broader context of innovation and economic performance should be considered (Porter and van der Linde 1995). In this view government policy plays a critical role within economies to encourage innovation and growth, and is important as a means for choosing the direction of change (Stoneman 1995; Foray 2009).

For the first time in history more than half of the world's population lives in cities. Cities now account for 75% of energy consumption (UN-Habitat 2011) and 75% of carbon emissions (Clinton Foundation 2010). Congestion, pollution and poorly provisioned services affect the productivity and health of all but fall particularly hard on the urban poor. With approximately 50% of the global population now living in emerging economies (World Bank 2010) that are rapidly urbanizing and developing the need for green city planning, infrastructure, and transportation is paramount.

2 Why the Green Economy

A Green Economy is one in which growth in income and employment is driven by public and private investments that reduce carbon emissions and pollution, enhance energy and resource efficiency, and prevent the loss of biodiversity and ecosystem services. This will require a substantial increase of investments in economic sectors that build on and enhance the Earth's natural capital or reduce ecological scarcities and environmental risks. These sectors include renewable energy, low-carbon transport, energy-efficient buildings, clean technologies, improved waste management, improved freshwater provision, sustainable agriculture and forest management, and sustainable fisheries, supported by national policy reforms and the development of international policy and market infrastructure (UNEP 2010a:3).

The green economy system is also a complex phenomenon which aims to achieve a low-carbon economy, life cycle analysis, and resource efficiency. Green Economy theory also acknowledges the vulnerability of human welfare, due to the widespread application of an unsustainable model of economic development. The UNEP urges cooperative efforts to address bringing economy and environment together, under the notion that if environment is where we live, development must attempt to improve our lot within that abode. The two are inseparable (UNEP 2007a).

In its simplest terms the green economy does the following:

(a) Produce low greenhouse gas emissions;
(b) Use resources more efficiently;
(c) Continue to generate growth, income and jobs; and
(d) Observe social equity and inclusiveness.

The 'Common Steps' that transition towards a global Green Economy (UNECA 2011) requires are:

(a) To measure the monetary value of the environment and its resources, often called "ecosystem services", so that they can be treated as a form of capital ("natural capital") on the same plane as physical-technical, human and financial capital, and which can similarly be depleted, or made productive;
(b) To prove the viability and profitability of enhancing this natural capital and related small-footprint technologies as a "new engine of growth" so that it can replace "business as usual" while also satisfying social goals; and
(c) To create the enabling conditions, such as policies and market mechanisms for "public and private investments to incorporate broader environmental and social criteria" (UNECA 2011).

The Green Economy is presented here as a corrective of the crises of recent years while ensuring long-term growth.

3 Green Economies and Sustainable Urban Development

Citizens, governments and the United Nations increasingly embrace the Green Economy as new paradigm for development. The Green Economy is pro-growth and pro-jobs as well as pro-environment. While some countries have attained high levels of human development this is often at the expense of their natural resource base, the quality of their environment, and results in high levels of greenhouse gas (GHG) emissions. Proponents of sustainable urban development however seek to develop their economies along pathways of higher, more equitable growth at lower carbon, energy and resource intensity. Increasing concern about climate change and emissions has seen the costs of renewable energy technologies decreasing, while resource-intensive conventional energy technologies are becoming more expensive. As an urban center grows the range and number of the functions that it supports generally increase and this can also yield superior results when it comes to creating a comprehensive model for urban development. The typically urban conditions of proximity, density and variety deliver increased productivity for firms and help stimulate innovation and new job creation.

Moving towards a green development path is almost certainly a means for attaining welfare improvements across a society, but it is also often a means for attaining future growth improvement. This is because a shift away from basic production modes of development based on extraction and consumption and towards more complex modes of development can be a good long-term strategy for growth.

Urbanization brings simultaneous challenges and opportunities for green cities. Challenges include the rapid pace of urbanization and related pressure on the environment and social relations, especially where this continues on the same trajectory (the 'business-as-usual' or BAU model). In addition, cities of different wealth levels impact the environment differently. Local environmental threats are most severe in poorer cities and relate to issues such as fresh water, sewage, health and the degradation of the living environment.

Opportunities for sustainable urban development include the possibility to design, plan and manage physical structure in ways that are environmentally advantageous, advance technological innovation as well as profit from synergies that exist between the constituent elements of complex urban systems. Urban from and density, land-use, integrated design strategies, technologies and the construction of buildings, energy, water and waste systems can be improved to reduce resource and energy consumption. Social opportunities can arise as cities start to benefit from cleaner urban environments and improved living conditions, lower exposure to resource scarcities, fewer price fluctuations and costs from energy wastage and more accessible public transportation.

The case for sustainable urban development can be made in terms of inter-linked economic, social, and environmental benefits. Economically the benefits include agglomeration economies, lower infrastructure costs and reduced congestion cost while reducing carbon emissions and other environmental pressure. Socially the benefits include employment creation, poverty reduction and improved equity and quality of life including improved road safety and community cohesion, among others. Environmental benefits are embedded in most of the economic and social benefits.

4 Green Cities Could Save Money

4.1 Green Cities

With the majority of the world's economic activity and more than 50% of its population concentrated in urban areas, cities have a central role to play in the realization of a Green Economy. Already the cities on this "Urban Planet" account for 67% of world primary energy consumption and contribute round 80% of total global greenhouse gas emissions. But while societies are dependent upon the performance of urban economies, the urban economy is dependent on ecosystem services and resources. Cities therefore have a crucial role to play in the creation of economies that are pro-environment as well as pro-growth and pro-jobs, and will be pivotal to the emergence of a green economy. Local governments need to be recognized as crucial actors towards greening the local and urban economy thereby contributing to a global green economy. They can do this by improving economic competitiveness, being strategic in spatial planning and planning around landscape ecologies. As the boundaries of each city's reach are difficult to define, such activities tend to focus on the city region, i.e. the area most associated with economic activity and resource flows.

Buildings currently account for 40% of energy use in most countries (IEA 2010b), with projections that demand in this sector will increase by 60% by 2050 (IEA and OECD 2010). This is larger than the transportation or industrial sector; the IEA and OECD (2010) estimate the building-sector carbon emissions will need to be reduced from the 15.2 Gt per year currently projected for 2050 to approximately 206 Gt per year as part of a strategy to successfully address climate change. There is no doubt that improvements in urban design, housing stock, public transit and waste management are crucial components of a strategy to combat climate change and improve the physical health of urban dwellers.

The good news is that many cities are already on the front lines of climate change mitigation and environmental conservation. The city of Boston for example implemented a Green Tech Initiative in 2008 which focuses on enhancing the city's growing clean technology sector, investing in existing clean technology companies as well as in attracting such businesses to the area. The GreenTech Initiative provides clean technology businesses interested in locating in Boston with assistance in site selection, hiring, workforce training, competitive financial incentives and navigating the city's business permitting and licensing process. Additionally the city is supporting the next generation of clean technology companies by offering networking opportunities for entrepreneurs and an incubator network for companies with similar industries and supply chains. Boston has made itself a natural fit for clean tech industries through progressive policies on issues such as green building and reduction of greenhouse gas emissions, as well as through programs like the city's Sustainable Business Leader Program which helps small to mid-size businesses leaders green their business practices.

Despite a recession that has adversely affected many industries nationwide the clean technology sector in Boston is thriving. Green business will be introduced in the building sector by integrating a green certificate system promoting CO_2 reducing green home and green buildings. About one million green homes will be constructed in the new town while another one million will replace current homes in the future. Green homes use renewable energies such as solar light, thermal heat and wind power as power sources. Public buildings will be transformed into energy saving green buildings and construction of new public buildings over a certain size will be mandated to use energy efficiency measures, and environmentally friendly materials are to be used when remodeling or replacing old facilities. In addition, around 20% of lights will be replaced by light-emitting diode (LED) lights to save energy in public buildings. Ponds, green space and rainfall use facilities will be supplied to schools where green space is insufficient and energy efficiency is low. They will be upgraded to green schools with the help of environmentally friendly paint and highly efficient lighting facilities (Geon Cho 2009).

Several cities in developing countries – particularly in South America – have also branded themselves as green cities. As early as 1968 authorities in Curitiba, Brazil introduced policies to integrate land use and transport planning and by the 1970s the city was equipped with an innovative bus rapid transit system (Economist Intelligence Unit 2010). Singapore introduced the world's first road-charging scheme in the 1980s and it is now at the forefront of sustainable policies on waste, water and the greening of the environment (Phang 1993; Suzuki et al. 2010).

Greening the global building stock will however require continued and considerable investment in new technologies, sustainable building materials and in design and engineering expertise. This will increase the upfront cost of building construction relative to continuing with "business-as-usual". Three principal green building strategies can be differentiated: design, technology, and behavior related. Particularly in a developing world context, passive design solutions to improve environmental performance are by far the most cost-effective approaches.

4.2 Economic Benefits

Greening cities can produce multiple economic and social benefits. First, as well as lowering per capita carbon emissions, densification as a green city strategy tends to enhance productivity, promote innovation, and reduce the capital and operating cost of infrastructure. Densification can also raise congestion and the local cost of living, but green city strategies and interventions to subsidize housing costs can help to mitigate these. Second, much of a green economy is service-based and will tend to cluster in urban areas where consumer markets are largest. Third, an emphasis on public transport, cycling, and 'walkability' allow social considerations to be fully integrated into the design of green cities.

Agglomeration economies exist in both developed and developing countries. Empirical studies in developed countries find that doubling the employment density of an urban area typically raises its labor productivity by around 6% (for a summary of the literature see Melo et al. 2009). The same basic patterns are found in developing countries, with strong evidence that urbanization boosts productive efficiency by lowering transport costs and widening trade networks (Duranton 2008; Han 2009). Agglomeration economies can also be achieved by connecting several cities as in China's Pearl River Delta region (Rigg et al. 2009), with the additional benefit of addressing inequality between leading and lagging regions within countries (Ghani 2010).

In developing countries however, urbanization may not provide the same kind of economic gains across cities and firms. For example, Brülhart and Sbergami (Brülhart and Sbergami 2009) found that within-country agglomeration boosts GDP growth only up to national income levels of US$10,000 per head.

The main reason for this is that very rapid – and sometimes chaotic – urbanization can outstrip national and city governments' ability to provide adequate infrastructure and services (Cohen 2006). Congestion could eat up the benefits of higher density as in the case of cities like Shanghai, Bangkok, Manila and Mumbai (Rigg et al. 2009).

Comparing smart growth areas and dispersed, car-dependent developments, Todd Litman suggests direct cost savings between US$5,000 and US$75,000 for building road and utility infrastructure per household unit; such as schools and medical clinics (Litman 2009a). A recent study of Tianjin concluded that infrastructure cost savings as a result of compact and densely clustered urban development reach 55% compared with a dispersed scenario (Webster et al. 2010).

The greening of the energy sector moving away from carbon-intensive energy sources and improving efficiency is a rapidly growing business. Global investment in renewable power and fuels set a new record in 2010, and the margin over totals for previous years was wide. Investment totaled US$211 thousand million in 2010, up 32% from US$160 thousand million in 2009, and nearly five and a half times the 2004 figure. For the first time, new investment in utility-scale renewable energy projects and companies in developing countries surpassed that of developed economies (UNEP 2011).

While urbanization has helped to reduce absolute poverty, the number of people classified as urban poor is on the rise (Ravallion et al. 2007). Between 1993 and 2002, there was an addition of 50 million poor in urban areas whiles the number of rural poor declined by 150 million (Ravallion et al. 2007). Urban growth puts pressure on the quality of the local environment, which disproportionately affects poorer people, such as the lack of adequate access to clean water and sanitation. This results in a huge disease burden that further affects their livelihood options. Improving the urban environment by measures such as traffic calming and promoting 'walk ability' can help foster a sense of community (Frumkin 2003; Litman 2006). Such changes are often designed to counteract instances of 'community severance', as identified by Bradbury et al. (2007).

4.3 *Investment in Green Buildings*

Greening cities is not cost free. There are tradeoffs and switching costs, creating both winners and losers. Consumer preferences are not always green. Cities may face financial, structural and technological constraints. And fragmented governance may lead to perverse outcomes of policy, if projects are not coordinated from the local to the international level. The rebound effect, where energy-saving innovations actually raise total energy consumption, illustrates how many of these issues come together. These factors suggest it is critical to look at both national and urban policy levers; and at the conditions that will enable cities in different parts of the world to make the transition to green economy models. In practice, green cities will require a coalition of actors across public, private and civil society sectors and multilevel governance models that allow these actors to come together effectively.

Climate change and GHG emissions being an overriding concern for the building sector, related to this are key environmental challenges such as water scarcity, land use, waste and sanitation. Climate change both impacts and is impacted by these. The social and economic dimensions are addressed in terms of how a more efficient use of resources in the building sector and a reduction of its GHG emissions can contribute to energy savings, health and productivity gains, as well as job creation. Overall, green building investment needs are primarily driven by climate and resource scarcity or efficiency imperatives. The building sector should be central to any attempt to use resources more efficiently. Buildings consume a large proportion of the global energy supply but opportunities to improve efficiency are huge and the sector has the greatest potential – more than any other covered in this report- to reduce global GHG emissions.

Governments can set a leadership example by using public procurement in the construction and management of their facilities to drive the greening of the building sector. Experience from Mexico and China has shown how energy-efficiency improvement programs in the public sector can also be boosted by the immediate pressure of high energy prices and energy shortages. Public assets, be they in the form of government buildings, hospitals or schools, hold wide-ranging opportunities of greening measures that result in a more efficient use of resources, reduced GHG emissions, improved productivity and avoided illness resulting from indoor air pollution.

Greening the global building stock will require considerable investment in new technologies, sustainable building materials as well as in design and engineering expertise. This will increase the upfront cost of building construction relative to continuing with "business-as-usual". The IEA and OECD (2010) estimate that a 12.6 Gt reduction by 2050 could be achieved with an average investment of US$308 billion per year between 2010 and 2050. A higher estimate of US$ one trillion per year on average between 2010 and 2050 was obtained in a separate study by the Peterson Institute for International Economics (Houser 2009).

This reduction of 12.6 Gt CO_2 emissions by 2050; published in the Energy Technology Perspectives 2010 (IEA and OECD 2010) revises earlier estimates that CO_2 emissions from buildings would need to be reduced by 8.2 Gt from a projected 20.1 Gt in 2050 to 11.9 Gt (IEA 2008). The earlier estimates formed a reference point for other analysis, including by the Peterson Institute for International Economics (Houser 2009). The 2010 estimates also include reductions achieved by fuel-switching and electricity de-carbonization, whereas the earlier estimates were limited to efficiency measures.

Net present value is calculated by subtracting the additional up-front operation and maintenance cost required for the more-efficient investment from the expected energy cost saving over the lifetime of the more-efficient investment. Energy cost savings are discounted by 6% annually.

5 Conclusions

Cities are where some of the world's most pressing challenges are concentrated: unsustainable resource and energy consumption, carbon emissions, pollution, and health hazards; but cities are also where hope lies. They are magnets attracting hundreds of millions of rural migrants in search for economic opportunities. Moving towards a green economy must become a strategic economic policy agenda for achieving sustainable development. Developing countries must take the same attitude in order to move forward towards sustainability. In that regard developed countries must encourage, advice and support the developing countries in order to enhance their capacity to take up the Green Cities concept.

There are many economic benefits to implementing the Green Cities attitude including employment creation, poverty reduction and improved equity and quality of life through such mechanisms as improved road safety and community cohesion. The investment in Green Cities represents a long term approach to saving money and for the mitigation and adoption to climate change impacts in many sectors.

References

Bradbury et al (2007) Investing in energy and resource efficiency Cities. http://www.unep.org/greeneconomy/Portals/88/documents/ger/GER_12_Cities.pdf

Brinkhoff T (2011) The principal agglomerations of the world. Available via: http://www.citypopulation.de. Cites 16 Aug 2011

Brown L (2011) World on the edge: how to prevent environmental and economic collapse. W. W. Norton & Company, New York

Meadows D, Brugmann J (1999). http://content.yudu.com/A1vce6/GECity/resources/a8.htm

Brülhart F, Sbergami F (2009) Agglomeration and growth: cross-country evidence. J Urban Econ 65:48–63

C40 Cities (2010d) Investing in energy and resource efficiency Cities. http://www.hbs.edu/environment/docs/HBS-Investing-in-Cities-of-the-21st-Century_Intro.pdf

Clinton Foundation (2010) Annual Report 2010

Cohen P (2006) Worldwide carsharing growth: an international comparison. University of California, Berkeley

Duranton G (2008) Cities: engines of growth and prosperity for developing countries? International trade and development in the Department of Economics, University of Toronto, Toronto

Economist Intelligence Unit (2010) Report on business 2010, embracing the challenge of change

EIA (1998) A look at commercial buildings in 1995: characteristics, energy consumption, and energy expenditures. Energy Information Administration, Office of Energy Markets and End Use, U.S. Dept. of Energy, Washington, DC

EIA (2003) Households, Buildings, Industry & Vehicles end-use energy consumption data & analysis. [Online]. U.S. Energy Information Administration, Independent Statistics and Analysis, U.S. Dept. of Energy, Washington, DC. Available via: http://www.eia.doe.gov/emeu/consumption/index.html. Cited 11 Jan 2011

EIA (2010) International Energy Outlook – highlights. U.S. Energy Information Administration, Office of Integrated Analysis and Forecasting, U.S. Dept. of Energy, Washington, DC. Available via: http://www.eia.doe.gov/oiaf/ieo/highlights.html

EMCDDA MONOGRAPHS, Harm reduction: evidence, impacts and challenges 0 1. http://eur-lex.europa.eu/SECMonth.do?year=2008&month=11
Frumkin H (2003) Environmental health: from global to local
Foray (2009) Introduction, setting the stage for a green economy transition, green economy
Geary C (2011) Sustainable connections: linking sustainability and economic development strategies
Geon Cho N (2009) The potential of the green economy, green cities, new approaches to confronting climate change. OECD workshop proceedings, Las Palmas De Gran Canaria, Spain, 11 June 2009
Ghani E (2010) The service revolution in India
Han (2009) Moving beyond deliberative control of impulses, the effect of construal levels on evaluative associations in self-control conflicts. http://www.ncbi.nlm.nih.gov/pubmed/19493322.
Hitchin R (2008) Can building codes deliver energy efficiency? Defining a best practice approach. A report for the Royal Institution of Chartered Surveyors by the Building Research Establishment, Watford
Houser T (2009) The economics of energy efficiency in buildings. Peterson Institute for International Economics, Washington, DC. Available via: http://www.www.piie.com/publications/pb/pb09-17.pdf
ICLEI - Local Governments for Sustainability http://www.iclei.org/index.php?id=global-themes
http://www.unep.org/search.asp?q=UNEP+Green+Economy+Initiate percent2C+January+2011&cx=007059379654755265211 percent3Ajkngxjgnyii&cof=forid percent3A11&sa.x=12&sa.y=10
Hutton G, Rehfuess E, Tediosi F, Weiss S (2006) Evaluation of the costs and benefits of household energy and health interventions at global and regional levels. World Health Organization, Geneva
ICLEI (2011) http://www.iclei.org/index.php?id=1487&tx_ttnews percent5Btt_news percent5D
IEA (2001) Dealing with climate change: policies and measures in IEA member countries. International Energy Agency, Paris
IEA (2008) Energy technology perspectives 2008: scenarios and strategies to 2050. International Energy Agency, Paris
IEA (2009a) World energy outlook 2009. International Energy Agency, Paris
IEA (2009b) Key world energy statistics. International Energy Agency, Paris
IEA (2010a) World energy outlook 2010. International Energy Agency, Paris
IEA (2010b) Policy pathways: energy performance certification of buildings. International Energy Agency, Paris. Avalable via: http://www.iea.org/publications/free_new_Desc.asp?PUBS_ID=2295
IEA and OECD (2010) Energy technology perspectives 2010 scenarios and strategies to 2050. International Energy Agency and the Organization for Economic Co-operation and Development, Paris
ILO (2001) The construction industry in the twenty-first century: its image, employment prospects and skill requirements. TMIC, ILO, Geneva
ILO (2009) Empregos Verdes no Brasil: Quantos são, onde estão e como evoluirão nos próximos anos. Organização Internacional do Trabalho, Escritorio no Brasil
IPCC (2007) Climate change 2007: mitigation of climate change. Contribution of working group III to the fourth assessment report of the intergovernmental panel on climate change. Cambridge University Press, Cambridge/New York
Kats GH (2003) Green building costs and financial benefits. Massachusetts Technology Collaborative, Boston. Available via: http://www.nhphps.org/docs/documents/GreenBuildingspaper.pdf
Kats G (2010) Greening our built world: costs, benefits, and strategies. Island Press, Washington, DC. Available via: http://www.cap-e.com/Capital-E/Resources_ percent26_Publications.html
Keivani R, Tah JHM, Kurul E, Abanda FH (2010) Green jobs creation through sustainable refurbishment in the developing countries. A literature review and analysis conducted for the International Labour Organization (ILO). International Labour Office, Geneva. Available via: http://www.ilo.org/public/english/dialogue/sector/papers/construction/wp275.pdf
Khor M (2011) The 'green economy' debate: a sustainability perspective. Presentation at the UN meeting on Rio Plus 20, panel on green economy, New York, 10–11 Jan 2011
Laustsen J (2008) Energy efficiency requirements in building codes, energy efficiency policies for new buildings. International Energy Agency, Paris. Available via: http://www.iea.org/g8/2008/Building_Codes.pdf

Litman T (2006) The future isn't what it used to be changing trends and their implications for transport planning
Litman T (2009a) Smart transportation economic stimulation, infrastructure investments the support economic development. Transport Policy Institute, Victoria
Loftness V, Hartkopf V, Gurtekin B (2003) Linking energy to health and productivity in the built environment. Available via: http://www.usgbc.org/Docs/Archive/MediaArchive/207_Loftness.pdf
Luhmann HJ (2007) Smart metering als neue Energie-(effizienz) quelle. Energ Manage 6:356–360. http://www.unep.org/greeneconomy/Portals/88/documents/ger/9.0_Buildings.pdf
Malhotra M (2003) Financing her home, one wall at a time. Environ Urban 15(2):217
Martinez-Fernandez C, Hinojosa C, Miranda G (2010) Greening jobs and skills labour market implications of addressing climate change. OECD Publishing, Paris
McDonough W, Braungart M (2002) Cradle to cradle: remaking the way we make things, 1st edn. North Point Press, New York
McGraw Hill (2009) Green building retrofit and renovation: rapidly expanding market opportunities through existing building. Smart market report. McGraw Hill Construction, Bedford. Available via: http://construction.ecnext.com/coms2/summary_0249-323452_ITM_analytics
Melo et al (2009) Bridging trade theory and labour econometrics: the effects of international migration. ICLEI - Local Governments for Sustainability
NHHP (2007) National urban housing and habitat policy 2007. Government of India Ministry of Housing & Urban Poverty Alleviation, New Delhi. Available via: http://mhupa.gov.in/policies/owingpa/HousingPolicy2007.pdf
Nicholas Stern, What is the Economics of Climate Change?, World Economics, 6(2). http://www.the-.eans.de/Presse/PMitt/2006/061030c76.pdf
NSF/IUCRC (2004) Guidelines for high performance buildings. Available at: http://cbpd.arc.cmu.edu/ebids/pages/home.aspx
OECD (2009) Declaration on Green Growth, adopted at the Meeting of the Council at Ministerial Level on 25 June 2009, [C/MIN(2009)5/ADD1/FINAL]. Cited 10 Oct 2011 at http://www.oecd.org/dataoecd/58/34/44077822.pdf
Oregon Department of Energy (2010). Business energy tax credits. Available via: www.oregon.gov/energy/docs/Tribal/2010-energy.pdf
Pearce DW, Markandya A, Barbier E, Dept. of the Environment (UK) (1989) Blueprint for a green economy. Earthscan, London
Phang (1993) Economic development and the distribution of land rents in Singapore: a Georgist implementation. http://www.jstor.org/discover/10.2307/3487624?uid=3737928&uid=2129&uid=2&uid=70&uid=4&sid=21101199905617
UN Population Division (2006) Available via: http://www.un.org/esa/population/publications/wpp2006/English.pdf
UN Population Division (2010) Available via: http://www.un.org/esa/population/publications/concise2005/Popdev.pdf
Porter ME, van der Linde C (1995) Toward a new conception of the environment-competitiveness relationship. J Econ Perspect 9(4):97–118
Ravallion et al (2007) A comparative perspective on poverty reduction in Brazil, China and India. Available via: Papers.ssrn.com/sol 3/papers.cfm?
Ravetz J (2008) State of the stock – what do we know about existing buildings and their future prospects? Energ Policy 36(12):4462–4470
Ries C, Jenkins J, Wise O (2009) Improving the energy performance of buildings: learning from the European Union and Australia. RAND Corporation, Santa Monica, CA
Rigg et al (2009) Working Papers Series in Economics and Social Sciences; Economic geographers and the limelight: the reaction to the 2009 World Development Report
Satterthwaite D (2011) Cities' contribution to global warming: notes on the allocation of greenhouse gas emissions. International Institute for Environmental Development. Environ Urban 20:539
Satterthwaite D, Moser C (2008) Human Settlements Discussion Paper Series, theme: climate change and cities – 3, climate change and cities discussion paper 3, IIED, London, October 2008

Simpson R (2010) A green economy for cities. http://www.stakeholderforum.org/sf/outreach2010/index.php/intersesh1-item1
Stern Review (2006) The economics of climate change. Available via: http://siteresources.worldbank.org/INTINDONESIA/Resources/226271-1170911056314/3428109-1174614780539/SternReviewEng.pdf
Stoneman (ed) (1995) Policy design and intervention in the innovation diffusion process: the case of China's communication sector. Available via: http://pubs.e-contentmanagement.com/doi/abs/10.5172/impp.2006.8.1-2.113?journalCode=impp
Suzuki et al (2010) Technical change, growth and trade: new departures in institutional economics. Available via: http://dspace.mah.se/handle/2043/9844
TEEB (2010) Ecosystem services and green growth
TEEB (2011) The Economics of Ecosystems & Biodiversity, A quick guide to TEEB for local and regional policy makers. Available via: http://www.scribd.com/mark_schauer_1/d/55266961-TEEB-D2-Quick-Guide
Thormark C (2000) Environmental analysis of a building with reused building materials. Int J Low Energy Sust Build 1:6–7, 12–13. Available via: http://dspace.mah.se/handle/2043/9844
Thormark C (2006) The effect of material choice on the total energy need and recycling potential of a building. Build Environ 41(8):1019–1026
UN Statistics Division (2008) http://unstats.un.org/unsd/pubs/gesgrid.asp?mysearch=environment
UNECA (2011) Final report, workshop on institutional and strategic frameworks for sustainable development in Africa, ECA/FSSDD/03/11
UNEP (2007) Available via: http://www.unep.org/PDF/AnnualReport/2007/AnnualReport2007_en_web.pdf
UNEP (2009a) Annual Report of 2009, seizing the green opportunity. UNEP, Nairobi. Available via: http://www.unep.org/pdf/UNEP_2009_ANNUAL_REPORT.pdf
UNEP (2009b) Global green new deal: an update for the G20 Pittsburgh Summit. United Nations Environment Programme, Pittsburgh. Available via: http://www.unep.org/greeneconomy/LinkClick.aspx? fileticket=ciH 9RD7XHwcpercent3d&tabid=1394&language=en-US
UNEP (2009c) Energy efficiency in the finance sector: a survey on lending activities and policy issues. United Nations Environment Programme, Finance Initiative, Geneva. Available via: www.unepfi.org/fileadmin/documents/Energy_Efficiency.pdf
UNEP (2010) GREEN economy driving a green economy through public finance and fiscal policy reform. Green Economy Initiative, UNEP, Nairobi. Available via: http://www.unep.org/greeneconomy/Portals/30/docs/DrivingGreenEconomy.pdf
UNEP/GRID (2009) Greening the world economy. Environment Times No. 6, UNEP/GRID-Arendal, Norway
UNEP (2011) Towards a green economy: pathways to sustainable development and poverty eradication
UNEP SBCI (2007a) Buildings and climate change: status, challenges, and opportunities. United Nations Environment Programme, Sustainable Buildings and Construction Initiative, Paris
UNEP SBCI (2010a) The 'State of Play' of sustainable buildings in India. United Nations Environment Programme, Sustainable Buildings and Climate Initiative, Paris. Available via: http://www.unep.org/sbci/pdfs/State_of_play_India.pdf
UN-Habitat (2011) Urban patterns for sustainable development: towards a green economy. UN-Habitat draft working paper, Jan 2011
WBCSD (2011) Energy efficiency in buildings. Business realities and opportunities. World Business Council for Sustainable Development, Geneva. Available at: http://www.wbcsd.org/DocRoot/JNHhGVcWoRIIP4p2NaKl/WBCSD_EEB_final.pdf.
Webster PJ et al (2010) Beyond the spring barrier? Nat Geosci 3:152–153
World Bank (2010) The Annual Report 2010
World Humanity Action Trust (2000) Governance for a sustainable future reports, reports of the commissions of the World Humanity Action Trust. World Humanity Action Trust, London. Available via: http://www.earthsummit2002.org/es/issues/Governance/whatgov1.pdf

Working in Green Cities: Improving the Urban Environment While Creating Jobs and Enhancing Working Conditions

Edmundo Werna

Abstract This paper analyzes the relationship between the urban environment and labor. Cities and towns will not be sustainable if the livelihoods of their inhabitants are not adequately addressed. At the same time, a well-trained labor force working in proper conditions is a vital asset for improving the urban environment. The paper shows that urban environmental problems have a negative impact on the lives and productivity of workers, and hence limitations for social and economic development. Improving the urban environment will not only counteract this impact, but also create businesses and employment. Initiatives to improve the urban environment require the active involvement of workers and enterprises. Therefore, the paper also notes the need for promoting training on green technologies, as well as the need for securing decent working conditions in the transition to greener cities and towns. The paper concludes by advocating an integrated approach to labor in urban areas, with particular considerations for the role of local authorities.

Keywords Urban labor • Decent working conditions • Urban environment • Business and employment generation • Social dialogue

1 Introduction

The existing environmental problems of urban areas have been extensively analyzed (Gallego et al. 2009; Gangolells et al. 2009; Keivani et al. 2010; Khatib 2009; Moavenzadeh 1994; Sarsby and Meggyes 2010; The Environment Agency in

---

E. Werna (✉)
Natural Resources and Industries, Department of Sectoral Activities,
International Labour Organization, Geneva, Switzerland
e-mail: werna@ilo.org

R. Simpson and M. Zimmermann (eds.), *The Economy of Green Cities: A World Compendium on the Green Urban Economy*, Local Sustainability 3,
DOI 10.1007/978-94-007-1969-9_7,

England and Wales et al. 2011; Werna 2011; Werna and Saboor forthcoming). They derive from possible hindrances related to the way cities and towns are planned or the lack of, how they are build and expanded, and their production and consumption patterns. There is ample evidence that major action needs to be taken throughout the world to address the existing challenges. At the same time, there is a substantial body of literature on the technical aspects of urban environmental improvement, while the role of labor in greening cities and towns has received less attention.

This chapter analyzes the greening of urban areas from the perspective of labor. It highlights that current urban environmental problems have a negative impact on the lives, wellbeing and productivity of urban workers, posing limitations for social and economic development. Conversely, initiatives to improve the urban environment will not only counteract the aforementioned impacts, but also have the potential to generate new businesses and employment. At the same time, considering that actions to improve the urban environment require the active participation of workers and enterprises, the chapter emphasizes the need for scaling-up skills training on green technologies, as well as the need for securing decent working conditions in the development of greener cities and towns.

The chapter starts with basic clarifications on the importance of labor in urban sustainability and introduces the concept of decent work used by the International Labor Organization (ILO) as a framework for the 'world of labor'. The concept is useful for analyzing the relationship between the urban environment and the different aspects of labor. The following section concentrates on how urban environmental problems affect the health of urban workers and explains their impact on livelihoods and productivity. The chapter then moves from negative impacts to net gains by examining the potential of initiatives to generate employment and income to green the urban economy. The potential will be illustrated with examples from selected sectors: construction, waste management, urban forestry, and agriculture. The chapter highlights concrete actions that can be taken to enable a decent work transition to a green urban economy. The chapter concludes with a call for an integrated approach to labor in urban areas to support environmental improvements, with specific considerations for local authorities.

2 Presenting the Urban Labor Framework

The concepts of sustainable development in general and urban sustainability in particular have been widely analyzed. They comprise of ecological, economic and social dimensions. Labor is an important element of the social dimension with spin-offs to the economic and ecological dimensions. As noted by Werna and Saboor (forthcoming), urban areas will not be sustainable if the livelihoods of their residents are not properly addressed. Urban poverty is extensive, especially in developing countries, and is increasingly overtaking rural poverty (UN-Habitat 2010, 2011). In addition to its direct consequences on the lives of a large number of urban inhabitants, poverty also limits the ability of urban workers to fully contribute to the economy and the greening of cities and towns. Initiatives related to urban infrastructure provision, such as slum upgrading, inner-city regeneration and the like, will not be sustainable if the livelihoods of the urban workforce are not properly addressed (Werna and Saboor forthcoming). Conversely, a well-trained labor force working in adequate conditions is a crucial asset for improving the urban environment. In addition, those workers and entrepreneurs trained in green techniques are prone to become champions of the environment, as they learn about its importance and are also motivated to use the skills they have learned (Werna 2011).

The ILO uses the concept of 'decent work' to examine and take action on the different facets of labor. The concept comprises of four pillars: employment creation, workers' rights, social protection and social dialogue. Each component will be briefly explained below:

Employment creation: Employment creation entails not only promoting 'job-rich' methods of production, but also increasing employability via for example skills training, enterprise development, and special initiatives for disadvantaged groups. Both developing and developed countries have unemployment and underemployment deficits, especially as half of the planet's population is now living in urban areas, and the rate and absolute number of the urban vis-à-vis the rural population continues to increase (UN-Habitat 2010, 2011). Growth of the urban population has overtaken employment opportunities. The lack of quality jobs in sufficient numbers has induced urban poverty and insecurity, particularly in a large number of cities and towns in developing countries (Lawrence and Werna 2009; Werna and Saboor forthcoming). As the chapter will later elaborate, the greening of the urban economy can be an important source for employment creation to address this challenge.

Workers' rights: With regard to workers' rights examples for existing urban challenges include informality, casual work, specific difficulties faced by migrant workers, child labor, bonded labor, discrimination against female workers.

Social protection: Deficits in social protection also prevail in the urban economy, particularly, although not only, for casual and informal workers. The urban poor face a plethora of problems including sickness, disability, accidents, premature death, loss of assets, and crime, among others. Exposure to these numerous risks is high and the low-income populations are least protected. Large numbers of urban workers and self-employed entrepreneurs do not have access to proper healthcare, protection against loss of revenue due to job loss, illness, accident or old age, and to other types of social protection. Under such circumstances, injury or illness of one income earner in the household can lead to destitution, child labor or debt.

Social dialogue: Social dialogue, in turn, has been a key process for workers, employers and governments to discuss and negotiate labor related issues. Socially viable cities and towns need to have a fair representation of workers in decision-making. However, large numbers of urban workers, and also enterprises, are not sufficiently organized and are subject to the mainstream social and economic dialogues that take place (Ghai 2006; ILO 2004, 2006; Lawrence et al. 2008; Lawrence and Werna 2009; Van Empel and Werna 2010; Werna and Saboor forthcoming).

These four pillars of decent work have an impact on the greening of urban areas. Before delving into the analysis of this relationship, the chapter will examine how existing urban environmental problems impact the health of urban workers with consequences on their livelihoods and productivity. Such an analysis supports the argument in favor of greening urban areas.

3 Urban Environmental Problems Are Bad for Business and Employment

There is a rationale for greening urban areas even if only to prevent the consequences of existing environmental problems to workers. This section presents such problems and their impact on health. Diseases have a direct effect on the lives and productivity of workers. The data below is based on a compilation by Comaru (2011) based on WHO (2011) and Campbell-Lendrum and Corvalan (2007).

Lack of access to safe water and sanitation: This is by and large the major environmental contributor to ill health in urban areas. Vast numbers of low-income workers face this risk, which leads to cholera, diarrheal diseases and other waterborne diseases.

Floods: Drainage systems have not kept pace with the scale of urban expansion, and many urban areas are prone to flooding, especially in the South, which often occurs on a periodic basis. The workers who live or/and work in low-income neighborhoods are exposed to such risk. As such, neighborhoods are often built in flood-prone areas, such as river banks and low lands, with no protection. Floods are a medium for the spread of waterborne diseases (mentioned before), and may also lead to accidents including fatal ones; in addition to other consequences which are not related to health but still are vital for livelihoods (such as loss of domestic or/and economic assets).

Sound pollution: Many cities or parts of cities have an excessive level of noise leading to cardiovascular problems, increased annoyance response, and sleep disturbances.

Overcrowding: Frequent in low-income settlements and many inner-city areas especially in the South, overcrowding facilitates the spread of airborne diseases.

Uncontrolled urban growth: Facilitates the spread of vector-borne diseases.

Traffic accidents: Traffic injuries are among the top ten causes of death in urban areas. Developing countries account for more than 85 % of all the fatalities and over 90 % of Disability Adjusted Life Years (DALYs) lost due to road traffic injuries (WHO 2004), and road injuries affect the poor more than the affluent in developing countries.

Outdoor air pollution: Can generate cancer of the lung and upper respiratory tract, acute myocardial infarction and arrhythmias, chronic bronchitis and asthma. For example, it has been estimated that outdoor air pollution in Sao Paulo, Brazil, leads to a reduction in 1.5 life years per person on average and causes some 4,000 premature deaths per year.

Indoor air pollution: Exposure to this risk takes place when the worker is at home during vacant time. Also, there are a burgeoning number of urban workers who work at home, and are therefore exposed for a longer period of time to this risk. Indoor smoke from fuel combustion is the eighth most important risk factor among the burdens of disease. In 2004, indoor air pollution was responsible for almost two million deaths. Indoor smoke occurs due to the use of inefficient stoves or/and lack of ventilation. It has been associated with a high risk of acute lower respiratory illnesses including pneumonia. Chronic obstructive pulmonary disease from indoor smoke causes one million premature deaths a year. Other health problems have also been reported, especially when coal is used as a source of energy. They include lung cancer, asthma, and cardiovascular diseases, among others.

Inappropriate building materials: A number of materials frequently used in buildings, especially low-income ones, entail health risks. For example, asbestos, which is still widely used, is carcinogenic and may also cause mesothelioma. Mold and moisture may lead to asthma. Volatile organic compounds may lead to poisoning, cancer, asthma and other conditions.

Extreme thermal conditions: Poorly insulated, poorly heated or poorly ventilated buildings lead to exhaustion, heat cramps, heat stroke, and cardiovascular or respiratory health problems.

The above are examples of the burdens of disease caused by urban environmental problems. Figures 1 and 2 provide general data. Such diseases have a significant toll on productivity at work, number of work days lost from sick leave, as well as premature retirement and premature death. According to the World Economic Forum (2008), a study which compared the competitiveness of 31 industrial and developing countries clearly showed that better safety and health standards equal better national productivity. The results indicate that a low number of fatal accidents and high competitiveness are directly proportional making the most competitive economies register the lowest number of fatal accidents. Countries such as Switzerland, with just over three fatal occupational accidents per 100,000 employees, are not only amongst the countries with the highest standards of health and safety, but are also amongst the most competitive national economies in the world.

In conclusion, improvements in the urban environment will support social and economic development by reducing and eventually eliminating the aforementioned burdens of disease. Such improvements also increase the comfort and security of workers and entrepreneurs, as well as combat the risk of climate change and improve adaptation, while providing benefits to social and economic development. As is explained in the next section, such improvements can also contribute to generating needed employment and business opportunities for urban workers and enterprises.

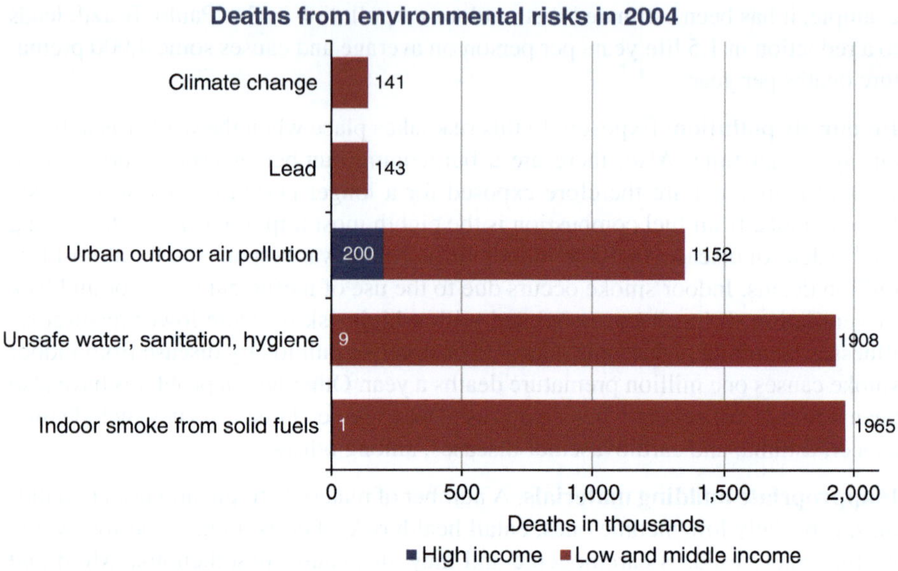

Fig. 1 Deaths from environmental risks in 2004, WHO (2012)

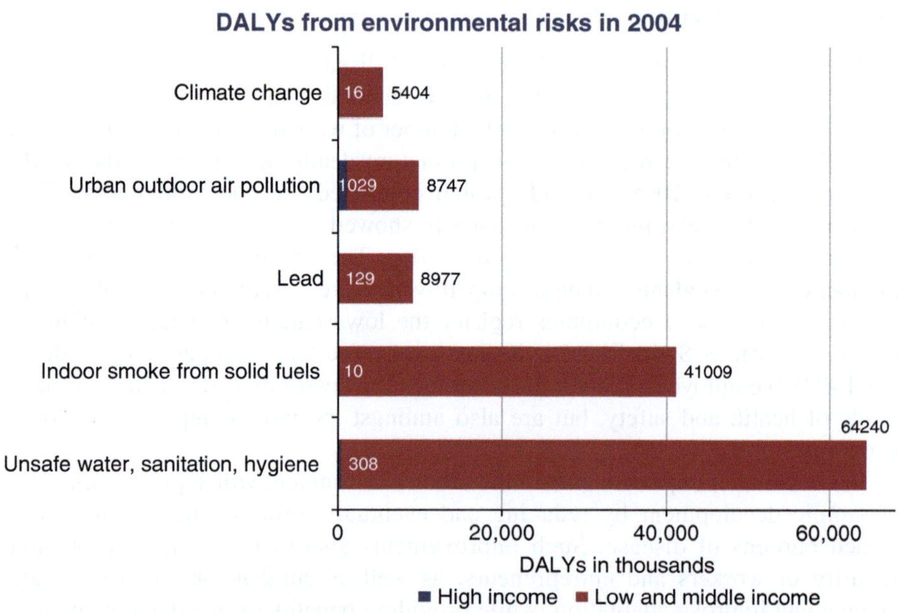

Fig. 2 Disability adjusted life years from environmental risks in 2004, WHO (2012)

4 Reversing the Trend: Improving the Urban Environment Is Good for Business and Employment

The urban environmental problems presented in the previous section call for action for their elimination. This section analyzes possibilities in selected sectors: construction, waste management, urban forestry and agriculture.

4.1 Construction

The built fabric of cities and towns has considerable impacts on the environment. A number of issues mentioned in the previous section are specifically related to construction (e.g. indoor air pollution, inappropriate building materials, extreme thermal conditions). Further impacts relate to decisions regarding the construction sites, the methods of production, and the selection of construction equipment (Werna 2011). It has been estimated that as much as 40 % of energy used in the world is related to the process of construction and demolition, and activities in buildings (WBCSD 2009). Therefore, a large magnitude of action needs to be taken in order to improve the environmental impact of construction, generating employment and business opportunities. They can entail:

- Improvements in the construction process (to address air emissions, noise pollution, waste disposal, water discharges).
- Construction of new green buildings.
- Construction of durable buildings (to postpone demolition).
- Retrofitting old buildings.
- Increased production of green construction materials, appliances and components.
- Energy-efficient operations and maintenance of buildings.
- Recycling of construction waste and demolitions (Werna 2011).

LEED is a certification process used by the World Council for Green Buildings. Data from LEED buildings alone shows incremental employment creation of about 60 jobs per million square feet (92,903.04 m^2) of green building constructed, including direct, indirect and induced jobs. Retrofitting of existing buildings also show potential for employment growth. Studies have estimated that every US$1 million invested in efficient retrofits creates 10–14 direct jobs and 3–4 indirect jobs. Data from the USA alone shows that retrofitting of 40 % of the building stock in such a country would generate 6,250,000 jobs over a period of 10 years. A large-scale retrofitting project of the German Alliance for Work and the Environment led to around 140,000 new or saved jobs. A study of ten new EU member states projected that 50,000–185,000 jobs would be created from the retrofitting of the existing residential building stock. In addition, it has been estimated that an investment of US$3.5 billion to modernize appliance standards would result in 29,876 jobs. In India, the introduction of fuel-efficient biomass cooking stoves in nine million households could produce 150,000 jobs (UNEP 2008; Werna 2011 based on data from UNEP).

Above are just some examples that illustrate the employment potential of greening construction. Many more examples abound (e.g. ILO 2010, 2011a; Keivani et al. 2010; Ofori and Abdul Aziz 2010; UNEP 2008; Van Wyk et al. 2010; Waste Concern Consultants 2010). While the data presented above are specific to employment, there are also ample opportunities for entrepreneurs. The construction industry nowadays entails a large-scale level of subcontracting, with several layers of enterprises, including a sizeable number of small and medium enterprises. The expansion of green construction requires new enterprises coming on board.

4.2 Waste Management

Urban areas generate a large amount of domestic and office waste. Sao Paulo, Brazil, alone generates approximately 15,000 ton of waste per day (Comaru 2011). A large number of cities and towns, especially in the South, still do not have a proper system of waste management, and their inhabitants face the health and wellbeing burden of large amounts of uncollected litter. This is particularly significant in low-income neighborhoods.

An initiative of the ILO in Dar-es-Salaam, Tanzania, implemented in the previous decade has shown that it is possible to clean low-income settlements and at the same time provide business/employment opportunities through a system of waste management involving micro-enterprises. Residents of low-income neighborhoods have shown willingness to pay for waste collection if such a service is provided.

New businesses and employment can also be generated through waste recycling. A large amount of domestic and office waste, even when collected, is still not recycled.

Existing waste management services include waste pickers and recyclers. It has been estimated that there are some 15 million pickers in developing countries (Comaru 2011). However, by and large such pickers have very low returns and low productivity. Furthermore, income can be generated by improving the skills of such workers.

In addition to domestic and office waste, other sectors also offer opportunities for recycling. For example, the construction process generates a large amount of waste, and it has been estimated that 50 % of such waste is not recycled (Zordan 2011). Also, material from demolitions is often not recycled. There is potential for expanding the recycling of construction waste, which would contribute to significantly reducing the impact of the construction industry on the environment. Another sector that is attracting growing attention, both from the perspectives of the environment and employment, is electronic waste.

4.3 Urban Forestry and Agriculture

While urban forestry (e.g. parks, gardens and green verges) has been on the agenda of public authorities for a long time, vast numbers of cities and towns are still short of such green spaces, especially in the South. This is due to the pressure to develop land and widen roads. While there is a rationale to increase urban density ("compact cities"), this has to be balanced against the need to have urban forestry. Lack of such spaces has resulted in the rise of heat islands and dust levels. For example, an increase in tree canopy cover by 10% reduces air temperature by 1.4 °C during daytime. Urban forestry also absorbs rainwater reducing the pressure on the drainage system and the risk of floods (ILO forthcoming; Uddin 2006).

Employment can be generated through the creation and maintenance of urban forestry, as the cities and towns, which already invested in such spaces show (see Uddin 2006). This can and should be extended to the large number of cities and towns, which still lag behind in urban forestry. Income can also be generated through the utilization of the products of trees (e.g. selling of fruits and medicine) as shown for example in one study on Dhaka, Bangladesh (Uddin 2006).

Urban agriculture, in turn, can expand the extent of green spaces, adding to the environmental benefits of urban forestry. The literature on urban agriculture has noted the benefits on the income of poor families, which are able to save money through planting or raising livestock for self-consumption. Recent studies by the ILO show that the income-generating potential of urban agriculture can go beyond such self-help schemes, as shown for example by an on-going investigation in Harare, Zimbabwe. In Caracas, Venezuela, the government has used urban agriculture to provide employment for young people in low-income settlements (ILO forthcoming). Businesses and employment can be generated not only through the selling of urban agricultural products, but also through the management of its waste.

4.4 Further Opportunities

The urban economy is vast and diverse, and other sectors can also contribute to the improvement of the environment, generating green employment and business opportunities, in addition to the sectors analyzed before. UNEP's *Green Economy Report* and ILO's background note to such report have each a specific article on cities with an overview of the urban economy (ILO 2011a; UNEP 2011). At the same time, it is important to note that employment and business opportunities are not just a passive consequence of green economic development. Positive action needs to be taken.

5 Positive Actions to Ensure a Transition Towards Decent Work in a Green Urban Economy

Provision of training is a fundamental requirement for the attainment of employment and business opportunities. Changes in processes of production require new working and managerial skills. This is confirmed, for example, by recent studies of the ILO on green construction (e.g. ILO 2010, 2011a; Keivani et al 2010; Ofori and Abdul Aziz 2010; Van Wyk et al. 2010; Waste Concern Consultants 2010). Expansion of existing services such as waste management and urban agriculture and forestry also require new workers and enterprises with adequate skills.

There are different ways through which required training can be achieved. It is necessary to adapt the curricula of existing training schemes to the requirements of green processes of production or/and to have specific green courses, depending on the situation in each city or town. One example of such potential is the 'green plumbers' initiative, which is in place in countries such as Australia and South Africa. Green plumbers have particular skills on water saving, with techniques to collect rainwater, and to use grey water to flush toilets. This attracts environment-conscious clients as well as those interested in saving costs on water consumption, therefore generating businesses/employment for the plumbers (Werna 2011).

Existing training schemes need to be complemented to change the urban economy. In the construction industry, for example, the training of master craftsmen as trainers will improve the on-the-job apprenticeship system that is found in several countries. In addition, the involvement of subcontractors, labor contractors and intermediaries in joint training schemes with cost reimbursement, seem to be essential if these schemes arc to be effective in meeting the skill needs of the industry (ILO 2001). These general recommendations could be used for the development of green construction (Werna 2011).

Urban forestry is a sector in which government authorities should be involved, either through direct provision or contracting out, because such green spaces are typically public. The promotion of new areas, such as for example waste recycling, may also require the involvement of public authorities in training, either directly or through substantial support to other actors to implement it. The expansion of urban agriculture as a source of income also requires technical training, as it often still takes place as a self-help activity without a business approach. All in all, different sectors may require different training schemes, depending on specific characteristics of each sector. Public authorities play an important role, either to provide training or to promote its provision by other actors.

In addition to skills and business training, concrete actions also need to be taken to ensure improvements in workers' rights, social protection and social dialogue. This chapter has already noted deficits in these pillars for decent work. Werna and Saboor (forthcoming) provide an account of such deficits in specific relation to the greening of the urban economy, followed by suggestions to address such deficits, as summarized below.

Deficits in workers' rights and lack of social protection affect working conditions and productivity, and hence the capacity of workers to fully engage in the development of a green urban economy. To give one example, waste management, which is vital for a green urban economy, currently involves a substantial percentage of people working under unhealthy conditions (ILO 2011a).

The transition to a green economy also entails changes with new implications for working conditions, derived from the introduction of new technologies, new substances and work processes, changes in the structure of the workforce and work organization. One example is occupational health and safety. Activities related to the production of new/green products and the assembly of new equipment (e.g. solar panels and wind generators) needs specific provisions to protect the workers.

Finally, considering the existing deficits in social dialogue, efforts to green cities and towns may not be sustainable in the long run. Workers and employers should be involved in the discussions concerning the greening of urban areas together with the government, as they are the ultimate actors who will deliver the necessary green goods and services.

There are possibilities to address the plight of different groups of workers as well as deficits in social protection, based on ILO's conventions, recommendations and other instruments (ILO 2011a, b). Such ILO instruments can be complemented with particular initiatives of governments, the private sector, NGOs and other actors. One example is ILO's International Programme for the Elimination of Child Labour (ILO 2011b), which has been widely supported through local actions. Poor people have also now and then put together their own resources and organized their own risk protection via mutual health protection schemes and community surveillance. There are also cases of partnerships involving local governments, the private sector, and local communities.

With regard to social dialogue, it is important to seek new roles for trade unions and other actors. Where there are legal restrictions on the rights of sections of the workforce to organize, trade unions can campaign for their removal. There are also cases of collaboration between trade unions and employers to improve working conditions (ILO 2001). Informal urban workers can also organize themselves, as happened for instance in the construction industry in Dar-es-Salaam, Tanzania (Jason 2008). Such and other forms of organization and participation can be specifically linked to actions to improve the urban environment. On a broader scale, there is a growing discussion on the 'right to the city', where workers are important stakeholders. 'Right to the City' was also the theme of the latest World Urban Forum (2010), which included workers' unions and federations as participants. There are also early examples of specific coalitions of urban workers, such as in Brazil. At the same time, there are specific initiatives related to the rights of urban citizens to a better environment (Berg 2011; DePhillis 2011; Kousis 1999; Losito 2010; Network for Improved Policy in South Asia 2011; Tree Media Group 2011; UNEP 2005) Synergies among such initiatives should be explored, bringing together labor and environmental issues.

6 Conclusion

This chapter supports the argument that improvements in the urban environment are good for business and for employment. First, the chapter noted that current urban environmental problems affect the health of workers with spin-offs to productivity and working time. Therefore, improvements in the environment will benefit the urban economy by addressing such health-related issues. Furthermore, the chapter noted how improvements in the environment generate green businesses and employment, with examples from selected sectors. At the same time, such opportunities cannot be just a passive outcome of environmental improvements. Following, the chapter described actions that need to be taken to ensure that workers and enterprises are prepared, and decent working conditions are in place, all with a positive effect on the transition to a green urban economy.

As shown before, labor has different components, encompassed under the framework of decent work. All components are important, and each one has an impact on the greening of cities and towns. At the same time, there are mutual influences among the different decent work components. Therefore, it is important that urban actors, including workers, employers and governments, adopt a comprehensive and integrated approach to address decent work, together with improving the urban environment. The goal of sustainable cities and towns needs to encompass such an approach.

Local authorities play a vital role in greening cities and towns through investments, regulation, as well as incentives to the private sector. With regard to labor, while many local actions are under the coordination of the ministry of labor in each country, local authorities also have an important role. For example, they can ensure decent work conditions in the sectors or activities under their direct administration, such as parks, and through procurement regulations for the ones contracted out. They can also promote local social dialogue. Local social dialogue has the advantage of addressing labor issues that are specific to each municipality.

In addition, local governments can bring on board actors that by and large do not take part in social dialogue at the national level, such as unregulated workers and enterprises and community-based organizations. Marikina (Philippines) and the Municipal Decent Work Programmes in Belo Horizonte and a number of towns in the metropolitan region of Sao Paulo (Brazil) are examples of such local social dialogue. This dialogue has the comprehensive and integrated approach noted above, and can become an important instrument for greening cities and towns.

References

Berg N (2011) Grassroots environmental movement sprouts in China. Available via: http://www.planetizen.com/node/30782. Cited 24 June 2011

Campbell-Lendrum D, Corvalan C (2007) Climate change and developing country cities: implications for environmental health and equity. J Urban Health Bull N Y Acad Med 84:115–122

Comaru F (2011) Informal Urban Workers and Health: vulnerabilities, solutions, and perspectives. Paper produced for a post-doctoral research at the ILO and submitted for presentation at the international conference on informal economy, vulnerabilities and employment, University of Geneva, Geneva, 09–10 February 2012

DePhillis L (2011) The 'livability movement': successor to the environmental movement. Available via: http://www.washingtoncitypaper.com/blogs/housingcomplex/2011/01/14/the-livability-movement-successor-to-the-environmental-movement/. Cited 24 June 2011

Gallego E, Roca X, Perales JF, Guardino X (2009) Determining indoor air quality and identifying the origin of odour episodes in indoor environments. J Environ Sci 21:333–339

Gangolells M, Casals M, Gassó S, Forcada N, Roca X, Fuertes A (2009) A methodology for predicting the severity of environmental impacts related to the construction process of residential buildings. Build Environ Int J Build Sci Appl 44:558–571

Ghai D (ed) (2006) Decent work: objectives and strategies. International Institute for Labour Studies, International Labour Organization, Geneva

ILO (International Labour Organization) (2001) The construction industry in the twenty-first century: its image, employment prospects and skill requirements. Report for an ILO tripartite meeting on the construction industry. International Labour Organization, Geneva

ILO (International Labour Organization) (2004) Cities at work: employment promotion to fight urban poverty. SEED & EMP/INVEST Report. International Labour Organization, Geneva

ILO (International Labour Organization) (2006) A strategy for urban employment and decent work (Fact-sheet). International Labour Organization, Geneva

ILO (International Labour Organization) (2010) Accelerating action against child labour. IPEC Report. International Labour Organization, Geneva

ILO (International Labour Organization) (2011a) ILO background note to UNEP's 'Towards a green economy: pathways to sustainable development and poverty eradication'. Green Jobs Programme Report. International Labour Organization, Geneva

ILO (International Labour Organization) (2011b) Children in hazardous work – what we know, what we need to do. IPEC Report. International Labour Organization, Geneva

ILO (International Labour Organization) (forthcoming) Formulating projects and studies concerning labour issues in greening the sectors of the built environment. Guidance Manual of the Sectoral Activities Department. International Labour Organization, Geneva

Jason A (2008) Organizing informal workers in the urban economy: the case of the construction industry in Dar es Salam, Tanzania. Habitat Int 32:192–202

Keivani R, Tah JH, Kurul E, Abanda H (2010) Green jobs creation through sustainable refurbishment in the developing countries. Sectoral Activities Department Working Paper 275. International Labour Organization, Geneva

Khatib JM (ed) (2009) Sustainability of construction materials. CRC Press LLC, London

Kousis M (1999) Environmental protest cases: the city, the countryside, and the grassroots in Southern Europe. Mobilization Int Q 4:223–238

Lawrence R, Werna E (eds) (2009) Labour conditions for construction: building cities, decent work, and the role of local authorities. Wiley-Blackwell, Oxford

Lawrence RJ, Gil MP, Fluckiger Y, Lambert C, Werna E (2008) Promoting decent work in the construction sector: the role of local authorities. Habitat Int 32:160–171

Losito B (2010) Citizenship education in a changing urban environment: opportunities and challenges. Available via: http://www.bpb.de/files/BIOAQO.pdf. Cited 24 June 2011

Moavenzadeh F (1994) Global construction and the environment: strategies and opportunities. Wiley, Oxford

Network for Improved Policy in South Asia (2011) SHEHRI-citizens for a better environment. Available via: http://www.nipsa.in/shehri-citizens-for-a-better-environment/. Cited 24 June 2011

Ofori G, Abdul Aziz AR (2010) Energy efficiency and green jobs in the building industry in Malaysia. Internal report prepared for the International Labour Organization, Geneva

Sarsby R, Meggyes T (2010) Construction for a sustainable environment. Taylor & Francis Group, London

The Environment Agency in England and Wales; SEPA in Scotland and the Northern Ireland Environment Agency (2011) How can construction and building trades affect the environment.

Available via: http://www.netregs.gov.uk/netregs/businesses/construction/62311.aspx/. Cited 20 Jan 2011
Tree Media Group (2011) Urban roots. Available via: http://www.treemedia.com/treemedia.com/Urban_Roots.html. Cited 24 June 2011
Uddin MN (2006) The relationship between urban forestry and poverty alleviation. Doctorate Thesis Degree Project within Urban Forestry and Urban Greening (P0401). Dept. of Landscape Management & Horticultural Technology, Swedish University of Agricultural Sciences, Alnarp
UNEP (United Nations Environment Programme) (2005) Urban environmental accords. Available via: http://archive1.globalsolutions.org/programs/health_environment/urban_environmental_accords.pdf. Cited 24 June 2011
UNEP (United Nations Environment Programme) (2008) Green jobs: towards decent work in a sustainable, low-carbon world. United Nations Office at Nairobi, Nairobi
UNEP (United Nations Environment Programme) (2011) Towards a green economy: pathways to sustainable development and poverty eradication. United Nations Environment Programme, Nairobi
UN-Habitat (United Nations Human Settlements Programme) (2010) Taking forward the right to the city. Available via: http://www.unhabitat.org/downloads/docs/Dialogue1.pdf. Retrieved 23 June 2011
UN-Habitat (United Nations Human Settlements Programme) (2011) Cities and climate change: Global report on human settlements 2011. Earthscan for UN-HABITAT, London/Washington, DC
Van Empel C, Werna E (2010) Labour oriented participation in municipalities: how decentralized social dialogue can benefit the urban economy and its sectors. Sectoral Activities Department Working Paper 280. International Labour Organization, Geneva
Van Wyk L, Kolev M, Osburn L, de Villiers A, Kimmie Z (2010) Research on the employment aspects of energy-related improvements in construction in South Africa. ILO Report. International Labour Organization, Geneva
Waste Concern Consultants (2010) Assessment of green jobs in construction sector. Internal report prepared for the International Labour Organization, Geneva
WBCSD (World Business Council for Sustainable Development) (2009) Transforming the market: energy efficiency in buildings. World Business Council for Sustainable Development, Geneva
Werna E (2011) Green jobs in construction. In: Ofori G (ed) Contemporary issues in construction in developing countries. SPON Press, Abingdon, pp 408–441
Werna E, Saboor A (2012) The role of decent work conditions in greening the urban environment. In Kennet M. & Ka-Ming Mak W. (eds.), Green economics and climate change. Green Economics Institute, Reading
WHO (2004) World report on road traffic injury prevention: summary. WHO, Geneva. Available via: http://www.who.int/violence_injury_prevention/publications/road_traffic/world_report/summary_en_rev.pdf. Cited 5 Dec 2011
WHO (2012) Deaths from environmental risks in 2004, DALYs from environmental risks in 2004. Available via: http://www.who.int/quantifying_ehimpacts/global/envrf2004/en/index.html
WHO (World Health Organization) (2011) Health in the green economy – health co-benefits of climate change mitigation – housing sector. WHO, Geneva
Zordan SE (2011) Entulho da Industria da Construçao Civil. Department of Civil Construction Engineering, University of Sao Paulo, Brazil. Available via: http://recycled.pcc.usp.br/entulho_ind_ccivil.htm. Cited 13 Jan 2011

Sustainable Development for Cities and Citizens: Green Housing, Employment and Transport

Asif Kabani and Maliha A. Kabani

Abstract This chapter focuses on three key urban management initiatives by which smart growth and climate change adaptation are mobilized towards the production of a sustainable urban environment. Significant headway has been made and opportunities for further progress revealed in building energy retrofits, green workforce development and transport-oriented development. In their identification as local priorities, such urban management initiatives have proven a new resource for sustainable development. While such initiatives are encouraging and indicative of the opportunities that sustainable urban development provides, the authors also contend that cities must prioritize the needs of their low-income population as part of their green strategies and programs.

Keywords Climate change • Smart growth • Green economy • Sustainable development

---

A. Kabani (✉)
Founder and Hon. VP, International Sustainable Development Resource Centre, Geneva

Head of Research and Project Management, Ministry of Information and Broadcast, Govt of Pakistan
e-mail: kabani.asif@gmail.com

M.A. Kabani
Founder and Hon. President, International Sustainable Development Resource Centre, Geneva

Manager, Education - IKEA, Save the Children International (SCI), Country office, Pakistan
e-mail: kabani.maliha@gmail.com

R. Simpson and M. Zimmermann (eds.), *The Economy of Green Cities: A World Compendium on the Green Urban Economy*, Local Sustainability 3,
DOI 10.1007/978-94-007-1969-9_8,

1 Introduction

On one hand, recent developments in urban sustainability have been encouraging as most cities of the world begin to seize the opportunities that sustainable urban development provides. We have found that few cities prioritize the needs of their low-income population as part of their green strategies and programs in making policies, plans, programs and projects for sustainable development.

Over the past decade climate change has moved from a scientific theory to a reality all over the world from Global context to local reality. Governments and communities around the globe are moving quickly to cut greenhouse gas emissions in the hope of warding off the most devastating impacts of climate change. Most cities do not always need to wait for action from central or regional government to begin turning themselves into green "laboratories," testing ways to lower greenhouse gas emissions, and building healthier, cleaner, more sustainable environments. Many cities have or will soon have detailed plans on how they will reduce greenhouse gasses; many are calling for emission cuts of between 10 and 20% in the next 5–10 years. Cities, it seems, have seen the future and are embracing it – sustainability has become a priority.

It is no longer a question of "if" the nations of this world will begin the challenging transition to a greener economy but "how" we will get there. As this chapter reveals, the answer to that question lies in large part within cities around the world. Cities are on the vanguard of developing climate change solutions. It will be vital that as cities' green efforts are scaled up, it is also ensured that all residents, including the poorest, have access to the benefits of urban development initiatives especially developing countries.

2 Green Buildings

Cities have made green buildings a priority amongst other early efforts towards sustainability, but often require help with retrofitting current structures (Kabani 2011). Roughly half of all greenhouse gases emitted are produced in order to build, heat, cool and power the structures in which we live, work, shop and play (Pew Center on Global Climate Change 2005). It is no surprise therefore that many cities have devoted considerable resources in attempting to make buildings more efficient. The green building industry has grown immensely and green building techniques progressed substantially in the last 20 years. This is partly because, in the face of rising energy costs and with the aid of city-level authority over municipal and local building codes, greening buildings is often one of the first steps taken by city officials in order to "green" their cities. Many cities have made considerable progress with green buildings and have put tough new codes in place particularly with new city structures and commercial projects. This is a significant accomplishment and has made a serious dent in building related emissions. These efforts must not only target commercial and institutional buildings however but also include the residential sector. In this regard a particular effort must be made to include low-income households, who may lack resources and require extra help in order to insulate houses and replace appliances as part of the drive towards reducing emissions.

Such efforts can build on the experience of central and regional government as well as various agencies and their local partners which run programs helping the poor to weatherize their homes. Green retrofitting cannot only help cities achieve real and meaningful reductions in greenhouse gas emissions, it can also create new "green" jobs for blue-collar workers. By addressing the low- and mid-skilled job losses which have resulted from the current global financial crisis, investments in retrofitting can produce immediate economic impacts, a key consideration for policymakers and "green" initiatives (Satterthwaite 2010).

Cities will need to take great care to ensure that low-income residents have access to energy efficiency upgrades. Navigating the complexities of the various loans, utility bills and tax credits involved with retrofitting a home is a difficult task even for middle-income households. Households faced with particular financial difficulty may find retrofitting a home virtually impossible. While low-income households may face the biggest challenge in creating greener dwellings, these households are proportionally best placed to benefit from such changes and from their resulting cost and energy savings.

Many large cities are mandating that new public buildings should be built to green standards. Cities such as Karachi and Singapore go even further, requiring the same for private constructions (Kabani 2011). This focus on new buildings is however limiting especially where urbanization is slow or stagnant. Mass retrofits of existing structures are therefore key to significant greenhouse gas reductions. While many cities would like to encourage mass retrofits, paying for them is another enormous challenge, especially given that cities will need to ensure that such programs reach low-income neighborhoods.

3 Cities and Green Jobs

Smart growth in green cities will only be made possible if green jobs are made a priority. Cities are poised to help implement policy in this direction but the field is still in its inception. The notion that a vibrant "green jobs" sector can revitalize the economy has garnered significant attention in recent years especially from the media, politicians and NGOs. Even as economic recovery is underway, the green-collar job movement is in need of programs and leadership that can turn this promise into a reality. The shift to a green economy will test cities, as most economic and workforce development systems are outmoded, fragmented and unprepared to adequately respond to new opportunities. The green economy and green jobs have the capacity to become a real vehicle for engineering vital improvements in these sectors and for ensuring that underemployed low-income residents become a strong focus of new workforce efforts.

City officials and planners are well aware of the promise that a green jobs' sector and green collar jobs and industries hold. Most however have only recently begun to take preliminary measures, aided by partnership with surveying companies, universities and other interested parties. Such collaborative efforts have led to progress in at least some cities. For example, to help retrofit existing affordable housing units, Chicago recently instituted the Chicago Energy Efficiency Building Retrofit Program, a public-private partnership to provide financing and technical assistance to owners of affordable multiunit rental properties. The program will provide loans to building owners for energy and water efficiency measures that will be repaid from the operating savings those improvements make possible (Satterthwaite 2010). Despite this promise however, the number of green businesses and jobs remains modest.

As this chapter has indicated, one solution particularly in the short term would be to provide employment through green building and mass retrofits, thereby employing laid-off workers from the construction sector and simultaneously making buildings more efficient. Effective finance structures will be key to unlocking this opportunity. While most cities are currently exploring the potential of green jobs, nascent programs have led to just a handful of jobs so far. The task of developing and enabling a green jobs' sector to thrive will test both economic and workforce development, two arenas in which cities have traditionally struggled. It will however present an opportunity to re-engineer and adapt current systems especially in the employment sector and particularly regarding low-income workers and the unemployed who desperately need skills.

4 Transportation

Perhaps the most significant step a city can take towards becoming a green city is improving public access to greener forms of transportation. Traffic is a massive source of greenhouse gas emissions. in recent years, cities across the world are now started to have invested in multiple public transport projects, from the expansion of light rail networks to the addition of bus lines and cycle paths. The drive to extend and improve transport is not only informed by the desire to tackle climate change but also to revitalize neighborhoods, improve air quality and help the local economy. These policies have simultaneously made urban areas more efficient and more desirable places to live. A range of public transport projects are underway across the world in both developed and developing countries, even in the cities where highways have traditionally been prioritized. Despite these positive trends however, public transport remains a complex fiscal and logistical puzzle for many cities and regions.

Cities have also struggled to address questions of equity in their planning for public transport. It is likely that most new projects are not reaching poorer neighborhoods, limiting their ability to benefit from such schemes. Poorer neighborhoods are also more likely to be exposed to car fumes due to a lack of public transport in these areas, exposing residents to a higher risk of asthma and other illnesses linked to car and truck traffic. New transport systems can also act as a magnet for local development which can drive up retail and property prices. This can adversely affect poorer neighborhoods when lower-income residents and small business owners can no longer afford rents and are forced to move away.

In recent years rising energy costs have dramatically increased the popularity of public transport in virtually every city. As Kelly Rayne, policy advisor to the mayor of Shelby County, US, observes, "Before, the public viewed mass transit as something poor people take but Once gas prices [started rising], people had standing room only. That has done wonders for perception" (Rayne 2011). The popularity of public transport has also increased in areas of urban sprawl as city dwellers incur the cost of travelling long distances by car. Such areas often lack developable land for the provision of public services, and residents may need to travel long distances for access to such services. As a result areas of urban sprawl can become particularly congested. "We've seen that that kind of spread-out development isn't feasible... There has been a cry from the community for walkable neighborhoods and alternative transportation" (Rayne 2011).

As cities have begun to question their love affair with the car, they are also changing the way transportation serves cities. Cities are studying, expanding, and investing in public transport, particularly inter-city transport and rail networks. This is a huge task for any city, and requires building consensus with regional partners to get such large scale projects built. While one of the main sources of funding cities' transportation needs are federal grants and taxes, these funds are rarely awarded to individual cities directly. They are instead often given to planning bodies which typically include representatives either from city areas or from other municipalities and government.

Governance structures are rarely a major obstacle to the implementation of public transport systems, especially when partnerships between local governments and stakeholders are successful. About one in four cities do report difficulties however, especially when cities and government do not have adequate consultation with stakeholders and as a result different transport systems are poorly coordinated, making it hard to live without a car in much of the region. "People want a coordinated transport service and are ready to work together at every level, but right now we don't have clear leadership or institutional support to help get us to that next level of collaboration. A new light rail system was even more costly due to the need for extensive tunneling. Officials have made transport a priority, and while money is not always forthcoming a million in new revenue each year is creating a fresh and reliable income source for public transport and roads."

It remains to be seen which counties will levy a tax towards such initiatives and how funds will be spent. Finding the funds to expand public transport will be especially challenging for individual cities given that there is national competition in this area – cities across the world are investing in public transport and other strategies to reduce car use in order to spur neighborhood development. Where federal transportation policies complicate regional coordination and funding allocation, cities struggle to implement successful transport systems. As an additional challenge, cities also need to make sure that new transport systems benefit low-income communities and to work against the displacement of local residents, thus ensuring that sustainable development benefits all residents in adjoining neighborhoods as well as protecting the environment.

5 Conclusions and Recommendations

Green Cities are well positioned to lead the way towards an economic recovery that will launch the green economy. A growing number of cities are on the vanguard of addressing climate change issues, creating new mechanisms to simultaneously reduce energy waste, cut carbon emissions and create new economic opportunities. Despite this determined approach, such efforts at the city level are limited in isolation. Financing for public transport and building retrofits is exceedingly complicated and difficult to obtain. Initiatives aimed at providing green jobs and assisting low-income or unemployed workers are still very much in the nascent stage. Cities can, however, make far greater progress with the support of active and engaged stakeholders, of administration committed to both transport and energy retrofit programs, and by targeting workforce development funds specifically towards green-collar jobs. The agenda on energy and environmental issues must include a commitment to cap greenhouse gas emissions, deploy "smart grid" technologies and invest in sustainable growth and transportation. While cities should be able to count on assistance from central, regional and local government, they must continue to innovate and evolve in themselves. To make this transition successful, cities need to work with a wide range of local partners including business, service providers and charitable organizations.

Sustainability issues are inherently interconnected, and any thoughtful approach also requires cities to think across sectors (for example housing, transportation, economic development, education and workforce development, energy policy) and to act collaboratively in order to construct feasible sustainability plans. Consultation between experts, practitioners, and city and local administration is essential in order to advance greener buildings, greener jobs and smarter transport for low-income people. These recommendations do not purport to be comprehensive. Rather, we intend that these ideas help set the stage for cities – in partnership with local stakeholders and regional and central government, to forge a comprehensive set of policies that help to create greener and more equitable communities. In each of these areas we recommend a thematic approach to policymaking, as follows:

- To achieve the energy savings and green job opportunities possible through green buildings, cities must retrofit through systems that can achieve scale.
- To create green-collar jobs at scale, cities must re-engineer their local economic and workforce development systems.
- To spur more equitable transport-oriented development, cities need to reorient their local real estate markets.
- We are acutely aware that none of this will be easy. In many cases incremental progress and piecemeal solutions are the only option. We believe however that larger visions and sustained political commitment at the local level will be required to seize this moment of opportunity.

References

Brown MA, Southworth F, Stovall TK (2005) Towards a climate-friendly built environment. Pew Center on Global Climate Change, Arlington. Available via: http://www.pewclimate.org/docUploads/Buildings_FINAL.pdf

Kabani A (2011) Smart growth, green cities, with climate change and adaptation. Paper presented at the Pakistan urban Forum, Lahore. Available via: http://pakistanurbanforum.com/puf2011/Presentations/Tech%204b_1-SUE_Mr.AsifKabani.pdf

Rayne K (2011) Speech presented at Pakistan Urban Forum in Lahore

Satterthwaite D (2010) Green cities report – cities and climate change. Living Cities, New York

Cities and the Green Economy

Philipp Rode

Abstract This chapter is a condensed version of the Cities Chapter of UNEP's seminal publication 'Towards a Green Economy: Pathways to Sustainable Development and Poverty Eradication'. It analyses the particular role that cities play in the context of a transition to a Green Economy. Cities account for a substantial proportion of global greenhouse gas emissions, mostly due to the intense concentration of people, economic activity and transport in urban centers. But just as cities are the source of many environmental problems, they offer many of the solutions. Firstly, this chapter assesses the particular potential of cities for a transition to a Green Economy. Secondly, it argues the case for greening cities, in terms of the economic, social and environmental benefits that a Green Economy can bring to cities and their inhabitants. It then analyses various urban sectors and how these should be greened in order to create a green city and a greener urban economy. Keys to achieving this are the required enabling conditions which are imperative in order to overcome institutional and financial barriers. Setting the required enabling strategies and a regulatory framework is the foremost challenge for national, regional and local governments in terms of greener cities.

Keywords Green cities • Greening urban sectors • Governance • Planning and regulation • Civic engagement

*This is an abbreviated version of the cities chapter of the Green Economy Report, *Towards a Green Economy: Pathways to Sustainable Development and Poverty Eradication (2011)* by the United Nations Environment Programme (UNEP). A list of all contributors is available here: http://www.unep.org/greeneconomy/Portals/88/documents/ger/ger_final_dec_2011/12.0-CIT-Cities.pdf

P. Rode (✉)
London School of Economics and Political Science, London, UK
e-mail: p.rode@lse.ac.uk

R. Simpson and M. Zimmermann (eds.), *The Economy of Green Cities: A World Compendium on the Green Urban Economy*, Local Sustainability 3,
DOI 10.1007/978-94-007-1969-9_9, © UNEP 2013

1 Introduction

Green cities are defined as those that are environmentally friendly. They are cities that have adopted a green (urban) economy. Indicators measuring environmental performance can include: levels of pollution and carbon emissions, energy and water consumption levels, water quality, energy mix, waste volumes and recycling rates, green-space ratios, primary forests, and agricultural land loss (Meadows 1999; Brugmann 1999). Other indicators include urban density, motorization rates and modal share of transport. Furthermore, green cities are also those that have ambitious green policies, a range of green projects and a principal trajectory towards a better environmental performance such as Freiburg, Germany or Curitiba, Brazil. The latter has implemented policies to integrate land-use and transport planning, whereas Freiburg has long had a long tradition of sustainable building and investment in recycling, and has reduced CO_2 emissions per capita by 12% between 1992 and 2003 (Dünnhoff and Hertle 2005).

2 Challenges and Opportunities

2.1 Challenges

The global population living in urban areas is now over 50% and is expected to reach 69% by 2050 (UN Population Division 2006, 2010). In some regions, cities are expanding rapidly, while in others, rural areas are becoming more urban. In developing economies, rural urban migration and natural population growth is swelling cities and in rapidly urbanizing countries, particularly evident in China and India, the struggle to develop infrastructure, mobilize and manage resources has negative consequences for the environment.

Cities of different wealth levels impact the environment differently. Local environmental threats are most severe in poorer cities and relate to issues such as fresh water, sewage, health and the degradation of the living environment. As cities become more prosperous, with wider and deeper patterns of consumption and production, their environmental impacts are increasingly felt at the global level. Urban areas in prosperous economies concentrate wealth creation as well as resource consumption and CO_2 emissions. Globally, with a population share of just above 50% but occupying less than 2% of the earth's surface, urban areas concentrate 80% of economic output, between 60 and 80% of energy consumption, and approximately 75% of CO_2 emissions (Kamal-Chaoui and Robert 2009; UN Population Division 2010). Particular activities, for example buildings, transport, and industry – which are constituent components of cities and urban areas – contribute 25, 22, and 22% respectively of global GHG emissions (Herzog 2009).

Cities per se are neither drivers of climate change nor the source of ecosystem degradation; certain consumption and production patterns as well as certain population

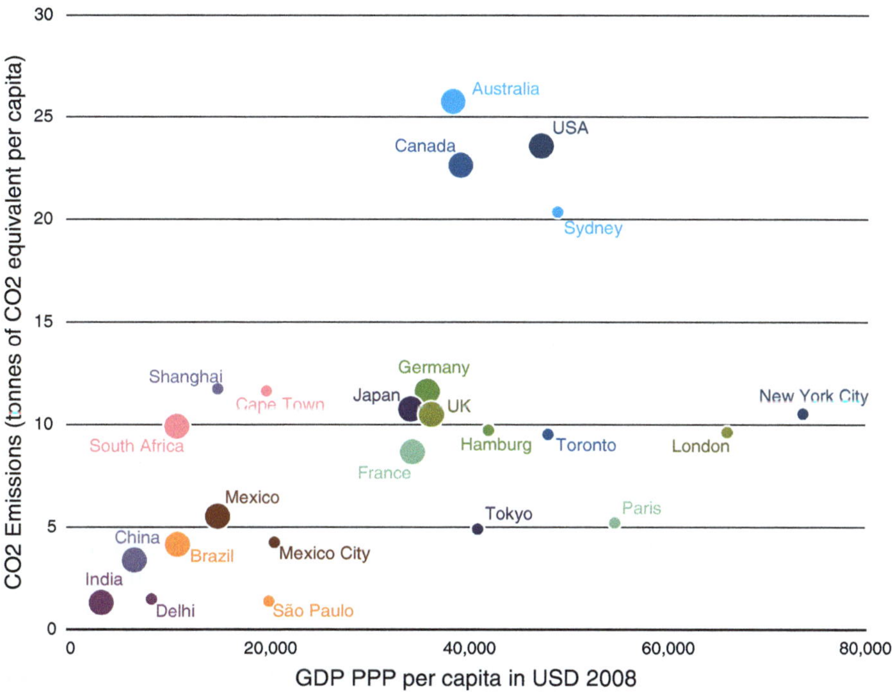

Fig. 1 Carbon emissions and income for selected countries and cities

groups within cities are. Carbon emissions are directly related to income. Per capita incomes are generally higher in cities than in rural areas, generating higher average per capita demand in major emissions sources. But this is the case only up to a certain income level, after which cities typically become more carbon-efficient compared with the average as can be seen by the relatively low levels of CO_2 emissions produced by high income cities like Tokyo or Paris (Fig. 1).

Patterns of urbanization in many areas also raise important social challenges. The traditional model of urban development – typical of rapidly urbanizing areas – is characterized by uncontrolled, often even incentivized, horizontal expansion. This leads to urban sprawl of affluent populations with lower development densities and increased dependency on the private car and to peripheralization of the urban poor, decreasing their access to the city and its workplaces, services and infrastructure. Typical developments further include the emergence of socially divisive neighborhoods in the form of gated communities, shopping centers and business districts, and a significant increase in the level of informal development with large swathes of slum housing with no access to basic services, infrastructure and sanitation. At a general level, the rapid growth of many cities coupled with insufficient resources and poor management compromises fresh water and electricity supply, waste treatment, transport, and other infrastructure provision, affecting the urban poor most.

2.2 *Opportunities*

2.2.1 Structural Capacity

The environmental performance of cities is dependent on a combination of effective green strategies and physical structure – urban form, size, density and configuration. They can be designed, planned and managed to limit resource consumption and carbon emissions. Or, they can be allowed to become voracious, land-hungry, all-consuming systems that ultimately damage the delicate global energy equation. More compact urban forms, reduced travel distances and investment in green transport modes lead to greater energy efficiency.

2.2.2 Technological Potential

Cities are incubators of innovation due to the close interaction of their residents and workers who benefit from the exchange of ideas and opportunities. In particular, they benefit from the concentration of diverse yet specialized skill-sets in research institutions, firms and service providers that can pilot and scale new technologies in an already highly networked environment. The OECD calculates for example, that there are ten times more renewable technologies patents in urban than rural areas and that 73% of OECD patents in renewable energy come from urban regions (Kamal-Chaoui and Robert 2009). The fast-growing 'cleantech' clusters in Silicon Valley and the North East of England are both examples of 'nursery cities', fostering innovative activity (Duranton and Puga 2001). Silicon Valley business leaders have been working for years to leverage the valley's 'innovation advantage' in the green economy (Joint Venture Silicon Valley Network 2009).

2.2.3 Urban Synergy and Integration Potential

Green cities can benefit greatly from synergies between their constituent parts. Recognizing, for example, the interrelationship of energy systems and city fabric can lead to particular synergies. An urban setting, which tends to support a diverse and compact pattern of production and consumption, is further advantageous to advance the notion of 'industrial ecology' (Lowe and Evans 1995). By optimizing and synergizing different industrial sectors and resource flows, outputs of one sector that become the input of another create a circular economy (McDonough and Braungart 2002). Principles of symbioses can also help cities minimize or recycle waste and improve overall resource efficiency.

3 The Case for Greening Cities

3.1 Economic Benefits

3.1.1 Agglomeration Economies

From an economic perspective, denser cities matter because they bring people and things closer together, help overcome information gaps, and enable idea flows (Glaeser 2008; Krugman 1991). It is for these reasons that 150 of the world's most significant metropolitan economies produce 46% of global GDP with only 12% of the global population (Berube et al. 2010). Agglomeration economies translate into productivity gains for firms and higher wages and employment rates for workers. Knowledge spillovers between firms and economic agents tend to be highly localized and die away within a few miles of the urban core (Rosenthal and Strange 2003).

There exists strong evidence that urbanization boosts productive efficiency by lowering transport and widening trade networks (Duranton 2008; Han 2009). Agglomeration economies can also be achieved by connecting several cities as in China's Pearl River Delta (Rigg et al. 2009), with the additional benefit of addressing inequality between leading and lagging regions in countries (Ghani 2010).

3.1.2 Lower Infrastructure and Operating Costs

Furthermore, densification reduces the capital and operating costs of linear infrastructure including streets, railway, water and sewage which can come at a considerably lower cost per unit in higher densities. Comparing smart growth areas and dispersed, car-dependent developments, Todd Litman suggests direct cost savings between US$5,000 and US$75,000 for building road and utility infrastructure per household unit (Litman 2009a) (Fig. 2).

Transport Infrastructure	Capacity [pers/h/d]	Capital costs [US$/km]	Capital costs/ capacity
Dual-lane highway	2,000	10m – 20m	5,000 – 10,000
Urban street (car use only)	800	2m – 5m	2,500 – 7,000
Bike path (2m)	3,500	100,000	30
Pedestrian walkway / pavement (2m)	4,500	100,000	20
Commuter Rail	20,000 – 40,000	40m – 80m	2,000
Metro Rail	20,000 – 70,000	40m – 350m	2,000 – 5,000
Light Rail	10,000 – 30,000	10m – 25m	800 – 1,000
Bus Rapid Transit	5,000 – 40,000	1m – 10m	200 – 250
Bus Lane	10,000	1m – 5m	300 – 500

Source: Rode and Gipp (2001), VTPI (2009), Wright (2002), Brilon (1994)

Fig. 2 Capacity and infrastructure costs of different transport systems

Cost savings are also derived from a shift away from car infrastructure towards public transport, walking and cycling. For example at similar capacity levels, bus rapid transit (BRT) offers significant cost savings compared to traditional metro and regional rail. Bogotá's TransMilenio infrastructure cost US$5.8 million per km, US$0.34 per passenger over 3 years compared with estimates for metro rail with US$101 million per km, US$2.36 per passenger (Menckhoff 2005).

3.1.3 Reduced Congestion Costs

Diamond (2005) suggest that the economic advantages of being in cities tend to mitigate even severe congestion problems and there also exist opportunities to deal with congestion through demand management, for example, via mechanisms such as congestion charges. London's congestion charge reduced congestion by 30% from February 2003 to February 2004 compared with previous years (Transport for London 2004a) and led to benefits such as the reduction in the number of trips by private vehicles entering central London (Transport for London 2004b) and a 19.5% drop in CO_2 emissions (Beevers and Carslaw 2005).

3.2 Social Benefits

3.2.1 Job Creation

Greening cities can create jobs on a number of fronts: (1) urban and peri-urban green agriculture; (2) public transport; (3) renewable energy; (4) waste management and recycling; and (5) green construction. Furthermore, green services will generally be more urban-orientated than green manufacturing or primary industry, although there will be some high-tech green manufacturing clusters in or close to urban cores, drawing on knowledge spillovers from universities and research labs. Overall, the green economy cannot be expected to create or destroy net jobs in the long run; the supply and demand for labor tend to equate in accordance with labor market conditions. In the short run, however, with unemployed resources, the net employment creation effect is likely to be larger.

3.2.2 Poverty Reduction and Social Equity

The World Development Report (2009) describes increasing economic density – one of the main features of a green city – as "a pathway out of poverty". Along similar lines, Nadvi and Barrientos (2004) assess the impact of clusters or agglomeration effects on poverty in several urban areas of developing countries. It is observed that these clusters are labor-intensive, informal in nature and also employ a lot of women as household-workers. Urbanization has reduced absolute poverty even though the number of people classified as urban poor is on the rise (Ravallion et al. 2007). Innovative approaches to urban planning and management can make urbanization inclusive, pro-poor and responsive to threats posed by environmental degradation and global warming. For example, enhancing public transport use can reduce inequality in access to public services and other amenities, on top of reducing carbon emissions (Litman 2002).

Switching to cleaner fuels for cooking, transport and power generation can minimize local pollution and reduce health inequality (Haines et al. 2007). Poor urban households in low-income nations have to spend a large proportion of their income on energy needs including food and cooking fuel (Karekezi and Majoro 2002). Introducing cleaner and more efficient sources of energy offers the potential to both reduce direct expenditure and to lower health costs connected to indoor-air pollution (Bruce et al. 2002). Retro-fitting older buildings in lower-income neighborhoods can improve energy efficiency and resilience, reducing the vulnerability of poorer communities when energy prices rise (Jenkins 2010). In addition, improving sanitation and fresh water supply can reduce persistent poverty and the adverse impacts of water-borne disease (Sanctuary et al. 2005).

3.2.3 Improvement in Quality of Life

Improving the urban environment by promoting "walkability" and introducing public green spaces can help foster a sense of community (Frumkin 2003; Litman 2006). Kuo et al. (1998) observed that the more trees and greenery form part of inner-city public spaces, the more these spaces are used by residents. The study also found that, compared with residents living near barren spaces, those closer to greenery enjoy more social activities, have more visitors, know more of their neighbors, and have stronger feelings of belonging. A further dimension in the quality of life surrounds road safety. Road traffic accidents are the leading cause of death among young people between 15 and 19 years, according to a report published by the WHO in 2007 (Toroyan and Peden 2007).

There also exists a relationship between green cities and cities with a high quality of life. Among the top 20 'quality of living cities' identified by Mercer in 2009, at least half have particularly strong green credentials such as Vienna, Zurich (strong focus on city transport) (Ott 2002) and Vancouver. Similarly, the integration of green space and natural elements within the city significantly enhance the quality of living which is also essential for attracting companies and a workforce to a city.

3.3 Environmental and Health Benefits

3.3.1 Reducing Pollution and Improving Public Health

Air pollution in cities remains a major public health burden, particularly in the developing world. Furthermore, there is a broader set of public health issues around healthier lifestyles in cities. It is estimated that physical inactivity accounts for 3.3%

of all deaths globally and for 19 million disability-adjusted life-years (Bull et al. 2004). Green urban transport is a unique opportunity to link physical activity and emissions reduction by promoting walking and cycling.

It is no coincidence that cities with a long tradition of applying land-use planning, public transport strategies and a focus on public green space are among the healthiest cities in the world. Portland was rated number one of the 100 largest USA cities in meeting Healthy People 2000 goals (Geller 2003), Vancouver is first amongst the Canadian cities (Johnson 2009), Copenhagen and Munich rank amongst the top ten healthiest and safest cities and Melbourne among the healthiest and safest in Australia (Sassen 2009).

3.3.2 Ecosystem Services and Risk Reduction

Urban greenery and vegetation represent a range of 'ecosystem services' with significant wider welfare effects (TEEB 2010). Ecosystem services further play a critical role in risk reduction measures. Restoration of urban ecosystems is part of the city greening effort, which can reduce the impact of freak weather conditions. Coastal regions in particular can benefit both in terms of lives and money. Mangrove replanting in Vietnam, for example, saves US$7.3 million annually on dike maintenance while it costs only US$1.1 million (International Federation of the Red Cross and Red Crescent Societies 2002. More generally, an increase in the amount of green cover in urban areas not only increases a city's ability to reabsorb CO_2 but also ameliorates the urban heat island effect (McPherson et al. 1994).

4 Greening Urban Sectors

4.1 Transport

The most familiar green transport strategies in cities primarily focus on reducing car use. Instruments such as the congestion charge in London and Singapore contributes to reducing car use, therefore congestion, pollution and GHG emissions. In Central London, for example, the congestion charge reduced daily vehicles trips by 65,000–70,000 (Transport for London 2004b) and CO_2 emissions by 19.5% (Beevers and Carslaw 2005).

Across a diverse range of cities, emission standards and car sharing schemes (Schmauss 2009; Nobis 2006) have reduced car dependency while low-emission zones and timed delivery permits have helped reduce congestion and pollution (Geroliminis and Daganzo 2005). Provided reliable and efficient public transport options, particularly surface transit such as bus rapid transit and light rail; cities can further reduce private car use, as can the promotion of Eco Mobility options – walking and cycling as the greenest forms of transport.

4.2 Buildings

Tackling the energy demand of existing building stock is a priority for cities, and urban green building strategies also include more efficient use of other resources such as water and materials. Three principal green building strategies can be differentiated: design, technology, and behavior-related. Particularly in a developing world context, passive design solutions to improve environmental performance are by far the most cost-effective approaches. For example, housing projects on the coast in Puerto Princesa City, the Philippines, have been designed to reduce energy demand through increased natural light, improved ventilation, the cooling effect of the roofing material, and strategic planting (ICLEI, UNEP and UN-HABITAT 2009).

Stringent building codes, mandatory energy certificates, tax incentives and loans, have had a measurable impact on energy demand in a number of European and US cities (C40 Cities 2010a). Berlin requires a solar-thermal strategy for all new buildings and Freiburg's energy efficient housing standard has reduced average household energy consumption for space heating by up to 80 percent (von Weizsäcker et al. 2009). Furthermore, municipal authorities are able to 'set an example' by implementing green strategies on their own public building stock.

4.3 Energy

Cities have the potential to either dissipate the distribution of energy or optimize their efficiency by reducing energy consumption and adopting green energy systems including renewable micro-generation, district heating, and combined heat and energy plants (CHP). In Freiburg, PV systems, encouraged by Germany's generous feed-in tariff, now supply 1.1% of the 470 Cities city's electricity demand. A biomass CHP system and wind turbines provide for a further 1.3 and 6% respectively of the city's energy needs (IEA 2009).

Oslo and São Paulo have harnessed power generated by nearby hydro-electric facilities to gain a relatively high share of renewable energy. Grid-based, decentralized energy system, with district heating systems can provide space and water heating for large urban complexes (like hospitals, schools or universities) or residential neighborhoods. They can significantly reduce overall energy demand. Their efficiency further improves with combined heat and power energy generation systems. Copenhagen's district heating system, for example, supplies 97% of the city with waste heat (C40 Cities 2010b).

4.4 Vegetation and Landscape

Parks, protected green space and gardens, street trees and landscaping provide vital ecosystem services, acting as "green lungs" absorbing and filtering air pollution or as acting as filters for waste water (TEEB 2010). They also provide a habitat for wildlife and offer recreational benefits to city dwellers.

Green landscaped areas help regulate natural processes, including the mitigation of local temperature extremes. Vegetation and "soft" open space also play a role in decreasing storm-water volumes, thus helping cities to manage the consequences of heavy rainfall etc. Measures for protecting green areas are of particular importance along the city fringe, where urban growth boundaries in cities such as Portland and London restrict development. In Stockholm, thanks to the protection of green areas, almost the entire population lives within 300 m of parks and green areas (City of Stockholm 2009).

New design strategies have pioneered the use of green roofs and facades on buildings, to add to the quantity of natural (as opposed to man-made) surfaces in cities and to reduce cooling energy demand. For example, Itabashi City in Tokyo is promoting climbing plants as "Green Curtains" around public buildings and private homes to avoid buildings overheating in summer and to reduce the use of air conditioning (ICLEI 2009b).

4.5 Water

Cities require significant transfers of water from rural to urban areas with water leakage being a major concern. Upgrading and replacement of pipes has contributed to net savings of 20% of potable water in many industrialized cities. Over the last 10 years alone, Tokyo's new water system has reduced water waste by 50% (C40 Cities 2010c). Volumetric charging has proven most effective in incentivizing more efficient water use. Many cities are introducing water meters and are shifting away from simple water-access fees.

To further reduce water consumption and provide alternatives to piped water supply, rain can be harvested and used as drinking and non-drinking water. To counter severe water shortages in Delhi, the Municipal Corporation made rainwater harvesting a requirement for all buildings with a roof area above 100 m^2 and a plot area greater than 1,000 m^2. It is estimated that 76,500 million liters of water per year will be made available for groundwater recharge (ICLEI, UNEP and UN-HABITAT 2009).

4.6 Food

The "food footprint" of a city has significant impacts on its green credentials, especially if one takes into account the energy use generated by transporting food from remote locations to urban marketplaces (Garnett 1996). For example, the food supply of European cities accounts for approximately 30% of their total ecological footprint (Steel 2008). Some cities do benefit from the location and have strong links with their regions; regarding food production and consumption.

Approximately 15–20% of the world's food is produced in urban areas. The extensive role of food production in cities is a common feature of many developing-world cities. Estimates suggest that 35% of households of Nakuru, Kenya were engaged in urban agriculture in 1998 (Foeken 2006; David et al. 2010). Furthermore, successful urban agriculture projects are scattered across some Western cities, albeit usually on a small scale, making use of communal gardens, roof spaces and unused urban spaces.

4.7 *Waste*

A waste economy plays a dominant role in a city's ecological footprint. Yet, cities can and have demonstrated considerable resilience in finding green solutions that reduce overall waste through new forms of environmentally friendly treatment of unavoidable waste. In developing world cities, which typically suffer from insufficient formal waste collection, informal recyclers and declaimers provide an essential service that often remains unrecognized.

In many European cities, recycling levels are in the region of 50%, while Copenhagen only sends 3% of its waste to landfills (C40 Cities 2010d). Composting is a further critical component for greening waste. Positive examples range from Dhaka's decentralized composting to San Francisco's municipal food composting programs (Zurbrügg et al. 2005).

4.8 *Infrastructure and Digital Technology*

A growing body of knowledge suggests that cities are the natural sites for investments in smart infrastructure to deliver more sustainable environments. The digital infrastructure of the internet and data centers creates an 'intelligent' infrastructure that connects people to people, people to city systems and city systems to each other, allowing cities and their residents to respond to changing circumstances by adapting in near real-time and to recognize patterns to help make informed decisions.

In addition, smart transport systems are being used to tackle congestion, facilitate road user charges or supply real-time information on traffic problems. Examples include Stockholm's congestion tax and Singapore's electronic road pricing. They also facilitate bike sharing schemes that are increasingly becoming an important component of urban transport worldwide, Amsterdam currently trials smart work centers that allow workers to use local office facilities rather than commuting to their main office (Connected Urban Development 2008).

5 Enabling a Green City

5.1 Barriers

- **Fragmented governance** – lack of coordination between policy frameworks that promote green economy measures at supra-national, national, regional and metropolitan levels;
- **Affordability** – even cost-effective green measures may be out of the reach of poorer cities, leaving them saddled with more wasteful urban infrastructure;
- **Lack of investment** – private and public sector have not prioritized green investment in basic city infrastructure (such as green planning, public transport and housing strategies);
- **Negative tradeoffs** – without effective policy intervention and infrastructure investment, green city strategies can lead to greater congestion, higher land values and costs of living;
- **Consumer preferences** – when given a choice consumers may not be willing to adopt new models of urban living that require changes in individual and collective patterns of consumption (e.g. high-density apartment living, public transport use);
- **Switching costs** – high short-term transition (welfare and capital) costs for businesses that shift from brown to green, leave many companies without adequate compensation to make the investment;
- **Vested business interests** – industry dynamics in construction, road-building and infrastructure are resistant to change that challenges existing business models and threatens the potential of short-term return on investment;
- **Risk aversion** – individuals, corporate and government organizations are resistant to any change that does not demonstrate immediate improvement in economic well-being, quality of life or enhanced status within the community; and
- **Behavioral response and the rebound effect** – consumers may respond to reduced energy costs (generated by energy efficiency measures) by either increasing per (Fig. 3)

5.2 Enabling Strategies

Overcoming this set of barriers and constraints requires a multi-faceted response across different sectors, which are addressed in turn, from governance and planning to incentives and financing. Enabling conditions, institutional strength and democratic maturity illustrates the breadth of policy instruments and tools that can promote investment in greening cities. Importantly, it correlates their effectiveness over time in relation to the strength of local institutions and the strength of the democratic system in different urban contexts. By plotting the enabling conditions available in systems with both 'strong' and 'weak' institutions against weaker and more

Fig. 3 Enabling conditions, institutional strength and democratic maturity

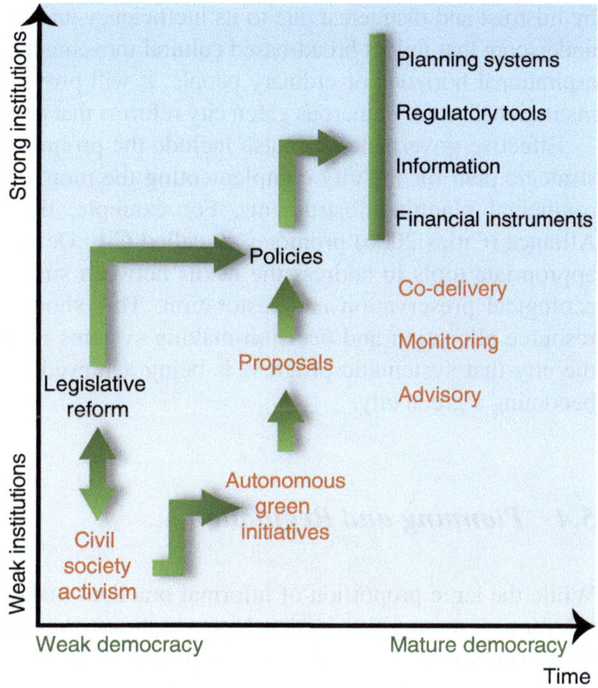

mature democracies, it suggests that the process of change is in most cases a long one, and requires the development of 'mature' institutions before long-term change can be implemented, whilst recognizing that civil-society, activism and autonomous green initiatives can be effective in the short-to-medium term, especially in weaker institutions and less mature democracies.

5.3 *Governance*

Governance encompasses the formal and informal relationships linking the various institutions involved in the urban system – the local, metropolitan, regional, state, civil society and private-sector actors – and its quality depends on the depth of reciprocity, trust, and legitimacy. These are enhanced by mechanisms and opportunities to facilitate meaningful dialogue, and by well-structured organizations in civil society, the business sector and the relevant government level. The practical imperatives of debating trade-offs and priorities in pursuing green city development can contribute to the maturing of governance relationships.

In contexts with strong local government it is possible to envisage a range of planning, regulatory and financing instruments to advance a multitrack approach to greater urban sustainability, however due to issues of mistrust and corruption, this may prove difficult for local governments. In countries where local government is weak or marked by mistrust and disinterest due to its inefficiency and/or corruption, it is important to underscore that unless broad-based cultural movements are fostered that can shift the aspirational horizons of ordinary people, it will prove very difficult to promote and institutionalize the numerous green city reforms that are usually proposed.

Effective governance can also include the promotion of the idea of a long-term strategic plan for the city complementing the more conventional spatial and environmental planning instruments. For example, the internationally-based Cities Alliance (Cities 2006) promotes so called City Development Strategies (CDS), as appropriate tools to address the nexus between sustainable economic growth and ecological preservation and restoration. This should be backed up by effective resource allocation and decision-making systems that demonstrate to everyone in the city that systematic progress is being achieved towards the long-term goal of becoming a green city.

5.4 Planning and Regulation

While the large proportion of informal practices make planning and regulation less relevant in some cities in developing nations, they are the most common policy instruments that shape urban development in more complex and mature political environments. Urban planning can divert development to a desired course whereby, denser urban forms allow for economies of scale and sustainable service provisions.

To maximize synergies across different urban sectors, integrated planning that combines land use and urban development with other policies and cuts across the urban functional region of cities is critical in achieving greater environmental performance, as well as economic viability. The recently launched World Bank Eco2 Cities program, for example, demonstrates why planning, finance and infrastructure imperatives are inextricably linked in a low-carbon world (Suzuki et al. 2010). This program argues for a one-system approach to: "realize the benefits of integration by planning, designing and managing the whole urban system." On a practical level this implies that all cities need to understand their urban form and the nature and patterning of material resource flows through the urban system.

The intersections of infrastructure and the dynamics, resilience or vulnerability of urban form are crucial. As described previously, it is not untypical for poor people to live without access to various infrastructure networks in the most climate-vulnerable areas of a city (Moser and Satterthwaite 2008). Possible impacts on urban form and resource flows need to be considered when planning infrastructure investments, especially given the enormous sums required for capital expenditure in rapidly urbanizing areas. More than anything else, urban sustainability will depend upon how these sums are going to be allocated.

A combined understanding of urban form and resource flows helps isolate effective actions to achieve greater overall resource efficiency. It also forces a longer-term horizon for understanding trends, the most strategic intervention points, and how to weigh up trade-offs between various spaces of an urban region. If it is based on sound data, it will hold the potential to provide a shared basis for understanding what is going on in a city, where it may be leading and what needs to be done to change the efficiency of the overall system (Crane et al. 2010). It is only when this kind of analysis and political discussion becomes commonplace, that one can achieve a broad-based commitment to effective long term strategic planning.

5.5 *Information, Awareness and Civic Engagement*

Effective planning and governance across different administrative levels requires high-quality information to raise awareness amongst urban residents to promote behavior change, especially in terms of consumer choice. In addition, given that cities contain large consumer markets which are potentially valuable to producers of green goods and services, information is also an essential tool to influence consumer choice. But consumer preferences, in developed and developing nations, are not always green. For example, very dense urban development is not always popular in many parts of the UK and Europe (Cheshire 2008) and the North American propensity for suburbanization is well documented. At the same time, information and active communication on the potential benefits of greener lifestyles in cities can enable consumers to make more informed decisions. For example, in Munich new residents are given an information package on green mobility opportunities.

5.6 *Incentives*

Information alone is insufficient to change behavior patterns; it needs to be supplemented by incentives to bring about lasting change. For example, firms and workers in 'brown industries' may face higher prices as cities shift their industrial structures towards greener models. National and city-level policy makers need to compensate these short-term losers while recalibrating urban economies. Incentives may be within the tax system (e.g. tax breaks or taxing environmental "bads"), other types of charges (e.g. road pricing) or payments (e.g. targeted subsidies). Subsidies were successfully used as part of the policy mix in the Bavarian example where Munich now has the largest green tech cluster in Germany. At the same time, full cost pricing (internalizing external environmental costs), whether as taxes or user charges is essential for inducing behaviors to be consistent with green city criteria.

5.7 *Financing*

Finance, particularly in times of government austerity can be a significant barrier and in some countries, national fiscal policy prevents local authorities from

Taxes	Cities need to be able to raise local taxes and service charges as they are the main revenues sources that can be used for public green city strategies
Cost recovery	Introduce user fees of municipal services to help greening these services and supporting the development of greener alternatives
Land value capturing	Financing public transport based on integrated "transport-property" development models
Micro-financing	Critical financing opportunity where micro-enterprises are involved in green city strategies, e.g. recycling developing country cities
Profit-making public companies	Cities to hold shares of profit making companies, e.g. utilities to allow for long-term green investments
Purchasing pools	Cities can also work together to purchase technology thereby bringing down the cost
Carbon credits	Clean Development Mechanisms (CDM) already pay for a range of green city projects in Bogotá, São Paulo and Dhaka

Fig. 4 Selected financing instruments

accessing the relevant funds. There are three key imperatives central to advancing on green city finance.

1. Understanding the existing financial position in terms of potential revenue, where a comparison with similar sized cities internationally is helpful.
2. City governments need to initiate a partnership with local business and community organizations, which is essential for leveraging private sector finance.
3. Horizontal and vertical networks are required especially for cross-municipal cooperation and regional and international participation in various local government policy forums.

Such examples of green financing include London and Paris, where urban bike hire schemes are paid for privately in return for prime advertising space. Figure 4 illustrates a breakdown of financing opportunities. A priority in any green urban planning is investment in cost-effective public transport infrastructure particularly over investment in road construction that further promotes private car use. Surface public transport such as bus rapid transit needs to play a central role particularly in lower income contexts. Non-motorized transport has to be recognized as basis of any transport system and requires greater shares of overall transport budgets.

6 Conclusions

Greening cities results in a wide range of social and economic benefits. As well as lowering carbon emissions, densification as a green city strategy tends to enhance productivity, promote innovation and reduce the capital cost of infrastructure. Densification can also raise congestion and the local cost of living, but green city strategies and interventions to subsidize housing costs can help to mitigate these.

In most countries cities will be important sites for the emerging green economy. Cities' basic offer of proximity, density and variety delivers productivity benefits for firms, and helps stimulate innovation and new job creation – for example in high-tech clusters, as are already emerging in urban regions like the Silicon Valley. Much of a green economy is service-based, and will tend to cluster in urban areas where consumer markets are largest.

Numerous instruments for enabling green cities are available and tested but need to be applied in a tailored, context-specific way. In contexts with strong local government it is possible to envisage a range of planning, regulatory, information and financing instruments to advance green infrastructure investments, green economic development and a multitrack approach to greater urban sustainability. City governments need to coordinate policies and decisions with other levels of government, but more importantly, they need to be equipped with strategic and integrated planning capacities, including the capacities to choose regulatory tools and economic incentives to achieve locally appropriate green city objective.

References

Beevers S, Carslaw D (2005) The impact of congestion charging on vehicle emissions in London. Atmos Environ 39:1–5

Berube A, Friedhoff A, Nadeau C, Rode A, Paccoud A, Kandt J, Just T, Schemm-Gregory R (2010) Global metro monitor: the path to economic recovery. Metropolitan Policy Program. The Brookings Institution, Washington, DC and LSE Cities, London School of Economics and Political Science, London

Bruce N, Perez-Padilla R, Albalak R (2002) The health effects of indoor air pollution exposure in developing countries. World Health Organization, Geneva

Brugmann J (1999) Is there a method in our measurement? The use of indicators in local sustainable development planning. In: Satterthwaite D (ed) Sustainable cities, the Earthscan Reader. Earthscan, London

Bull F, Armstrong T, Dixon T, Ham S, Neiman A, Pratt M (2004). Physical inactivity. In: Ezzati M, Lopez A, Rodgers A, Murray CJL (eds) Comparative quatification of health risks. World Health Organization, Geneva

C40 Cities (2010a) C40 cities: best practices – energy. Available via: www.c40cities.org/bestpractices/energy/. Cited 10 Dec 2010

C40 Cities (2010b) Copenhagen, Denmark: 97% of Copenhagen city heating supplied by waste heat. Available via: www.c40cities.org/bestpractices/energy/copenhagen_heat.jsp. Cited 10 Dec 2010

C40 Cities (2010c) Tokyo, Japan – world leader in stopping water leakage. Available via: www.c40cities.org/bestpractices/water/tokyo_waterworks.jsp. Cited 10 Dec 2010

C40 Cities (2010d) Copenhagen, Denmark – Copenhagen's waste plan 2008: Copenhagen puts only 3% of waste into landfill. Available via: www.c40cities.org/bestpractices/waste/copenhagen_landfill.jsp. Cited 10 Dec 2010

Cheshire P (2008) Reflections on the nature and policy implications of planning restrictions on housing supply. Discussion of 'Planning policy, planning practice, and housing supply' by Kate Barker. Oxford Rev Econ Policy 24(1):50–58

Cities Alliance (2006) Guide to city development strategies. Improving urban performance. Cities Alliance, UNEP and ICLEI, Washington, DC

City of Stockholm (2009) The city of Stockholm's climate initiatives. Environment Administration, City of Stockholm. Available via: www.stockholm.se/international. Cited 10 Dec 2010

Connected Urban Development (CUD) (2008) Smart work centers: will they work? CUD blog [blog] 3 December. Available via: http://www.connectedurbandevelopment.org/blog/?p=22. Cited 10 Dec 2010

Crane W, Swilling M, Thompson-Smeddle L, De Witt M (2010) Towards urban infrastructure sustainability. In: Pieterse E (ed) Counter-currents: experiments in sustainability in the Cape Town region. Jacana Media, Johannesburg

David S, Lee-Smith D, Kyaligonza J, Mangeni W, Kimeze S, Aliguma L, Lubowa A, Nasinyama G (2010) Changing trends in urban agriculture in Kampala. In: Prain G, Karanja N, LeeSmith D (eds) African urban harvest: agriculture in the cities of Cameroon, Kenya and Uganda. Springer and Ottawa IDRC, New York

Diamond J (2005) Collapse: how societies choose to fall or survive. Penguin, London

Dünnhoff E, Hertle H (2005) Ergebnisse der CO_2 – Bilanzierung für die Stadt Freiburg 1992 bis 2002/2003. IFEU, Heidelberg

Duranton G (2008) Viewpoint: from cities to productivity and growth in developing countries. Can J Econ/Revue canadienne d'économique 41(3):689–736

Duranton G, Puga D (2001) Nursery cities: urban diversity, process innovation and the life cycle of products. Am Econ Rev 91(5):1454–1477

Foeken D (2006) To subsidize my income – urban farming in an East African town. Brill, Leiden/Boston

Frumkin H (2003) Healthy places: exploring the evidence. Am J Public Health 93(9):1451–1456

Garnett T (1996) Growing food in cities: a report to highlight and promote the benefits of urban agriculture in the UK. National Food Alliance and SAFE Alliance, London

Geller AL (2003) Smart growth: a prescription for liveable cities. Am J Public Health 93(9):1410–1415

Geroliminis N, Daganzo CF (2005) A review of green logistics schemes used in cities around the world. UC Berkeley Center for Future Urban Transport: A Volvo Center of Excellence. Institute of Transportation Studies, UC, Berkeley

Ghani E (2010) The poor half billion in South Asia: what is holding back lagging regions? OUP, New Delhi

Glaeser E (2008) Cities, agglomeration and spatial equilibrium. OUP, Oxford

Haines A, Smith K, Anderson D, Epstein P, McMichael A, Roberts I, Wilkinson P, Woodcock J, Woods J (2007) Policies for accelerating access to clean energy, improving health, advancing development and mitigating climate change. The Lancet 370(9594):1264–1281

Han Z (2009) A model of clustering process in low income economies. Int J Bus Manage 4(12):46–51

Herzog T (2009) World greenhouse gas emissions in 2005. WRI working paper. World Resources Institute, Washington, DC. Available via: http://pdf.wri.org/working_papers/world_greenhouse_gas_emissions_2005.pdf. Cited 10 Dec 2010

ICLEI (2009b) Itabashi: leader in green curtain movement. Available via: http://www.iclei.org/index.php?id=9853 and follow the document links

ICLEI, UNEP, UN-HABITAT (2009) Sustainable urban energy planning: a handbook for cities and towns in developing countries. UN-HABITAT, Nairobi

IEA (2009) Cities, towns and renewable energy: yes in my front yard. IEA Publications, Paris

IFRC – International Federation of the Red Cross and Red Crescent Societies (2002) World disasters report 2002. IFRC, Geneva

Jenkins DP (2010) The value of retrofitting carbon-saving measures into fuel poor social housing. Energ Policy 38(2):832–839

Johnson T (2009) Canada's healthiest cities 2009. Best Health Magazine Online. Available via: www.besthealthmag.ca/gethealthy/health/canadas-healthiest-cities-2009. Cited 10 Dec 2010

Joint Venture Silicon Valley Network (2009) Climate prosperity. A greenprint for Silicon Valley. TDA, San Jose

Kamal-Chaoui L, Robert A (2009) Competitive cities and climate change. OECD regional development working papers 2009/2. OECD, Public Governance and Territorial Development Directorate, Milan

Karekezi S, Majoro L (2002) Improving modern energy services for Africa's urban poor. Energ Policy 30(11–12):1015–1028

Krugman P (1991) Increasing returns and economic geography. J Polit Econ 99(3):483–99

Kuo FE, Sullivan WC, Levine Coley R, Brunson L (1998) Fertile ground for community: inner-city neighbourhood common spaces. Am J Commun Psychol 26(6):823–851

Litman T (2002) Evaluating transportation equity. World Transport Policy Pract 8(2):50–65
Litman T (2006) Cities connect: how urbanity helps achieve social inclusion objectives. Paper presented at Metropolis conference, Toronto, Canada, 14 June 2006. Victoria Transport Policy Institute, Victoria. Available via: http://www.vtpi.org/citiesconnect.pdf. Cited 10 Dec 2010
Litman T (2009a) Understanding smart growth savings. What we know about public infrastructure and service cost savings, and how they are misrepresented by critics. Victoria Transport Policy Institute, Victoria. Available via: http://www.vtpi.org/sg_save.pdf. Cited 10 Dec 2010
Lowe EA, Evans LK (1995) Industrial ecology and industrial ecosystems. J Clean Prod 3(1–2):47–53
McDonough W, Braungart M (2002) Cradle to cradle: remaking the way we make things. North Point Press, New York
McPherson EG, Nowak DJ, Rowntree RA (eds) (1994) Chicago's urban forest ecosystem: results of the Chicago urban forest climate project. Gen. Tech. Rep. NW-186. U.S. Department of Agriculture, Forest Service, Northeastern Forest Experiment Station, Radnor, PA
Meadows D (1999) Indicators and information systems for sustainable development. In: Satterthwaite D (ed) The Earthscan reader in sustainable cities. Earthscan, London
Menckhoff G (2005) Latin American experience with bus rapid transit. Paper presented at the annual meeting of the Institute of Transportation Engineers, Melbourne
Moser C, Satterthwaite D (2008) Towards pro-poor adaptation to climate change in the urban centres of low- and middle-income countries. Human settlements working paper series climate change and cities, 2. IIED, London
Nadvi K, Barrientos S (2004) Industrial clusters and poverty reduction. United Nations Industrial Development Organization (UNIDO), Vienna. Available via: http://www.unido.org/index.php?id=o24736
Nobis C (2006) Carsharing as key contribution to multimodal and sustainable mobility behavior: carsharing in Germany. Transport Res Rec J Transport Res Board 1986:89–97
Ott R (2002) The Zurich experience. In: Greater London Authority, Alternatives to congestion charging. Proceedings of a seminar held by the Transport Policy Committee, GLA, London
UN Population Division (2006) World urbanisation prospects: the 2005 revision. Executive Summary, Fact Sheets, Data Tables. UN, Department of Economic and Social Affairs, New York
UN Population Division (2010) World urbanisation prospects: the 2009 revision. UN Department of Economic and Social Affairs, New York
Ravallion M, Chen S, Sangraula P (2007) New evidence on the urbanization of global poverty. Popul Dev Rev 33(b4):667–701
Rigg J, Bebbington A, Gough KV, Bryceson DF, Agergaard J, Fold N, Tacoli C (2009) The World Development Report 2009 reshapes economic geography: geographical reflections. Trans Inst Brit Geogr 34(2):128–136
Rosenthal S, Strange W (2003) Geography, industrial organisation and agglomeration. Rev Econ Stat 85(2):377–393
Sanctuary M, Tropp H, Berntell A, Haller L, Bartram J, Bos R (2005) Making water a part of economic development. Stockholm International Water Institute (SIWI), Stockholm
Sassen S (2009) Cities in today's global age. SAIS Rev 29(1):3–34
Schmauss A (2009) Car2go in Ulm, Germany, as an advanced form of car-sharing. European Local Transport Information Service (ELTIS). Available via: www.eltis.org/PDF/generate_pdf.php?studyid=2121&lan=en. Cited 10 Dec 2010
Steel C (2008) Hungry city. Chattoo and Windus, London
Suzuki H, Dastur A, Moffatt S, Yabuki N, Maruyama H (2010) Eco2 cities: ecological cities as economic cities. The World Bank, Washington, DC
TEEB (2010) The economics of ecosystems and biodiversity: mainstreaming the economics of nature: a synthesis of the approach, conclusions and recommendations of TEEB
Toroyan T, Peden M (eds) (2007) Youth and road safety. World Health Organization, Geneva
Transport for London (2004a) Congestion charging Central London: impacts monitoring. Second annual report. Transport for London, London

Transport for London (2004b) TfL publish C-Charge annual report. Transport for London [online]. Available via: www.tfl.gov.uk/static/corporate/media/newscentre/archive/4339.html. Cited 10 Dec 2010

Von Weizsäcker E, Hargroves K, Smith MH, Desha C, Stasinopoulos P (2009) Factor five. Earthscan, London

Zurbrügg C, Drescher S, Rytz I, Sinha M, Enayetullah I (2005) Decentralised composting in Bangladesh, a win-win situation for all stakeholders. Resour Conserv Recycl 43:281–292

Green Cities into Practice

Vito Albino and Rosa Maria Dangelico

Abstract This chapter aims to explain how green economy principles can be applied to cities to make them *green* cities. Toward this aim the green economy concept is first described. Then a matrix is employed to characterize green practices that are connected with the green economy and are adopted in the urban environment. The matrix will draw upon a sample of best performing cities. This matrix is built upon two analytical dimensions: environmental focus and impact. Specifically, three types of environmental focus are considered: material, energy, and pollution; and three types of environmental impact are considered: less negative, neutral, and positive. The sample city is based upon the winning cities of the European Green Capital Award from 2010 to 2013 (Stockholm, Hamburg, Vitoria-Gasteiz, and Nantes). This chapter could be particularly useful for policy makers, since it offers structured examples of best practices to make cities *green* along green economy principles.

Keywords Green economy • Green cities • Green practices • European Green Capital Award

1 Introduction

The economic model pioneered by today's industrial countries is not viable for the world as a whole. In fact today's economy is profoundly out of balance with the world's natural system with impacts on economic and social equity among nations.

V. Albino (✉)
Department of Mechanics, Mathematics and Management,
Politecnico di Bari, Bari, Italy
e-mail: v.albino@poliba.it

R.M. Dangelico
Department of Computer, Control and Management Engineering,
Sapienza – University of Rome, Rome, Italy
e-mail: rosamaria.dangelico@uniroma1.it

R. Simpson and M. Zimmermann (eds.), *The Economy of Green Cities: A World Compendium on the Green Urban Economy*, Local Sustainability 3,
DOI 10.1007/978-94-007-1969-9_10,

As stated in Chapter 4.3 of Agenda 21 (United Nations 1993), "the major cause of the continued deterioration of the global environment are the unsustainable patterns of consumption and production, particularly in industrialized countries, which is a matter of grave concern, aggravating poverty and imbalances".

Urban areas are the source of many of today's environmental challenges and can play an important role in achieving the objectives of the EU Sustainable Development Strategy (Council of the European Union 2006). Several diverse initiatives are being put in place by pioneering green cities. However, most European cities are far from realizing an urban environment that could be considered environmentally sustainable. This chapter aims at providing structured examples of best practices to make cities *green* in line with a green economy.

The chapter is structured as follows. We first describe the green economy. Then green practices which are connected with the green economy and adopted in the urban environment by a sample of best performing cities are characterized through the use of the Green Cities Practices Matrix. Finally we conclude by discussing the main findings and highlighting key implications for policy makers.

2 Green Economy

In a world characterized by climate change and financial crisis it has become clear that our global community has to adopt more sustainable patterns of production and consumption to reduce both the use of natural resources and pollution, and move towards low-carbon societies and a green economy (UNEP 2011). This is crucial in order to decouple economic growth from the exploitation and degradation of the environment in both developing and developed countries, while creating opportunities for the poor to meet their basic needs. Social inequality in many countries, often caused by economic disparities in the distribution of economic assets and income, is cause for discrimination and negatively affects people's sense of dignity. Moreover, large economic and social disparities can cause social instability and damage economic development.

In the last years the debate about the concept of a new and green economy has grown dramatically. Political and academic attention has been forced to consider new economic principles and values. Green economy has been defined by UNEP as "one that results in improved human well-being and social equity, while significantly reducing environmental risks and ecological scarcities". In its simplest expression, a green economy can be thought of as "one which is low carbon, resource efficient and socially inclusive" (UNEP 2011). A green economy not only generates increases in wealth, in particular a gain in ecological commons or natural capital, but also produces a higher rate of GDP growth. A transition to a green economy should eventually create new jobs at a rate which over time will exceed job losses caused by the "brown economy". There is also an important link to poverty eradication and the better maintenance and conservation of the natural environment, as the income poor often derive direct benefit from local natural capital.

In addition to recognizing the value of and investing in natural capital, alleviating poverty and enhancing social equity, a green economy entails several actions to create green urban economies. Specifically, as highlighted by UNEP (2011), a green economy substitutes fossil fuels with renewable energy and low-carbon technologies, promotes enhanced resource and energy efficiency, as well as delivers more sustainable urban living and low-carbon mobility.

3 A Characterization of Green Practices of Best Performer Cities

3.1 The Sample Cities

The sample cities represent best European performers in applying green practices, which are also connected to green economy principles. The selected cities are the winners of the European Green Capital Award 2010–2013: Stockholm, Hamburg, Vitoria-Gasteiz, and Nantes. The European Green Capital Award (EGCA) is an initiative developed by the European Commission to address today's environmental challenges and highlight the key role of urban areas. This award recognizes local efforts to improve the environment, the economy and the quality of life in cities and is awarded each year, since 2010, to a city leading in environmentally friendly urban living (European Commission 2012a).

Specific conditions to be satisfied by the winning city are:

- Having a consistent record of achieving high environmental standards.
- Being committed to ongoing and ambitious goals for further environmental improvement and sustainable development.
- Having the potentialities to act as a role model to inspire other cities.

3.2 The Green Cities Practices Matrix

The Green Cities Practices Matrix (GCPM) is an adaptation of the Green Option Matrix, developed by Dangelico and Pontrandolfo (2010), for urban areas. The GCPM can be employed to characterize green urban practices, connected to green economy principles. This matrix is built upon two analytical dimensions: environmental focus and impact. Three types of environmental focus are considered: material, energy, and pollution. Environmental impacts are distinguished by less negative, neutral, and positive.

By *environmental focus* we refer to the main category of environmental impact of a green practice, as the latter can improve its impact on the environment with emphasis on materials (including water), energy, or pollution (emissions and toxic waste) (Dangelico and Pontrandolfo 2010).

Once the respective main type of environmental focus of a green practice is identified, the *type of environmental impact*, as less negative, neutral, or positive can be assessed. A practice can be considered green, in terms of one environmental focus, if it has an environmental impact lower than conventional/commonly used practices, or if it has a neutral impact, or if it positively contributes to the natural environment i.e. enhancing it. Below we detail the definition with respect to each environmental focus.

A green practice with a focus on materials is, for example, a practice that:

- Allows using fewer materials than conventional/commonly used practices would (less negative environmental impact).
- Allows using only recycled materials or natural/biodegradable materials at a sustainable rate (neutral environmental impact).
- Fosters the recycling, reuse, disassembly and remanufacture, so reducing the environmental impact of products or practices that will not require the consumption of virgin materials (positive environmental impact).

Similarly, a green practice with a focus on energy is, for example:

- More energy efficient than conventional practices and thereby uses less energy, or if part of the energy used comes from renewable energy sources (less negative environmental impact).
- Uses only energy from renewable sources (neutral environmental impact).
- Produces energy from renewable sources, and in so doing reduces the need for energy from fossil fuels (positive environmental impact).

A green practice with a focus on pollution is, for example, a practice that:

- Is less polluting than conventional practices (less negative environmental impact).
- Does not pollute (neutral environmental impact).
- Reduces pollution (positive environmental impact).

As pointed out by Dangelico and Pontrandolfo greater attention needs to be directed to practices with positive environmental impact (Dangelico and Pontrandolfo 2010). While other practices can be considered green as they create less environmental problems than conventional/commonly implemented ones, practices with positive environmental impact in fact contribute to solving environmental problems.

In the next sections we use the GCPM to characterize the main green practices adopted by the selected cities.

3.3 The Green Cities Practices Matrix Applied to Stockholm

Stockholm is the winner of the European Green Capital Award 2010 (European Commission 2012d). Stockholm is an energy efficient city with cutting-edge environmental technology and a target of becoming free of fossil fuel use by 2050 (European Union 2010). In Table 1 the city's green practices are characterized with the GCPM.

Table 1 The green cities practices matrix applied to Stockholm

		Environmental focus		
		Materials	Energy	Pollution
Environmental impact	Less negative	• When purchasing goods and services, the committees and companies of the city must try to minimize the amount of waste created	• More than 70% of households have access to district heating, which is mainly powered by renewable energy • Installation of individual measuring and billing of heat and hot water in all housing and other premises let by the city • Co-generation plants	• Use of district heating (cut of emissions of CO_2, sulfur dioxide, and other hazardous substances) • The water is soft (the lowest dosage of detergent can be used) • All rail services are operated with certified renewable electricity, and all city buses run on renewable energies (biogas, ethanol or hybrid)
	Neutral	• Action plan for bathing waters (environmental quality norms and mandating bathing-water profiles) • Drinking water supplied by the Lake Mälaren and water quality tightly controlled by the Food Administration Authority	• Measures to reduce transport need (e.g. a fiber-optic network to allow e-services) • A well developed public transport system • All rail services are operated with certified renewable electricity, and all city buses run on renewable energies (biogas, ethanol or hybrid) • Use of eco-labeled electricity • Incentives to the use of bikes (expansion of cycling paths and lanes, safer local streets, rent stations, bicycle-pump points)	• Measures to reduce transport need (e.g. a fiber-optic network to allow e-services) • A well developed public transport system • Incentives to the use of bikes (expansion of cycling paths and lanes, safer local streets, rent stations, bicycle-pump points) • The transition to renewable energy sources and efficient conversion technology should make Stockholm free of fossil fuels (and polluting emissions) by 2050
	Positive	• Innovations in transporting waste (e.g. use of underground vacuum-controlled systems to transport waste, especially bio-waste)	• Heat in waste water used to generate district heating • Cold water from lakes and the sea, as well as the cooling effect from heat pumps that extract energy from seawater or waste water, provide for district cooling	• Around 40% of the city's land consists of parks or recreational areas and there are seven nature reserves within city limits; these enhance citizens' physical and mental health and well-being, help reduce noise, filter the air, and purify the water flowing through its wetland areas

(continued)

Table 1 (continued)

	Environmental focus		
	Materials	Energy	Pollution
	• Prohibition to send to landfill any organic waste • All separately collected food waste is recycled as biogas and fertilizer/compost • Household waste is incinerated for producing heat and electricity, reducing the amount of waste sent to landfill and replacing fossil fuels with combustible waste • Products and equipment must be reused, and information about available items must be distributed to all concerned • Campaigns to raise awareness about waste separation and recycling • Well-functioning integrated waste-handling and disposal system • Campaigns to raise awareness about waste reduction		

3.4 *The Green Cities Practices Matrix Applied to Hamburg*

Hamburg is the winner of the European Green Capital Award 2011 (European Commission 2012b). Hamburg has adopted an innovative policy for 'Responsible Growth' which has helped to make the city an example of environmental best practice. Hamburg has set ambitious targets for its future development including cutting CO_2 emissions by 40% by 2020 and by 80% by 2050 compared to 1990 emission levels. The city has already made important progress in environmental protection and intends to increase its efforts in the future (European Union 2011). In Table 2 the city's green practices are characterized with the GCPM.

3.5 *The Green Cities Practices Matrix Applied to Vitoria-Gasteiz*

Vitoria-Gasteiz is the winner of the European Green Capital Award 2012 (European Commission 2012e). Vitoria-Gasteiz is an environmentally advanced city thanks to impressive environmental policies developed over the years and to the commitment of a citizenry that represents an example for the entire European Community regarding recycling, mobility, and water consumption (Vitoria-Gasteiz City 2012). In Table 3 the city's green practices are characterized with the GCPM.

3.6 *The Green Cities Practices Matrix Applied to Nantes*

Nantes is the winner of the European Green Capital Award 2013 (European Commission 2012c). Nantes is actively protecting the environment and fighting climate change with a long-term commitment stretching across more than 20 years. The city aims to obtain a greenhouse gas emission reduction of 25% by 2020 compared to 1990 emission levels (Dantec et al. 2010). In Table 4 the city's green practices are characterized with the GCPM.

4 Discussion, Implications, and Conclusion

As illustrated in Tables 1 and 4 several green practices have been implemented by the selected cities. These practices appear to be almost uniformly distributed across the different cells of the GCPM. Attention is devoted to all types of environmental focus and through the implementation of practices with different types of environmental impact. With regard to the focus *materials* several practices are related to the reduction in material use and in waste production, as well as to their reuse and recycling. With regard to the focus *energy*, the practices related to energy efficiency, use of renewable energy sources (in buildings as well as in transport systems) and

Table 2 The green cities practices matrix applied to Hamburg

		Environmental focus		
		Materials	Energy	Pollution
Environmental impact	Less negative	• Measures to control per capita water consumption (metering, reducing leakage and high investment in infrastructures) • Partnership program, called 'Enterprise for Resource Protection' encourage voluntary investment in increasing resource efficiency in enterprises	• Energy-saving standards in public buildings • Partnership program, called 'Enterprise for Resource Protection' encourage voluntary investment in increasing energy efficiency in enterprises	• Increased use of cleaner means of transport (e.g. the world's largest fleet of hydrogen-fuelled buses) • In the port, containers are transported by rail for long distances and via automated "container taxis" between port terminals eliminating the use of trucks • Easy accessibility to the use of public transport • Financial incentives encouraging schools to cut their CO_2 output • The Hamburg ECO-Partnership, a joint venture between local government and the commercial sector, encourages companies to go beyond statutory requirements about CO_2 emissions
	Neutral	• Hamburg draws on high-quality, natural, local groundwater for 100% of its supplies	• Increased use of buildings rooftops for the generation of solar electricity • Promotion of enhanced construction of passive houses and climate-friendly renovation of buildings • Plans about building a new central terminal in the Steinwerder district with options for solar energy and kinetic energy recovery • The Hamburg Water Cycle® enables sewage disposal which is energy self-sustaining	• The world's largest fleet of hydrogen-fuelled buses • Pilot project to encourage restaurants, schools and retail stores to use regional products • Measures to incentive the use of bicycles (network of cycle path, additional bike lanes and cycle-hire schemes) • In the future, vessels will have the ability to plug in to less-polluting onshore electricity supplies • The Hamburg Port Railway is conducting a pilot project where a new geothermal railroad switch heating system is being tested with the total absence of emissions • The Hamburg Water Cycle® enables sewage disposal which is neutral for the climate

Positive	• Campaigns to encourage households and businesses to sort waste • Well-developed recycling and incineration processes • Precious raw materials are returned to the production cycle • Innovative processes to remove ammonia from sewage sludge	• 600 renewable energy businesses in and around Hamburg that build solar plants, design wind farms, and work on new ways to use biomass • Wind energy as the core of renewable energy business • Hamburg will soon be able to pull into Europe's biggest hydrogen filling station, in the port, producing hydrogen from solar or wind power generation • Innovative processes to recover energy from sludge incineration • Pilot schemes that allow for heat from wastewater to warm houses	• Project for removing the pollutants from the river Elbe • Covering of 4 km of the A7 highway to fight noise

Table 3 The green cities practices matrix applied to Vitoria-Gasteiz

		Environmental focus		
		Materials	Energy	Pollution
Environmental impact	Less negative	• Improvement of the water supply network in order to prevent leaking	• Improvement of the efficiency of public lighting (reduction in installed power, installation of header flow reducers, all incandescent or halogen traffic lights replaced by LEDs) • Installation of immotic systems for the management of interior lighting of public buildings and car parks • Bioclimatic design of buildings in the new areas of expansion of the city • Municipal buildings requiring energy providers to guarantee energy saving in heating, air conditioning and hot water under contract • Improvements in the interior lighting of municipal buildings • Free distribution to citizens of low consumption light bulbs • Efficiency improvement measures in new and existing buildings: – Energy refurbishment projects (improvement in insulations – cladding, windows, renewable energy installations- thermal and photovoltaic solar) in housing and public buildings – Mandatory energy certification in all public housing since 2000 (B class minimum) – No coal usage and diesel domestic hot water and heating systems gradually replaced – Installation of electricity and heat co-generation systems in the public hospitals of the city	• Improvement of the public transport network (introduction of two tram lines; restructuring of the network of public transport buses with an increase in frequencies and speed; construction and installation of infrastructures to give priority to public buses; introduction of flexible fuel vehicles in municipal fleets) • Use of biofuels in municipal fleet and urban transport

		• Installation of co-generation systems with natural gas in industries	
Neutral	• Use of biofuels in municipal fleet and urban transport	• Production of photovoltaic solar power on the roofs of public and private buildings • Production of solar thermal energy on the roofs of houses, commercial buildings and public buildings • Production of low-temperature geothermal energy for air conditioning and sanitary hot water in public buildings	• Measures to incentive the use of bicycles: free bicycle loan service; pedestrianizing streets and measures to limit traffic • "Municipal market garden" program for the local production of food
Positive	• Promotion of composting • Construction and start-up of a chemical-biological treatment plant	• Production of renewable electrical power from the biogas obtained in the management of urban waste and waste water	• Enlarging urban green areas • Protection and conservation of the forestry areas of the municipality
	• Construction and start-up of a construction and demolition waste treatment plant	• Production of wind energy (outside the municipal boundary there are three wind farms)	• Recovery for forestry use of land given over to agriculture • Ecological improvement actions in degraded areas
	• Fostering of selective separation of waste at source both in sidewalk containers and door-to-door	• Progressive degasifying of the landfill to take advantage of the biogas for the production of electricity	• Activities to raise awareness on mobility and the sustainable use of transport
	• Awareness campaigns on water saving	• Activities to raise awareness on the efficient use of energy	• Awareness campaigns on climate change

Table 4 The green cities practices matrix applied to Nantes

		Environmental focus		
		Materials	Energy	Pollution
Environmental impact	Less negative	• Houses equipped with individual meters intended to make consumers responsible for consumption of drinking water • Improvements in appliances and water savers to decreased water consumption in the domestic category	• The Green Buildings Program promotes the large-scale energy efficiency integration in new constructions and refurbishments as well as the mainstreaming of renewable energy systems • The local urban planning entails advanced energy standards and energy efficient refurbishments	• Bringing into conformity the two incinerator plants in the metropolitan area (reduction of industrial emissions) • Development of services that are cleaner in carbon (energy-performative computer equipment, e-services) • Reduction of agricultural emissions and non-agricultural pesticide emissions
		• Optimization of network structure to reduce potential leaks • Awareness-raising initiatives in favor of water savings • Monitoring of input-consumption indicators in the public green spaces • Water-recycling jetter truck prompted in the public operating body to reduce drinking-water consumption	• 1% of citizens are connected to district heating	
	Neutral	• Biodegradable waste is not landfilled • Removal of single-use waste collection bags	• The local urban planning entails renewable energy systems in new constructions and development of the district heating produced from renewable sources	• Actions to promote travelling by bicycle: an increase of 66% of cycle paths, more services (free bicycles distributed at 89 stations, long-term rentals for students, medium-term rental for individuals and enterprises), 5,600 bike racks at 700 locations • Increase of park and ride stations • Creation of the car-sharing service and the carpooling area creation

			• Extending of the coverage and frequency of transport network (modern trams, buses with a very high level of service, suburban rail and waterway links)
Positive	• Purification plants to treat waste water • Landfills replaced by two incinerators • Well-developed waste collection services (e.g. door to door collection; recycling bins in public areas; ecotox van routes for hazardous household waste; waste-collection centers for bulky waste; waste-collection centers for wood and WEEE sorting; individual composters or collective composters in vertical collective housing)	• Energy from incineration heats an urban heating network • Steam recovered from incineration is used in industrial processes • The remaining heat recovered from incineration is converted into energy for consumption on site and sold on to a company that sells energy • There is a scant local photovoltaic electricity production (<1%) and a small local wind-farm electricity production (<1%)	• Increased green public areas, parks and family gardens (100% of Nantes residents live within 300 m from a green area) • Measures to preserve and improve air quality (e.g. development of forecasting models, involvement in studies and action research projects) • Measures to inform the public (both inhabitants and tourists) of air quality levels (e.g. via websites or information screens) in order to raise public awareness • Adoption of an active, coherent policy in the area of noise management and of environmental sound management

production of renewable energy sources are very common. With regard to the focus *pollution*, the developed practices include reduction or elimination of emissions (e.g. CO_2 and noise), for example through environmentally friendly transport systems or noise management initiatives. It should be noticed that we included awareness campaigns among practices with a positive impact, since they can be considered as the premise to enhance the natural environment.

This chapter aims to be helpful for policy makers, since it offers structured examples of best practices to make cities *green* (from an environmental sustainability perspective) coherently with green economy principles. These examples can be used to set environmental policies and implement environmental actions within cities.

It should be noticed that, while the green economy does give an important role to the social dimension of sustainability, the European Green Capital Award only refers to its environmental dimension. It would therefore be useful if future research analyzed the green practices developed by best performer cities from a social point of view.

References

Council of the European Union (2006) Renewed EU sustainable development strategy. Available via: http://register.consilium.europa.eu/pdf/en/06/st10/st10917.en06.pdf. Cited 18 Jan 2012

Dangelico RM, Pontrandolfo P (2010) From green product definitions and classifications to the green option matrix. J Clean Prod 18(16–17):1608–1628

Dantec R, Marest P, Guillard M, Figureau C, Quero T, Vicenzotti A (2010) Nantes European Green Capital Nantes application. Presentation to the jury, 7 September 2010. Available via: http://ec.europa.eu/environment/europeangreencapital/wp-content/uploads/2011/04/nantespresentationpart1.pdf. Accessed 18 Jan 2012

European Commission (2012a) European Green Capital Award website. Available via: http://ec.europa.eu/environment/europeangreencapital/about-the-award/index.html. Cited 18 Jan 2012

European Commission (2012b) European Green Capital Hamburg website. Available via: http://ec.europa.eu/environment/europeangreencapital/winning-cities/2011-hamburg/index.html. Cited 18 Jan 2012

European Commission (2012c) European Green Capital Nantes application website. Available via: http://ec.europa.eu/environment/europeangreencapital/winning-cities/2013-nantes/nantes-application/index.html. Cited 18 Jan 2012

European Commission (2012d) European Green Capital Stockholm website. Available via: http://ec.europa.eu/environment/europeangreencapital/winning-cities/stockholm-european-green-capital-2010/index.html. Cited 18 Jan 2012

European Commission (2012e) European Green Capital Vitoria-Gastiez website. Available via: http://ec.europa.eu/environment/europeangreencapital/winning-cities/2012-vitoria-gasteiz/index.html. Cited 18 Jan 2012

European Union (2010) Stockholm European Green Capital 2010 (Luxembourg: Publications Office of the European Union). Available via: http://ec.europa.eu/environment/europeangreencapital/wp-content/uploads/2011/04/Stockholms-application-for-Europan-Green-Capital-revised-version.pdf. Cited 18 Jan 2012

European Union (2011) Hamburg European Green Capital 2011 (Luxembourg: Publications Office of the European Union). Available via: http://ec.europa.eu/environment/europeangreencapital/wp-content/uploads/2011/06/ENV-10_034_Hambourg_MEP_final.pdf. Cited 18 Jan 2012

UNEP (2011) Introduction. Setting the stage for a green economy transition. Available via: http://www.unep.org/greeneconomy/Portals/88/documents/ger/ger_final_dec_2011/1.0-Introduction.pdf. Cited 18 Jan 2012

United Nations (1993) Agenda 21: Earth Summit – The United Nations Programme of Action from Rio

Vitoria-Gasteiz City (2012) Application to the, European Green Capital Award 2012–2013. Available via: http://ec.europa.eu/environment/europeangreencapital/docs/cities/2012-2013/European%20Green%20Capital%20Award%202012-13%20nuevo%20estandar.pdf. Cited 18 Jan 2012

Emerging-Market Cities Could Set a New Standard for Sustainable Development

Jonathan Woetzel, Shannon N. Bouton, and Molly Lindsay

Abstract A new tool, the urban sustainability index, highlights five themes of sustainable development for cities in emerging economies: industrial restructuring linked to land renewal, green urban planning, use of transparent standards and charges, integrated large-scale recycling, and cross-departmental coordination.

Keywords Emerging markets • China • Sustainability index • Benchmarking • Growth

1 Introduction

Analysis recently conducted by the McKinsey Global Institute indicates that 423 emerging-market cities will generate more than 45% of global GDP growth from 2007 to 2025. The population of these cities will grow by an estimated 40% over this period, and the average income (measured in GDP per capita, adjusted for purchasing power parity) will more than double from $13,000 to $31,000. As a result, these cities will account for nearly 20% of the global population and about 30% of global GDP by 2025 (McKinsey Global Institute 2011).

Urbanization on this scale could dramatically raise productivity and standards of living in emerging markets. It also puts cities in these regions in a position to set a new standard for sustainable development—economic growth that improves lives without exhausting the environment or other resources—particularly because their rates of growth are expected to be so rapid as to be transformative.

China is a case in point. Over a billion people will live in China's cities by 2030, requiring construction on a scale never before seen. A tidal wave of urban migration

J. Woetzel (✉) • S.N. Bouton • M. Lindsay
McKinsey & Company, Shanghai, China
e-mail: jonathan_woetzel@mckinskey.com

R. Simpson and M. Zimmermann (eds.), *The Economy of Green Cities: A World Compendium on the Green Urban Economy*, Local Sustainability 3,
DOI 10.1007/978-94-007-1969-9_11, © Springer Science+Business Media Dordrecht 2013

on the order of 15–20 million long-term residents a year will continue to surge into Chinese cities, resulting in a 70% urbanization rate by 2030. To accommodate these migrants, China could continue to add 1,500 buildings that are more than 30 floors tall per year, equivalent to a new Chicago every year, resulting in more than 950 cities by 2030 (McKinsey Global Institute 2009).

This explosive growth gives China a unique opportunity to create and develop its cities in a way that supports economic growth, preserves the environment, and delivers the highest possible quality of life for its citizens. If handled correctly, these developments will not only benefit China but will also have profound implications for the rest of the twenty-first century world.

Yet today most emerging-market cities lag significantly on sustainability when compared with cities in the developed world, and they may have few opportunities to implement best practices from developed-market cities since the challenges they face are so different. Moreover, due to the absence of accepted frameworks for evaluating success in emerging-market cities, officials lack opportunities to discover and implement effective solutions.

The Urban China Initiative (a joint effort of Columbia University, The School of Public Policy and Management of Tsinghua University, and McKinsey & Company that is dedicated to finding solutions to China's most pressing urbanization problems) created a new metric, the *urban sustainability index*, to help policy makers in emerging markets identify approaches that will work in their cities.[1] The index is designed to measure the performance of cities in five sustainability categories: how well they are meeting their citizens' basic needs, resource efficiency, environmental cleanliness, built environment, and commitment to future sustainability.

By analyzing the policies and programs of some of China's most sustainable cities, we identified five common themes for achieving sustainability in emerging-market cities: industrial restructuring linked to land renewal, "green" urban planning, transparent standards and charges, integrated large-scale recycling, and cross-departmental coordination. We believe urbanization trends in China have great relevance globally and that these four themes can serve as guides to emerging-market cities around the world as they strive to lay the foundations for sustainable growth.

2 The Urban Sustainability Index

The urban sustainability index measures the performance of cities in the emerging market countries on a common set of sustainability categories. Our goal was to gauge not only the environmental sustainability of cities but also city officials' commitment to handling their growing urban populations in a sustainable way, and their efficiency in using resources.

[1] For the full report on which this article is based, see *The urban sustainability index: A new tool for measuring China's cities* (www.urbanchinainitiative.org).

A five-part index includes 18 indicators necessary for sustainable development in emerging economies.

Categories	Indicators	Description of the indicators
Basic needs	• Water supply • Housing • Health • Education	• Water access rate (%) • Living space (square meters per capita) • Doctors per capita • Student-teacher ratio (primary school)
Resource efficiency	• Power • Water demand • Waste recycling • % GDP from heavy industry	• Total electricity consumption (kilowatt-hour/GDP) • Water consumption (liters per capita) • Ratio of industrial waste recycled and utilized (%) • Heavy-industry GDP/total GDP (billion renminbi)
Environmental cleanliness	• Air pollution • Industrial pollution • Wastewater treatment • Waste management	• Concentration of SOx, NOx, PM10[1] (milligrams/cubic meter) • Industrial sulfur dioxide discharged/GDP (tons/renminbi) • Wastewater treatment rate (%) • Domestic waste collected and transported (10,000 tons per capita)
Built environment	• Urban density • Mass-transit usage • Public green space • Building efficiency	• Persons/square kilometer of urban space • Passengers using public transit (bus, trolley) • Public green space (square meters per capita) • Building heating efficiency
Commitment to future sustainability	• Green jobs • Investment in environmental protection	• Number of environmental professionals per capita • Amount of environmental sanitation funds/GDP

[1]Sulfur oxides, nitrogen oxides, and particulate matter.

Fig. 1 Eighteen indicators of urban sustainability, Urban China Initiative (Urban China 2010)

To that end, we created a comprehensive five-part definition of sustainable development, encompassing 18 individual indicators that are important in emerging economies and for which data are readily available (Fig. 1). For example, the index accounts for basic needs such as availability of drinking water, which, while nearly universal in developed economies, varies widely in emerging countries.

To determine elements that are critical to sustainability, we evaluated 112 cities selected by China's national government as the focus of sustainable development, using data for 2004–2008. We examined policy successes and failures among urban areas featuring similar financial constraints, policy environments, and experience.

The definition of sustainable development comprises five categories:

- *Basic needs*. Access to safe water, sufficient living space, adequate health care, and education are fundamental priorities for urban populations.
- *Resource efficiency*. A city's efficiency in such areas as the use of water and energy and the effective recycling of waste directly correlates to the quality of life of its citizens.
- *Environmental cleanliness*. Limiting exposure to harmful pollutants is fundamental to a city's livability.
- *Built environment*. Equitable access to green space, public transportation, and dense, efficient buildings makes communities more livable and efficient.

- *Commitment to future sustainability.* An increase in the number of employees and the level of financial resources devoted to sustainability suggests how vigorously city governments are committed to implementing national and local policies and standards.

An encouraging finding is that sustainability often proceeds hand-in-hand with economic growth. Indeed, almost a third of cities in our study group improved sustainability while increasing GDP from 2005 to 2008 at an above-average rate.

3 Five Themes for Sustainable Development

As a result of analyzing the policies and programs of some of China's best-performing cities, we identified five themes common to sustainable cities in the emerging world.

3.1 Industrial Restructuring Linked to Land Renewal

Rising costs and tighter national environmental standards—particularly for sulfur dioxide emissions—have made many city officials press heavy industries to shutter urban factories and build more modern plants in new industrial parks or in suburban development and economic zones. In Tianjin, for example, smokestack industries are moving east from the city center into some parts of the Binhai New Area, a development zone. Chengdu has removed almost all traces of heavy industry from its urban core and banned the industrial use of coal. This phenomenon is common to many growing markets in the industrialization phase.

Rising land prices has made locating facilities within city boundaries more expensive for industry, and many industries that relocated have invested money raised by selling land-use rights in urban cores to buy state-of-the-art technology and emissions-control equipment, as well as to cover their relocation costs. Although the trend is just beginning, indications are that consolidating heavy industry away from urban centers brings economies of scale large enough to offset the costs of sophisticated infrastructure retrofits and new equipment.

Equally crucial is "brownfield" redevelopment at the sites left behind. These sites provide large-scale opportunities for planning, because they are typically large plots of land in high-value inner-city locations. Chengdu launched plans to redevelop its eastern region when industry moved out, leaving a number of large factories behind. Most of the efforts involved cleaning up the sites so they could be used for housing and commercial activity. In some cases, factory buildings were repurposed for new uses that build on the area's history. For example, in the "East Music City" area, old factory sites were transformed for use as art galleries and music venues.

3.2 Green Urban Planning

Chinese cities that have successfully balanced sustainability and growth incorporate both objectives when they create mass-transit networks and urban amenities. Efficient and attractive mass transit takes cars off urban roads, cutting emissions and congestion. Urban forests and green areas serve as a net to filter dust particles caused by vehicles, industrial development, and other sources; they also absorb carbon dioxide, helping clean the air. The best-performing cities have recognized these benefits and included efforts to enhance mass transit and green space in their development programs.

Financing is perhaps the most difficult aspect of public transit. China's city officials are seeking ways to manage subsidies, expanding services with minimal public funds. Reasonable fares and a proper regulatory framework for private participation are essential for the long term. In Shenyang, for instance, the municipal public-transport company contracts with private operators and transport firms to supplement its own services.

3.3 Use of Transparent Standards and Charges

Our research indicates that cities that adopt clear goals, publicize their progress toward meeting them, and hold responsible parties accountable for their performance are more likely to achieve high standards of sustainability. For example, superior environmental supervision and strict monitoring of digital information pay off for cities such as Qingdao.

Part of Qingdao's consistent performance in wastewater treatment is the result of pressure from Shandong province officials, who publicly identified the region's 1,000 biggest polluters and set aggressive waste-reduction targets for each of them. By 2008, more than 1,000 companies and 170 wastewater-treatment plants in the province were being monitored. Each company on the list was required to provide digital data on its status regularly. Such policy enforcement at the provincial level in effect places cities in a healthy public competition that encourages improvements.

In Chengdu, the environmental protection bureau launches a campaign targeting one highly polluting industry each year. In 2011, for example, it focused on ensuring that factories producing lead-containing batteries were meeting environmental standards. Those found to fall short of standards were forced to halt operations until they were able to comply with regulations.

3.4 Integrated Large-Scale Recycling

The best-performing cities excel at creating efficient local linkages among industrial producers from different sectors. In the next 5 years, leaders of rapidly industrializing small and midsize cities must find ways to reduce the volume and increase the efficiency of resource consumption. One promising approach in China links manufacturing or utilities plants in a given locale. Tianjin's Binhai New Area, for example, started with two ambitious projects to transform itself into a desalinization center. Tianjin officials we interviewed pointed to its power plant near Beijing as an example of resource efficiency: the project links water, power, sea-salt production, waste reuse, and land conservation in an elegant desalinization system.

During the project's first phase, launched in 2005, the city invested 1.3 billion renminbi (around $160 million) to construct two 1,000-MW generators that would provide 200,000 ton of water a day for city residents as a by-product of power generation. In phase two, which began in 2010, two 1,000-MW clean, coal-fired generating units and saltwater-cooling towers will be added. The whole system is expected to provide 400,000 ton of fresh water a day, as well as 11 billion kilowatt-hours of power, 450,000 ton of salt, and 60,000 ton of minerals (Bromine, potassium chloride, magnesium chloride, and magnesium sulfate) each year. Fly ash and other waste will be sold cheaply to construction companies for building materials.

3.5 Cross-Departmental Coordination

Our interviews with urban officials in China indicate that success in executing sustainable development projects depends on coordination among city agencies and other bodies. For example, successful transit projects typically involve experts in urban planning, construction, and the environment. Efforts to increase environmental transparency involve representatives from industry and information management. And land-renewal projects involve experts in economic planning, land use, urban planning, and multiple industries.

To break down silos and facilitate cooperation, municipalities should establish processes to ensure projects meet coordination requirements before they are approved. In Shenyang, for example, all projects must be approved by a department directly affiliated with the state council, and officials must demonstrate that they have met the city's standards for coordinating with all relevant departments.

4 Concluding Remarks

The challenge facing rapidly growing cities in developing countries is enormous. Leaders in China and other emerging markets have recognized this and are already taking action to develop solutions to these challenges. Our ambition in creating the urban sustainability index is to provide a yardstick that cities can use to measure success and identify initiatives that they can implement to achieve sustainable development.

References

McKinsey Global Institute (2009) Preparing for China's urban billion. Available via: http://www.mckinsey.com/Insights/MGI/Research/Urbanization/Preparing_for_urban_billion_in_China

McKinsey Global Institute (2011) Urban world: mapping the economic power of cities. Available via: www.mckinsey.com/mgi/publications/urban_world/index.asp

Urban China Initiative (2010) The urban sustainability index: a new tool for measuring China's cities. Available via: www.urbanchinainitiative.org

Part III
Joining Forces to Accelerate Action

Introduction: Joining Forces to Accelerate Action

Phoebe Stirling

Keywords Collaboration • Actors • Introduction • Green urban economy

Swift and comprehensive action is necessary for addressing the global challenge of achieving sustainability. The previous part has set out a series of opportunities and challenges to optimize urban sectors and the urban economy by greening them. However any fundamental change in pathway, be it through the transformation of a mature urban economic system and urban infrastructure or the steering of rapid urban growth, requires the involvement and active engagement of a range of stakeholders. Such a spectrum encompasses public, private and civic actors, groups and institutions; each has a range of roles to play, but most importantly each has the ability to maximize their potential through collaboration, cooperation and coordination.

This embodies processes and structures to identify the benefits and incentives for collaboration and coordination towards mutually agreeable targets. For example, engagement between government at both the local and national level, civil society, the private sector and research and academic institutions not only fosters accountability, but allows for a more synergistic approach to developing policies and implementing decisions (Joss 2011).

Part III draws together contributions that detail considerations as to how this can occur. It offers practicable solutions, examples of best practice and policy approaches which enable such cooperation. The chapter focuses on some of the debates that are thrown up when multiple actors are brought to negotiation around the common issue of sustainability, and throws light on various paths for collaboration and for streamlining potentially divergent priorities towards a common goal. Most importantly the chapters focus on the significance of multi-stakeholder collaboration, and

P. Stirling (✉)
ICLEI – Local Governments for Sustainability, Bonn, Germany
e-mail: urban.research@iclei.org

R. Simpson and M. Zimmermann (eds.), *The Economy of Green Cities: A World Compendium on the Green Urban Economy*, Local Sustainability 3,
DOI 10.1007/978-94-007-1969-9_12,

the essential task of engaging a cross-section of actors capable of operating at a range of scales.

The first chapter by Bernd Hendriksen and Eric Copius Peereboom provides an extensive and comprehensive analysis of mechanisms available both to local and to national governments towards stimulating clean, low-carbon economic growth at the city level. They remind us that attracting private finance and enabling public private partnerships are vital for investment into infrastructure. At the same time governmental facilitation of stakeholders through the influence of procurement policies, financial instruments, a consistent regulatory framework and knowledge sharing are similarly important, amongst others.

Ulrich Mans and Sara Meerow for example illustrate the need for a stronger interaction between the public and the private in facilitation of local renewable energy clustering politics, through an analysis of 33 cities in developing countries. The chapter draws a distinction between 'green-conscious growth' through the pursuit of a green agenda, and 'green driven growth', in which cities can enhance competitiveness in the global economy by capturing market share in green business. The authors contend that green growth "cannot succeed as a purely public endeavor", concluding that the private sector and green energy sector in particular will play a vital role in a green urban economy.

Golam Rasul focuses on the role of the state, making an assessment of various regulatory and market-based mechanisms for enabling both the private sector and the public towards sustainable production and consumption. While advocating its effectiveness, Rasul contends that purely market-based incentivizing does not secure the necessary conditions for a green agenda. By themselves, such mechanisms can be blind to their effect on different socio-economic groups, can fail to deal with environmental externalities and cannot be used to guarantee access to non-exclusive public goods, and therefore require a regulatory backbone.

In her chapter *Local sustainability: driving green urban economies through public engagement* Rosalie Callway describes a strong role for local governments in creating enabling conditions for community actors. For Callway the role of citizens in greening the urban economy should be stimulated through open information, active partnership with community groups, market-based incentives and sustainable procurement, amongst others.

Sean Connelly, Sean Markey, and Mark Roseland similarly underscore the need for community engagement as a key process. They also argue that despite existing and extensive innovative and technical capacity "for meaningful, structural change", social change and positive decision making are still required in order to bridge the implementation gaps of "mobilizing and socializing sustainability".

In the final contribution to this chapter, Rula Qalyoubi appraises strategies available at both the local and nation level to move economies towards convergence on a green agenda. Qalyoubi advocates amongst others the use of financial instruments, investment into human capital and a "common language" to remove inherent ambiguities and to enable consistent decision making. The paper focuses on the ability of all actors, whether national or local government, NGO or the private sector, to "duplicate successful experiences around the world."

This compendium brings contributions together to argue that there are such opportunities to grow and transform urban economies in a green way, while directly and indirectly contributing to environmental and social objectives. Part III underscores that an ambitious green urban economy transformation and green urban growth requires a careful balance between economic, environmental and social goals, by which they support rather than undermine one another. As Rasul states "the effectiveness of policy instruments is largely dependent upon the involvement of diverse stakeholders and actors... it is necessary to strike a balance between environmental effectiveness, economic efficiency, equity, as well as administrative costs of implementation."

Such a balance is achieved through active collaboration between those actors who share a stake in and whose involvement impacts directly upon the environment. The development of knowledge sharing networks, of a "common language" and of behavior change are all part of what Hendriksen and Peereboom call "working together on mutual reinforcing goals". Shared interest in a strong economy and an equitable, productive society are thereby harnessed to forge consensus and to achieve a common objective. As Sean Conelly et al. contend, even with a solid basis in quantifiable and technological solutions, the drive towards sustainability still requires positive dialogue between all interested actors, in order to mobilize and socialize sustainability and to "enact structural change". By engendering citizen ownership of policy direction, public engagement of civil society can garner support and move towards comprehensive and more equitable urban sustainability initiatives. This requires the engagement of both private and public actors.

The joint objective of building and sustaining a strong economy and society in a way that is mutually beneficial to the divergent interests of different actors is no doubt a challenge. It is exactly this challenge on which the chapters of this part seek to start opening the debate.

Reference

Joss S (2011) Eco-city Governance: a case study of Treasure Island and Sonoma Mountain village. J Environ Policy Plan 13(4):331–348

Accelerating Green Urban Growth

Bernd Hendriksen and Eric Copius Peereboom

Abstract In 2007 we reached an urban tipping point when for the first time in history over 50% of the world's population was living in urbanized environments. According to estimates by the UN Population Division 2010, urban population is predicted to reach 69% by 2050. If managed ineffectively, rapid urban population growth can create severe environmental, economic and social problems – leading to loss of productivity, attractiveness, pollution, congestion, reduced quality of life and social exclusion. Rapid urbanization is not the only significant urban challenge; cities must adapt to changing attitudes towards the environment and social equity, and embrace new technologies that make transport and housing more energy efficient. These challenges require cities to aspire to be more integrated, less car-dependent and more mass-transit and pedestrian friendly.

In today's economic times, financial sustainability is equally as important as environmental and social concerns. To manage the pace of change requires unprecedented levels of investment at a time of tightening financial resources. Against this backdrop, however, many city leaders are finding creative ways to push ahead and – recognizing the clear economic, environmental and social benefits of green growth policies – are transforming their cities for the future.

A green growth model has emerged as a new paradigm that will identify policies and approaches that promote economic development while reducing greenhouse gas (GHG) emissions, minimizing waste and inefficient use of natural resources, and shift production and consumption towards a clean, low-carbon economy. In this article we present seven enablers that support the transition to a green urban economy.

Keywords Growth • Financing • Enablers • Greening and leadership

---

B. Hendriksen • E. Copius Peereboom (✉)
KPMG Advisory N.V., Amsterdam, The Netherlands
e-mail: Hendriksen.bernd@kpmg.nl; copiuspeereboom.eric@kpmg.nl

R. Simpson and M. Zimmermann (eds.), *The Economy of Green Cities: A World Compendium on the Green Urban Economy*, Local Sustainability 3,
DOI 10.1007/978-94-007-1969-9_13,

1 Introduction

In 2007 we reached an urban tipping point when for the first time in history over 50% of the world's population was living in urbanized environments. According to estimates by the UN Population Division 2010, urban population is predicted to reach 69% by 2050. If managed ineffectively, rapid urban population growth can create severe environmental, economic and social problems – leading to loss of productivity, attractiveness, pollution, congestion, reduced quality of life and social exclusion.

Rapid urbanization is not the only significant urban challenge; cities must adapt to changing attitudes towards the environment and social equity, and embrace new technologies that make transport and housing more energy efficient. These challenges require cities to aspire to be more integrated, less car-dependent and more mass-transit and pedestrian friendly.

In today's economic times, financial sustainability is equally as important as environmental and social concerns. To manage the pace of change requires unprecedented levels of investment at a time of tightening financial resources. Against this backdrop, however, many city leaders are finding creative ways to push ahead and – recognizing the clear economic, environmental and social benefits of green growth policies – are transforming their cities for the future.

A green growth model has emerged as a new paradigm that will identify policies and approaches that promote economic development while reducing greenhouse gas (GHG) emissions, minimizing waste and inefficient use of natural resources, and shift production and consumption towards a clean, low-carbon economy. In this

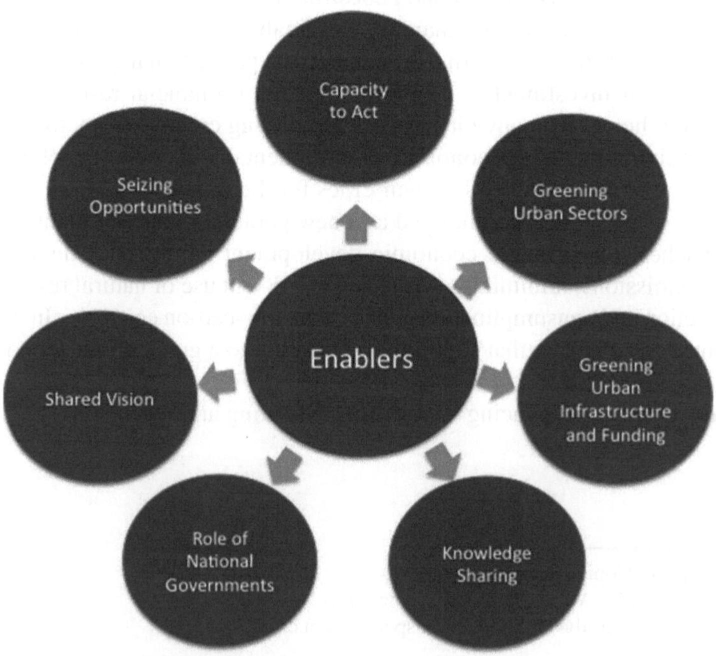

Fig. 1 Seven enablers that accelerate the green urban economy (Hendriksen et al. 2012)

article we present seven enablers that support the transition to a green urban economy, these include (Fig. 1):

There is a realization that sustainability is less a challenge of science and technology and more about human behavior change or, more specifically, choices. Cities have done this before, and many more are heading in this direction.

The vital link in transitioning to a green urban economy is leadership – a "firestone", i.e. a person or team with passion and energy, who are capable of negotiating multiple priorities and are willing to change.

2 Accelerating Green Urban Growth: Capacity to Act

The last decade has seen a rapid proliferation in the number of reports, studies and initiatives relating to sustainable cities. While city leaders have committed to several initiatives which demonstrates their political will and intent for local-level sustainability leadership, it is not clear whether they possess the capacity to act i.e. to create effective policies, develop economic models, raise adequate financing, manage and measure progress, and engage local stakeholders to affect behavior change. Furthermore their ability to act can be severely impacted by conflicting policies introduced at the regional and national level.

Advancing green growth strategies and policies will require local governments to overcome and respond to many of the challenges outlined above. The good news is that these challenges can be overcome. According to the OECD report entitled Cities and Green Growth: A Conceptual Framework (2011) cities are already demonstrating that they are active leaders *now* through the developing and sharing of information, projects, and solutions (Hammer et al. 2011).

2.1 Local-Level Sphere of Influence

Cities have through their jurisdictional power the ability to regulate land-use, service provision and master planning. For example London and Barcelona have enacted the introduction of solar technologies as a precondition for building permit approval, both of which have been widely credited for playing a key role in renewable energy deployment, notably Barcelona´s Solar Thermal Ordinance which has been a pioneer for solar regulations throughout the rest of Spain and Europe. Cities also play the role of enforcer for regional and national tiers of government in areas such as building codes, and can use their regulatory authority to fast track the permitting process for projects that can demonstrate environmental and/or societal benefits.

Greening public spending and procurement policies can accelerate green growth by providing early stage financing that can help boost market uptake. Furthermore cities can lead by example through the practical demonstration of green products and/or services.

Cities can also influence behavior and affect purchasing decisions through a wide range of financial instruments that either encourage or discourage certain actions. In the area of renewable energy, cities can work with national and regional governments to deploy financial incentives such as Feed-in Tariffs (FiTs) that can accelerate the uptake of renewable energy and can help build the market. Incentives can have a huge impact. FiTs are a case in point, but they need to be set carefully and consistently applied. In transport, mode switching from private vehicle to public transit can be achieved through financial disincentives such as congestion charging, fuel taxes, limited/high cost parking.

In the building sector the greatest obstacle to energy efficiency and/or renewable energy deployment remains the owner/tenant dilemma (split incentives). Although building owners want to protect their long-term investments they are normally reluctant to incur short-term costs as they are unlikely to benefit from energy reductions.

Box 1 Encouraging Building Owners to Go Ahead with Energy Renovations

In France a law has been passed that addresses this issue. The law passed allows for the redistribution of savings on energy costs to be split between the owner and tenant.

In the UK, the Department of Energy and Climate Change has introduced an energy efficiency proposal that helps to overcome the owner/tenant dilemma called the Green Deal. The Green Deal provides upfront finance for energy efficiency improvements (no upfront costs for owner) via a loan, which is attached to the energy use of the property and is paid back by the tenant(s) over a 20-year period.

In France, a new legislative text has recently allowed such a model (Article 119, Boutin Law 25 March 2009, Decree of 23 November 2009).

Likewise, tenants may also be reluctant to make longer-term investments into properties which they do not own. Therefore, the owner/tenant dilemma remains (Box 1).

The examples of France and the UK demonstrate how a major stumbling block (split incentives) can be overcome by encouraging owners to go ahead with renovation work. Increasing energy efficiency within existing buildings enables cities to positively address the three pillars of green growth (social, economic and environmental) at the lowest cost. Buildings can account for 30–40% of energy use within the urban area. Stimulating increased energy retrofits can therefore significantly reduce energy demand and costs, lower carbon emissions, address social concerns such as fuel poverty while creating local employment through labor intensive energy services.

Local, regional and national tiers of government must cooperate further to address the owner/tenant dilemma through low-interest loans for investments in distributed renewable energies or energy efficiency technologies, information and dissemination or by providing technical assistance to facilitate the use of Energy Service Companies (ESCOs).

To protect green spaces property taxes can be redesigned so that they support compact urban development by favoring brownfield development, which can help to reduce the impact of urban land sprawl, lower transportation costs, enhance social inclusion and improve quality of life through shorter journeys and increased green areas for recreation.

3 Accelerating Green Urban Growth: Greening Urban Sectors

There are significant challenges and opportunities arising from greening key urban sectors such as energy, transport, buildings and waste. Together these sectors may account for up to 80% of local energy use.

3.1 Energy

Cities are high-energy users. Many studies attribute up to 70% of global energy use to city regions. This figure is set to rise through increased urbanization, consumption trends, population growth and perceived climate change impacts.

Increasing the efficiency and effectiveness of the energy sector plays a key underlying role in the transition to a green economy, and can be achieved in two ways:

Energy efficiency Energy efficiency measures offer city governments the most cost-effective means to address economic, social and environmental issues. In addition to negative cost abatement potential, energy efficiency shifts expenditures from energy generation to energy efficiency services, which typically create more localized jobs.

Within an urban area that has a large residential building stock, an ambitious insulation program addresses the three pillars of sustainability (economic: low abatement costs/local jobs; environmental: energy and carbon reduction; social: fuel poverty/lower bills/increased comfort) more effectively than other interventions such as renewable energy systems, heating rebates or the development of a district heating network.

Renewable energy The renewable energy sector supplied an estimated 16% of the world's energy and 20% of its electricity in 2010 and has continued to perform well despite the economic slowdown and cuts in incentives (El-Ashry 2011).

Under the right conditions adopting renewable energy within the urban context can have positive economic and environmental impacts. Investments in renewable energy are considered to be far more labor intensive than investments to expand fossil fuel based energy generation.

3.2 Transport

Through their role as managers and planners of transport networks, cities can employ a range of policies that can contribute to economic development, reduce environmental impacts and foster social inclusion.

Road congestion is hugely expensive. In 'opportunity cost' the UK is expected to lose US$36 billion each year by 2025 (Eddington 2006). Reducing congestion is important for a number of other reasons. A clear winner is the environment, benefitting from fewer cars idling. Traffic congestion also impacts the quality of life of those living and/or working in the city by increasing travel times and air pollution. All of this has led to many urban areas to focus on renewing their development of mass transit systems. Many cities are also implementing road-use fees (congestion charging) as a way of discouraging private vehicle use. Hereto London and Singapore are examples of best practice.

3.3 Buildings

Buildings are often cited as being responsible for up to 30–40% of global carbon emissions (build/use) and for offering the lowest carbon abatement potential. Generally speaking there has been a prioritization towards new buildings, which account for only 1% of the building stock. However, in most cities the greatest potential lies in the retrofitting of the existing building stock. To capitalize on this opportunity cities can develop "Energy Saving Partnerships" as pioneered by the City of Berlin (Box 2).

Box 2 Berlin: Energy Saving Partnerships

The Berlin Energy Agency is responsible for Berlin's Energy Saving Contracting program (ESC). This program was introduced in 1996 with the objective to reduce the energy needs of public buildings. It is attractive for both private contractors and building owners since the owners have a guarantee of about 16–36% of cost savings and the contractors can refinance their investments through the savings within a period of 8–15 years. Berlin has 24 ESC. This means, more than 500 real estate properties with over 1,300 public buildings are currently managed by contractors. The contracting firms have already invested €70 million into energy-saving measures. So far, Berlin has already saved €30 million of energy and maintenance costs. At present the cumulated CO_2 reduction amounts to almost 70,000 ton per year.

City of Berlin submission to CDP Cities Project 2011

Greening residential buildings to increase energy efficiency will have a positive impact on low-income households who are victims to rising energy prices. In addition some schemes such as FiTs for Solar PV exclude low-income households as they are unable to make the upfront payment for installation or they do not own the required roof space. The job creation potential from greening buildings is also significant, which also tends to be local.

3.4 Waste

Urbanization and changing consumption patterns are rapidly increasing levels of waste generation, notably in developing countries where significant quantities of waste are either sent to landfill, incinerated or simply dumped. In China and India waste generation levels estimated for 2030 are expected to increase by 214 and 130% respectively, whereas emerging economies such as Brazil, Russia and Indonesia have already exceeded their projected 2030 estimates (Hammer et al. 2011).

To reduce the environmental and financial costs of increasing waste generation, pricing for waste services that reflect the true cost of their delivery and the limited availability of landfill space should be introduced. Partnerships between neighboring cities and regions provide win-win situations by increasing the volume of waste recovery (re-use, recycling or energy) plus enable administration structures to achieve economies of scale. Within cities, partnerships with local stakeholders can be developed to affect behavior change.

Growing evidence suggests that there are tremendous employment creation opportunities in the waste and recycling sector which can support green urban growth. By examining new and innovative methods that convert waste into energy for heat and power, cities can reduce energy and greenhouse gas emissions.

4 Accelerating Green Urban Growth: Greening Urban Infrastructure and Funding

Around the world, in the face of rapid urbanization, cities are striving to provide effective transportation systems, reliable energy and water networks, and efficient social infrastructure to meet the changing demands of a growing (and aging) population. The demands on urban infrastructure have reached unprecedented levels. In both the developing and developed world existing infrastructure is in need of revitalization. Already many global cities rely on poorly designed or badly maintained facilities that are, in many cases, more of an impediment to growth than a catalyst (Chism 2011).

Low-carbon infrastructure is incredibly expensive to design, build and operate (Smart Grids, Electric Vehicle Networks, District Heating/Cooling Networks) and cities cannot afford to get it wrong. To effectively address the demands of rapid urbanization, while recognizing and balancing economic, environmental and social concerns, requires greater collaboration across all tiers of government.

Taxes	Grants	User Charges	Lease Income
• Property taxes • Licence fees • Entertainment tax • Sales tax etc.	• From national, provincial or state governments	• Water • Sewerage and drainage • Tolls • Fares etc.	• Lease rental from land • Lease rental from markets • Lease rental from projects etc.

Fig. 2 Typical sources of funds for city governments (Brookes 2011)

4.1 Urban Growth Through Infrastructure

To lay the foundation for growth two types of basic urban infrastructure are essential. These are broadly classified as social and economic infrastructure.

Social infrastructure includes facilities such as hospitals and educational institutions. Economic infrastructure consists mainly of public utilities such as water, solid waste management systems, drainage, sanitation, flood protection systems, power and telecommunications.

The provision of economic infrastructure expands the productive capacity and the production possibility frontier, which enhances economic growth and social development. Benefits from economic infrastructure are easily valued and apparent over a relatively short timeframe. In contrast gains from social infrastructure such as healthcare and education are often difficult to quantify in the immediate term. However, empirical evidence has shown that economic and social infrastructure together lead to economic growth, as both play critical roles in laying the foundation for stable and self-sustaining development. Nevertheless, given the capital-intensive nature of economic infrastructure projects, they attract higher funding than social infrastructure projects.

Conventional financing of growth for cities is limited to taxes, grants, user charges and lease incomes. These are often insufficient to meeting the funding requirements of infrastructure projects. New approaches to growth financing include varying degrees of private sector participation through Public-Private Partnerships. Typically, cities fund infrastructural projects through four main sources as shown in Fig. 2 below (Brookes 2011). However, these typical sources are usually insufficient to meet the funding requirements for infrastructure projects.

An increasing gap exists between the funds available to city governments and the funds required for infrastructure development. This has prompted city governments to source non-public financing, typically through initiatives like public private partnerships (PPPs). Private sector participation has contributed immensely to the development of infrastructure projects and has helped governments access financing to implement much-needed projects.

Public-Private Partnership refers to the long-term partnering relationships between the public and private sectors in the delivery of services. It is a fairly new approach adopted by governments to increase private sector involvement in the delivery of public services.

Both the public sector and private sector have competitive advantages over one another in performing specific tasks and handling risks. The effective sharing of risks between public and private sectors creates more 'value for money' in PPPs. Private sector participation is largely in the form of PPPs, which are typically structured as special purpose vehicles (SPVs).There are various forms that PPPs can take, which vary widely depending on the level of private-sector involvement. The focus here will be on concessions, Build-Operate-Transfer (BOT) and other similar arrangements where private funding is involved.

There are several motivations for the public and private sectors to engage in PPPs to develop urban infrastructure and services. However, the principal reason is that PPPs are able to deliver better value for money than alternative approaches. From the perspective of city governments, PPPs offer advantages in the following broad categories:

- Financing
- Efficiency and Cost Savings
- Quality
- Change in Focus

Instead of relying on government and direct investments, the new approaches to growth financing are also complemented by alternative sources of funding such as Capital Markets, Institutional Investors, Multilateral Agencies, Domestic Financial Institutions and Asset Leverage.

Urban growth, essential for maintaining social stability and ensuring relevance, is supported by urban infrastructure development. Because infrastructure development is a complex process there have been many examples of failed attempts. For such projects to be sustainable a more collaborative approach should be adopted, whereby governments should open up infrastructure investment opportunities to private sector participation.

4.2 *Investing in Cities: Attracting Private Finance*

The combined effect of the financial crisis, concerns about scarcity and the pricing of natural resources and energy security, is changing the way private sector stakeholders (notably shareholders) value the management of environmental, social and governance (ESG) issues as a driver of long-term business value. This has led to the rapid growth and expansion of a number of investor-backed initiatives. For example, the United Nations Principles for Responsible Investment (UN PRI) has gone from 50 signatories in 2006 to 986 in 2011, with assets of over US$30 trillion (UN PRI 2011).

Mainstream investors are increasingly working with ESG information regarding business performance, where the integration of ESG considerations has become increasingly more important in investment decisions amongst mainstream investors.

Perhaps there is an opportunity for cities to explore the growing interest from the investment community, in particular pension funds, around ESG concerns. In developing countries, cities should look at Impact Investing, which is not only focused on financial or environmental returns but also considers wider societal benefits. A recent essay from the Hauser Center outlines a number of advantages to investors of sustainable cities as an investment vehicle such as scale, long-term horizons, positive externalities and risk mitigation (Woods 2011).

Further research is needed to explore the advantages of cities as an investment-grade vehicle for institutional investors and private investors.

5 Accelerating Green Urban Growth: Knowledge Sharing

Knowledge sharing between cities can help advance the green growth agenda. Current actions such as benchmarking the sustainability performance of cities can help to identify front-runners and leading practices. If these are shared, cities can benefit from the knowledge and experience of colleagues in other cities and that obviously benefits their sustainability policy. Benchmarking thus contributes to an increase in the effectiveness of the policy. However, it is important for cities to fully understand their baseline conditions so that they can prioritize policy interventions in order to maximize their investments.

Earlier research by KPMG (City typology as a basis for policy, 2010) found that there was an abundance of city benchmarks/rankings, each using a very specific set of indicators and approaches (Hendriksen et al. 2010). One of the key findings from the KPMG study was that a city´s baseline conditions were never taken into consideration. For example a historic city with a high-quality service economy has very different challenges than that of a port city with a large industrial sector. These differences are so great that the real effectiveness of existing benchmarks and to some extent knowledge sharing and demonstration projects must be put into serious perspective.

At present there is a lack of real insight into how cities perform with respect to a specific typology, although it offers invaluable insight to the definition of an effective green growth policy. The key question is: What is the best thing for a city to do in view of its natural and prevailing conditions? Answering this question requires a tailor-made approach.

We have developed a typology grid for European cities (see Fig. 3 below) on the basis of which cities can engage in internal discussions about where their optimum lies. A number of city characteristics are listed.

Figure 3 above provides insight into a number of elements that are crucial for green growth policies. These relate to the city's economic typology (the dominant economic sectors that are represented by the bars), the size of the city, the average

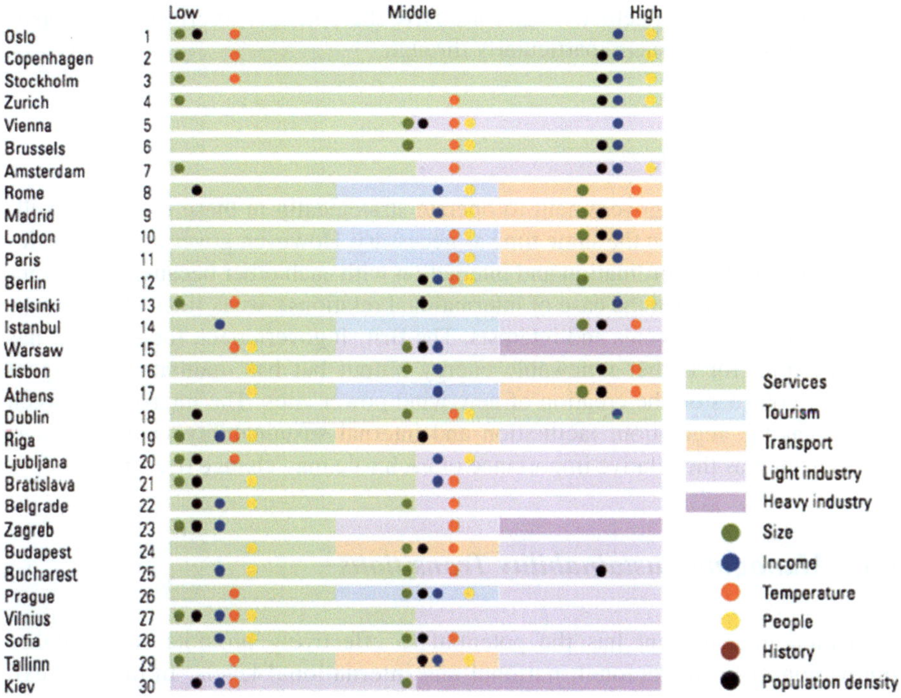

Fig. 3 European city typology grid (Hendriksen et al. 2010)

income of its population, the average temperature, the environmental awareness of its residents, and its history. Although it has no scientific basis it does provide an excellent way to focus one's thoughts when defining a climate policy. Further research needs to be conducted to explore how baseline conditions can impact green growth to support further best practice knowledge sharing between 'peer' cities.

6 Accelerating Green Urban Growth: Role of National Governments

In our view, national government has four distinct roles in addressing sustainability concerns: policy development; regulation; facilitation; and internal sustainability management. These roles need to be carried out coherently if they are to achieve their optimum effectiveness.

Implementation by government of coherent strategies to achieve sustainability goals requires an understanding of the different roles commonly played by government in relation to sustainability. Each of the policy making, regulation, facilitation and internal sustainability management roles of government has its own characteristics and success factors. Together, these roles have the potential to effectively support sustainability management, particularly through:

- Setting goals
- Driving change
- Leading by example

It is important for government to operate strategically in these roles in a well-designed sequence or at the same time as necessary. However, government agencies often lack internal coordination and interaction with each other because of jurisdictional discrepancies (in the case of international relations), weak lines of communication, or competing goals and interests. The British government, for example, has been pushing for higher renewable energy output but has maintained stringent planning controls on the creation of new wind power sites. The interplay between policy-making, regulation, facilitation and internal sustainability management is very dynamic as the relative importance of each role may change over time.

6.1 Managing Sustainability Transitions

Transition management has the potential to effectively promote sustainability innovation at local, regional, national and international levels. In this context, 'transition' is the conversion from the use of conventional to innovative and more sustainable methods. Transition may involve multi-level changes which fundamentally alter predominant societal patterns of behavior such as a shift from fossil fuels to renewable energies, or from private vehicles to public transport.

If carefully designed the facilitator role can be highly influential in long term transition management, because it relies on consultation and the ability of markets to come up with the most (cost) effective solutions rather than prescribing ways of pursuing policy targets. As a facilitator government will work towards realizing sustainability policy goals in cooperation with the private sector, research institutions and other stakeholders in addition to its regulatory, control and licensing functions.

6.2 Three Phases of Transition

There are three phases of transition management: pre-development; take-off and acceleration; and stabilization (Rotmans 2003).

Pre-development Identifying opportunities through extensive consultation, a preliminary cost-benefit analysis and an assessment of comparative advantages.
Take-off and Acceleration The public sector promotes projects which potentially have a high impact on sustainability and provide business opportunities, and seeks to mobilize other key players such as research institutions, companies and investors.

Stabilization Characterized by the increased market maturity of the new regime: at this stage government's role is largely limited to monitoring progress and promoting best practice. This ultimately leads to market self-regulation within the applicable regulatory and fiscal framework.

6.3 Criteria and Means of Successful Government Facilitation

Our experience indicates that successful government facilitation of sustainability transition depends on:

- Application of a set of criteria to determine whether and when government should act
- Ability and readiness of government to apply a range of methods simultaneously or in a carefully designed sequence, corresponding to the relevant phase of the transition process

We believe that general guidelines can be established irrespective of the specific political realities. These guidelines may assist decision-making processes within and between public sector agencies as well as consultation with interested stakeholders. The question of how far the public sector should go in facilitating sustainability transition is a controversial, and essentially a political one:

- When should government intervene and what means should it employ?
- When should markets be left to regulate themselves?

An understanding of the dynamics of sustainability transition processes is necessary to determine the right balance between market mechanisms and public sector facilitation. Whether and how government should actively facilitate sustainability transition depends on what stage a transition is in.

We believe that a better understanding of the potential, limitations and dilemmas of public sector facilitation is vital for government, business and society to collaborate effectively.

7 Accelerating Green Urban Growth: Shared Vision

To adequately respond to economic, environmental and social trends, cities will need the support of all stakeholders, particularly businesses that can help cities unlock opportunities and develop practical, cross-cutting solutions to support the transition to a low-carbon driven economy. Companies are important urban actors in the development of low-carbon cities because they provide a multitude of goods and services to city populations and directly influence urban carbon dioxide (CO_2) emissions (Whiteman et al. 2010).

The private sector is a key stakeholder in the transition to a green urban economy through employment, production of goods and services to city populations and through the ability to affect urban consumption and behavior through advertising that can directly influence CO_2 emissions. We propose four concepts on how cities and companies can work together towards a shared vision for green urban growth. These four concepts are:

- Regulatory concept
- Networking concept
- Green Business Zones concept
- Innovative financing concept

7.1 Regulatory Concept: Enabling Companies to Drive Urban Green Growth

Despite a complex array of challenges, many companies have recognized the imperative of sustainability and have created strategies and solutions to effectively respond to the issue. These companies have aligned sustainability to their operational strategies, and are starting to measure it in the same way as other investments. By viewing sustainability as an investment rather than a cost, they have adjusted their business models to drive long-term change and make them more competitive in the market.

The lack of a clear and consistent regulatory framework, however, makes it difficult for all companies to formulate business decisions that can have a long-term effect on sustainability. Some companies may lack an understanding of how to build an appropriate business case for sustainability programs as they grapple with sparse, inconsistent and often unreliable data. Therefore, in the absence of a clear international agreement on climate change, city leaders hold the keys in transitioning towards a low-carbon economy.

7.2 Networking Concept: Working Together on Mutual Reinforcing Goals

Cities and businesses are reliant upon each other in maintaining safe, resilient environments from which to operate (infrastructure, resources, people, services and revenues), therefore knowledge sharing and learning is critical for a long-term competitive advantage. The development of a Green Urban Growth Network can mobilize business and civic forces required to support mutual goals. In this network all participants would strive for a common goal beyond their own.

To fully engage the business sector the *Network* should focus on a specific mutual goal and determine a plan for action. We recommend one key consideration for cities, which is to install a central coordinator, a leadership team or person who can bring different government departments together so that businesses and other local stakeholders only have one point of contact, and also to ensure policies are consistent. The CDP Cities 2011 Global Report highlighted that a number of C40 cities are using a collaborative process during the design of their Climate Action Plans. For example the City of Toronto invites business and other stakeholders to participate in several Enviro-Action Working Groups: Commercial Fleets; Small Business; Good Neighbors Food Procurement; Trees and Greenspace; and Climate Change Adaptation (Riffle et al. 2011).

7.3 Green Business Zones Concept: In Search of the Green Urban Economy

Building on the successful model of Special Economic Zones (SEZs), cities could develop the concept of "Green Business Zones" (GBZs) i.e. living laboratories for experimenting with new energy efficient/green business models. Cities can work in collaboration with regional and national governments to create Green Business Zones (and/or Green Industrial Zones) to spur green activity and job creation. The premise of these zones includes financing mechanisms, government incentives and regulations, technology exchange and training, and provides businesses favorable conditions such as low taxes, low-carbon sourced energy, business support, low-carbon transport (electric vehicles or bike schemes) and advanced ICT technology (ultra-high speed broadband), which may help attract companies to relocate bringing to these places additional benefits such as job creation (Box 3).

Box 3

"We are trying to build innovative clusters in Stockholm that will attract investments. We did it with Information and Communication Technology (ICT). Now we are doing it with Green Tech."
Kristina Alvendal, former **Vice City Mayor of Stockholm**
Taken from interview with Kristina Alvendal 9th September 2011.

Just as SEZs provided developing cities with a laboratory to shape their participation in the global market economy, the GBZs could help global cities transition to a vibrant, low-carbon economy. The challenge for cities will be twofold: firstly GBZs must be endorsed at the regional and national level and will require transformational political leadership; and secondly, they require active participation from businesses. Therefore cities must utilize the political clout of their Mayor to gain support from regional and national governments, and to engage business leaders and media support to maximize the visibility of these zones (Box 4).

Box 4 Status on Green Business Zones (Low Carbon Economic Zones)

Across the world, many regions are establishing low-carbon zones to serve as demonstration areas that promote low-carbon economic transformation. For example in London there are ten low-carbon areas plus the Green 500 Initiative.

In response to the challenge of achieving a low-carbon economy in China, a consortium of Chinese and European institutes have recommended strengthening cooperation between the EU and China to establish a range of "low-carbon zones" across China. These will aim to stimulate transformational regional political leadership in a similar fashion to the special economic zones (SEZs).

Currently there exists an official EU-China work program to pilot low-carbon zones in China – in particular in Jilin City, where they are developing a methodology for a low-carbon zone. This example will be used as a best-practice case study throughout China and other developing countries.

Low carbon zones: Road to a green future – 5th May 2009, published on www.china.org.cn (author: Deputy Chair of UK´s Environment Committee).

7.4 *Innovative Financing Concept: Greening Urban Finances*

National government austerity measures together with reforms in the global financial markets are making the financing for carbon reduction projects increasingly difficult for both the public and private sector. Given recent government spending cuts, cities will not be able to rely on national governments to finance their own low-carbon investments. At the same time businesses vying for financing for relevant investments have been hit from both sides. Firstly, access to credit from global banks and financial institutions has been cut off, and secondly, the business case for many projects has been impacted by a temporary fall in many resource and energy prices (although the long-term trend is upward). Nevertheless, despite these constraints there have been many creative examples of how some financing approaches might work. For example energy services companies (ESCOs) have created various "pay as you save" products, which allows energy users (governments and businesses) to install energy efficiency measures with no upfront costs and with repayments generated from reduced energy costs.

Beyond existing fiscal instruments already at a city's disposal such as taxes and grants there are a host of new approaches being developed to channel public and private capital to promote low-carbon development. They include Green Investment Banks (UK, The Netherlands, US), environmental bonds, Public Private Partnerships (PPPs), Feed-In Tariffs (FiTs), and hitherto less developed by cities Carbon Finance. According to the OECD report Cities and Climate Change (2010), the possibilities for cities to use existing carbon finance instruments could be increased (Kamal-Chaoul and Robert 2010).

This article advocates that cities should work with national authorities especially on transport and buildings as key areas to reduce GHG emissions in the international negotiations to provide a rationale for involving urban areas in market mechanisms.

8 Accelerating Green Urban Growth: Seizing Opportunities and Climate Protection

Public policies that impose stringent green standards on urban development and economic activities can contribute to the development of a larger market for green services which can range from architecture and design firms, certification, construction/installation and financing to eco-tourism and urban farms. Growth of these services is based on the ability of cities to provide enabling conditions and to influence market demand through advertising and marketing campaigns.

Green urban growth presents cities and businesses with a strategic opportunity for innovation and competitive advantage. For companies this would be through the development of products and services to serve emerging new sectors and markets. For cities it provides opportunities for job creation, improvements to quality of life, and the enhancement of international branding and perception (tourism/hosting of major events).

In the publication Corporate Sustainability: A Progress Report (2011) by KPMG and Economist Intelligence Unit (EIU), sustainability was viewed as a source of innovation and new growth. Forty-four percent of business executives agreed that sustainability is a source of innovation, and 39% viewed it as a source of new business opportunities. This clearly points to a change in the way sustainability is being viewed by the business community.

Cities such as Seoul, London, Copenhagen, Tokyo, Yokohama and Rio de Janeiro are facilitating the development of industries related to climate change mitigation. Seoul is willing to spearhead a green policy paradigm shift that will take place over the next two decades. In order to become a leader Seoul is placing a critical value on the realization of the "Green Design City", which redesigns not only buildings, transportation and urban areas but also the functions of the city while reflecting upon low-carbon, low-energy and resource recovery aspects (Riffle et al. 2011).

9 Key Messages

A green growth model or green economy has emerged as a new paradigm that will identify policies and approaches that promote economic development while reducing GHG emissions, minimizing waste and inefficient use of natural resources, and shift production and consumption towards a clean, low-carbon economy. In this article we have presented seven enablers that support global cities in the transition to a green urban economy. They are:

- Capacity to act: creating enabling conditions and removing potential barriers.
- Greening urban sectors: the most cost-effective means to address economic, social and environmental concerns.
- Growth through greening urban infrastructure: examining the critical role of urban infrastructure and financing.
- Knowledge sharing: understanding the importance of knowledge sharing between peer cities.
- Role of national governments: facilitating and managing the transition.
- Shared vision: city to business partnerships – working together towards a shared vision.
- Seizing opportunities: green urban growth presents cities and businesses a strategic opportunity for innovation and competitive advantage, and also climate protection.

There is a realization that sustainability is less a challenge of science and technology and more about human behavior change or, more specifically, choices. It's easy to believe that perhaps your city/region doesn't have the ability, 'natural talent' or resources to transition itself to a low-carbon economy. The first thing to realize, however, is that other cities have done this before, and many more are heading in this direction.

The vital link in transitioning to a green urban economy is *leadership* – a "firestone", i.e. a person or team with passion and energy, who are capable of negotiating multiple priorities and are willing to change. All of these attributes are choices, and all of them are available to you today.

References

Alvendal K (2011) Direct quote during telephone conversation with Ms Alvendal, 9 September 2011

Article 119 Boutin Law 25 March 2009, Decree of 23 November 2009

Brookes G (2011) Financing the growth of your city. KPMG

Chism N (2011) Insight: urbanization. KPMG

Deputy Chair of UK's Environment Committee (2009) Low carbon zones: road to a green future, 5 May 2009. Available via: published on www.china.org.cn

Eddington R (2006) The Eddington Transport Study. UK Department of Transport, London

El-Ashry M (2011) Renewables 2011 Global Status Report. REN21, Paris

Hammer S, Kamal-Chaoul L, Robert A (2011) Cities and green growth: a conceptual framework. OECD, Paris

Hendriksen B, Copius Peereboom E, Jansen S (2010) City typology as a basis for policy. KPMG

Hendriksen B, Copius Peereboom E, Jansen S (2012) Seven enablers that accelerate the green urban economy. KPMG

Kamal-Chaoul and Robert (2010) Cities and Climate Change. OECD, Paris

Riffle C, Hendriksen B, Copius Peereboom E (2011) CDP Cities 2011 Global Report. CDP, C40 Cities & KPMG

Rotmans J (2003) Transitiemanagement: sleutel voor een duurzame samenleving. Koninklijke van Gorcum, Assen. * For reasons of similarity and simplicity we have modified Rotmans' model by combining take-off and acceleration phases into one
Senko et al (2011) Corporate sustainability: a progress report. KPMG & EIU
UN PRI (2011) PRI Progress Report 2011. UN PRI, London
Whiteman G et al (2010) Business strategies and the transition to low-carbon cities. Wiley/ERP Environment, Chichester
Woods D (2011) Making sustainable cities investable. Hauser Center. Available via: www.low2no.org

Linking Green City Politics with Green Business

Ulrich Mans and Sara Meerow

Abstract In the light of cumbersome national energy transitions, cities across the world are increasingly taking the lead in promoting renewable energy on the municipal level. Whereas cities in Europe and the United States are the most prominent pioneers of this trend, the need for renewable energy in Western cities is in fact less urgent. Energy savings often have a much greater priority for these city leaders. This is in contrast to cities in emerging markets, which are on a fast trajectory to become the megacities of tomorrow. Where population growth is imminent and economic development unimpeded, the living standards for many will increase. As a consequence, cities in many developing economies face the growing challenge to meet rising energy needs. For these cities, it is crucial to find adequate solutions for this additional demand despite the major costs involved. This chapter examines cities in developing countries to try to determine which of them have been able to stimulate green growth, in which the development of green energy businesses and green energy policies are mutually reinforcing. The authors assess current renewable energy trends across 36 cities and show that until today only a few cities have been able to create such synergies.

Keywords Renewable energy • Developing countries • Green-driven growth • Green energy business • Mitigation strategies

1 Introduction

In the first decade of the new millennium, renewable energy solutions have witnessed a major boost. They have grown exponentially since 2000 and there is a lively debate on how this development can be sustained in the form of a green

U. Mans (✉) • S. Meerow
Department for Geography, Planning and International Development Studies,
University of Amsterdam, Amsterdam, The Netherlands
e-mail: u.mans@uva.nl; sara.meerow@gmail.com

R. Simpson and M. Zimmermann (eds.), *The Economy of Green Cities: A World Compendium on the Green Urban Economy*, Local Sustainability 3,
DOI 10.1007/978-94-007-1969-9_14,

economy. Unfortunately, much of this discussion has focused on national economic policies – city developments are generally overlooked. This is despite the fact that the city-level is in fact the most challenging part of this debate. One the one hand, cities are the largest energy consumers and need to find new ways of creating a secure supply of sufficient energy for local industries, commercial entities and private households. And, on the other hand, because cities have limited authority over their energy infrastructure, as it is in most cases a national affair. Even though in many countries a general drive towards greater decentralization has led to more responsibilities being devolved to the city level, conventional energy generation is still largely beyond the scope of municipal decision-making.

As a result scholars and practitioners are keen to find best practices from around the globe to learn from municipal energy policy experiments: "we have learned that we need to learn from each other and have the courage to do things differently" (Jie 2010). However, most of these examples are limited to cities in the developed world. Illustrative of this imbalance is the 2011 REN21 report, where 79% (164 of 206) of the selected city cases are from the US, Canada, Europe, Australia, New Zealand and Japan (REN21 2011).

We observe this trend despite the fact that urban centers in developing countries are set to grow significantly in the medium term and will need much more energy in the future. The international Energy Agency (IEA) expects that "by 2030 over 80% of the projected increase in [energy] demand above 2006 levels will come from cities in non-OECD countries" (IEA 2009). This is due both to rural–urban migration trends and economic development. Urbanization rates in Asia are estimated at 2.7%, and Africa scores a phenomenal 3.3 % in 2010 – this compared to a global average of 2.1% (Euromonitor International 2010). With urban populations on the rise and continued economic growth, lifestyle standards are also expected to increase. With more money to spend, people in these increasingly populous cities are expected to have significantly greater energy needs.

2 The Role of City Economics in Renewable Energy

One of the bottlenecks for renewable energy uptake in cities in developing countries is the centralized system for energy generation and transmission. Locally generated electricity can only be fed into the grid if it is legally permissible, and if grid technology allows for independent power producers. This is an important barrier to increasing city-based renewable energy, and often is a key challenge for cities in industrialized countries too.

However, a much more urgent problem for cities in developing countries lies in the financing of renewable energy solutions. For example, the annual budget for Buenos Aires in 2011 amounted to US$6.3 billion compared to US$59 billion for the city of New York (New York Times 2009). Additionally, municipalities in developing countries tend to give priority to other socio-economic problems such as urban poverty or inadequate basic services infrastructure. When facing the choice between fixing an urgent waste collection problem and co-financing a wind farm, municipal authorities are still likely to give priority to the former.[1] Without the adequate monetary resources municipal leaders tend to limit energy-related innovations to cost-neutral initiatives, such as introducing regulations for energy-efficient lighting and building standards, or to projects that can be externally funded, for example through national subsidies for non-electric solar heating solutions on rooftops.

This is where green economic growth can make a difference. If a city sees monetary benefits from the uptake of renewable energy, its promotion is less problematic to agree upon during budgetary negotiations. Such an economic 'benefit' for the urban economy can come in various shapes and forms. In this context, the OECD speaks of a 'green growth strategy': "[a] growth strategy that accounts for increases in public and private investments and consumption leading to sustainable resource use, lower greenhouse gas emissions, and reduced vulnerability to climate change" (Kamal-Chaoui and Alexis 2009). However, this definition was designed for national economies and is therefore not city-specific. Nor does it acknowledge the difficulty of attempting to do two things at the same time: stimulate economic growth and use fewer resources.

When talking about a green urban economy it is therefore useful to divide the concept of green growth into two separate, interrelated aspects. On the one hand, economic growth can be achieved while limiting its environmental impact. On the other hand, it can be driven by the need for environmentally friendly solutions and products. The first could be called 'green-conscious growth', which encompasses a city's efforts to try to follow a green agenda and reduce its overall ecological footprint. The latter could instead be labeled 'green-driven growth', and stands for a more entrepreneurial vision for a green city. A city can create economic growth not despite being green, but rather because of it (see also Mans 2012). Naturally, this cannot succeed as a purely public endeavor; the role of private businesses is crucial to stimulating green-driven growth.

The recent boom of green, sustainable business has already reached many parts of the global economy. As part of this trend, the renewable energy business in particular is an illustrative example of major investments towards the green economy.[2] Renewable energy is a fast-growing business with a total market of US$210 billion in 2010 (Roland Berger 2011); and market analysts expect continued growth over the next couple of years (see also UNEP SEFI and New Energy Finance 2009). A large number of companies across the globe are part of this expanding business sector, such as India's Suzlon or China's Yingli Solar.

Municipalities are keen to attract such businesses, as it potentially creates additional sources of revenue. Thus, green cities should be at the frontline of this development. For those cities that follow a green agenda, it is particularly interesting to

[1] This was the case in Cape Town, where municipal authorities have to balance between major socio-economic challenges while trying to commit to their Energy and Climate Change Strategy (Mans 2010).

[2] The term green economy refers to all economic sectors that primarily produce products that support the reduction of CO_2 emissions. Some authors prefer to speak about clean economy. For the purpose of this chapter, the term green economy is used.

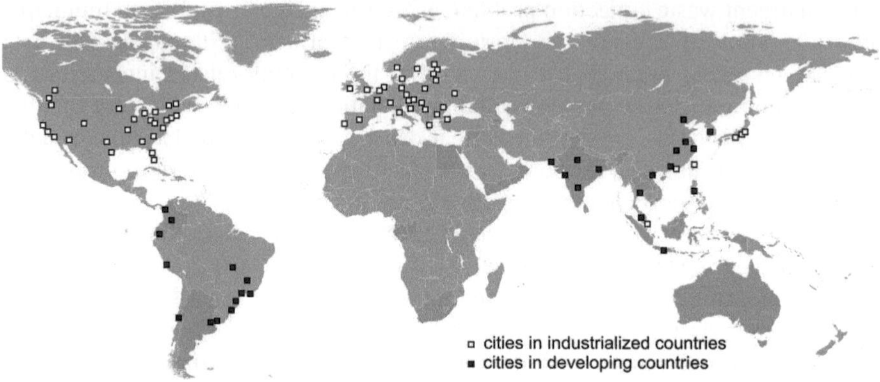

Fig. 1 The EIU green city index – global coverage (until November 2011)

have their own share of renewable energy business in town. First because renewable energy products would be locally available; second because a city can enhance its reputation as being 'green' through a business community that is supporting the green agenda; and third because green policies that come at the cost of taxpayers are easier to sell if they are perceived to create local jobs at the same time. In other words, in order to make 'green' an economically sustainable undertaking, municipalities and private actors could actively look for solutions that are mutually beneficial and manage urban centers in a way that green-conscious growth (the political agenda) meets green-driven growth (the business agenda).

In order to look for cities where this synergy is possible, this chapter uses the Economist Intelligence Unit's Green City Index (GCI), which is one of the first comprehensive efforts to collect data across cities within a comparable framework, and a dataset purchased from *Globaldata* on renewable energy business locations.[3] The GCI looks at eight aspects of a green city: energy and CO_2, transport, water, air quality, land use and buildings, waste, sanitation and environmental governance. For each aspect, a city is given a score from one to five, qualified as 'well below average', 'below average', 'average', 'above average' and 'well above average'. The total coverage spans 96 cities worldwide,[4] of which 33 are located in developing countries (Fig. 1).[5]

[3] It is important to note that this does not include a quantitative judgment on how much monetary benefit these businesses represent for a given municipality. This would depend on further research on the labor market, tax revenue, direct investment and other indirect positive effects on the urban economy.

[4] This chapter is based on the regional Green City Indexes for Asia, Latin America, US and Canada and Europe. The one done for Germany was not included in Fig. 1 for matter of readability.

[5] This means that Singapore, Osaka, Tokyo, Yokohama, Hong Kong and Taipei were not considered to be 'developing country cities'.

For the political (green) agenda, we created an additional ninth score to reflect each of the 33 cities' commitment towards renewable energy. We based this assessment on information on renewable energy initiatives from the GCI report and other public sources (Fig. 2). For the (green) business agenda we calculated a score for green-driven growth in every city in the sample using a proxy indicator: the number of renewable energy companies that have an office presence in the city. This makes it possible to correlate both indicators and to determine whether there are cities in developing countries that are able to combine green-conscious growth with green-driven growth.

3 Cities in Comparative Analysis

Looking at the top ten scores in terms of renewable energy business locations there are few cities that show a similarly high score for both indicators (Fig. 3). São Paulo (89 businesses) and Seoul (106) are the only cities within the top ten that have the highest green energy politics score (five). Another two cities score four on the index: Rio de Janeiro (43) and Bangkok (91). In comparison, the cities Jakarta (2; 87) and Kuala Lumpur (2; 108) share a middle position and score better than the remaining four cities in the top ten: Kolkata (1; 137), Manila (1; 97), Buenos Aires (2:56) and Santiago de Chile (1; 66). The overall correlation between political and economic renewable energy ambitions seems weak.

A more detailed look at the type of renewable energy businesses in each city reveals some interesting observations. Among the major business centers, Mumbai is dominated by one of the business categories. Out of 501 companies based in the city, 422 are active in the wind energy business. Other notable results include Shanghai, Beijing, Seoul, Lima, Jakarta, Guangzhou, Nanjing and Wuhan. In Seoul and the four Chinese cities Nanjing, Shanghai, Guangzhou and Wuhan, there are a relatively high number of solar photovoltaic businesses (61, 85, 71, 64 and 57% respectively). In Jakarta 44% of the businesses are active in the geothermal energy sector, and 71% of Lima's renewable energy businesses work in hydropower. In Bangkok, 45 companies of 91 in total are active in the biotech sector.[6] Figure 4 below shows an overview of the various renewable energy business foci for 13 selected top-ranking cities.[7] All other cities have less than 40 business locations.

The sub-sector data confirm that there is little correlation between green politics and green business. Among the cities in the sample, only three show a relatively clear link between municipal renewable energy priorities and the type of renewable energy businesses present in the city. In Seoul, for example, 70% of all companies (75 of 106)

[6] As many companies are involved in more than one sector (on average 1.75 sectors per company) these numbers do not present an 'exclusive' focus on a particular energy technology, rather it gives an indication about which sectors have the largest economic footprint in each city.

[7] Because many companies are active in more than one sector, the numbers do not match with the office count.

City	Energy and CO2	Land use and buildings	Transport	Waste	Water	Sanitarion	Air Quality	Environmental Governance	GCI overall	New RE Score
Bangkok	***	**	**	**	**	**	****	****	***	****
Beijing	**	***	***	***	****	***	**	***	***	**
Belo Horizonte	***	****	***	**	****	***	****	**	****	****
Bengaluru	***	***	***	***	***	***	***	***	**	***
Bogota	****	****	****	****	****	**	**	****	****	***
Brasilia	***	***	*	*	****	****	****	****	****	***
Buenos Aires	***	***	***	**	*	**	**	***	**	**
Curitiba	****	***	****	*****	****	****	*****	****	*****	****
Delhi	****	***	***	****	**	***	***	***	***	****
Guadalajara	**	***	**	***	*	**	***	*	*	*
Guangzhou	**	***	***	***	**	****	***	***	***	****
Hanoi	***	*	**	***	**	*	***	*	**	**
Jakarta	****	***	***	*	**	**	***	***	***	**
Karachi	**	**	*	**	***	**	*	**	*	**
Kolkata	**	**	*	**	***	**	**	*	**	*
Kuala Lumpur	**	***	****	*	*	**	****	***	***	**
Lima	***	*	***	**	**	**	**	**	*	**
Manila	***	**	**	**	**	**	****	***	**	*
Medellin	**	**	***	**	***	*****	****	***	***	***
Mexico City	****	****	****	***	***	**	**	*****	***	****
Monterrey	***	***	**	****	****	****	**	**	***	***
Montevideo	**	*	***	**	**	**	**	****	**	***
Mumbai	***	***	**	**	***	**	*	**	**	***
Nanjing	**	***	***	***	****	***	***	***	***	***
Porto Alegre	**	***	**	****	***	***	***	**	***	***
Puebla	**	***	**	****	***	***	***	***	***	**
Quito	***	**	****	****	***	**	****	***	***	**
Rio de Janeiro	****	****	***	***	**	***	***	*****	****	****
Santiago	*	***	*****	****	****	****	***	***	***	*
Saa Paulo	*****	****	****	****	****	****	***	***	****	*****
Seoul	****	****	****	**	****	****	***	****	****	*****
Shanghai	*	**	***	***	***	***	***	***	***	****
Wuhan	**	**	***	***	****	***	**	***	***	***

Fig. 2 Thirty-three cities and their respective scores for being 'green' (EIU 2010a, b and own estimates)

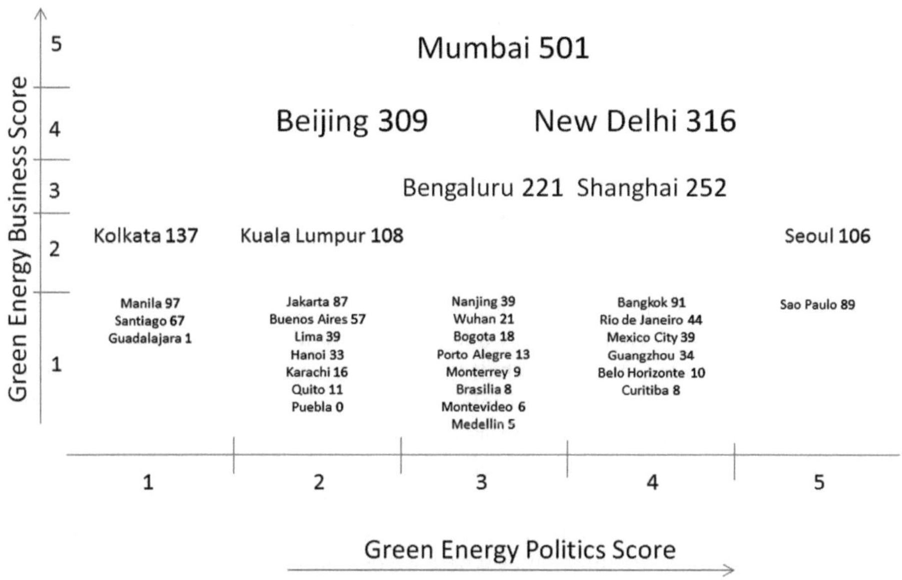

Fig. 3 Correlation between green business and green politics

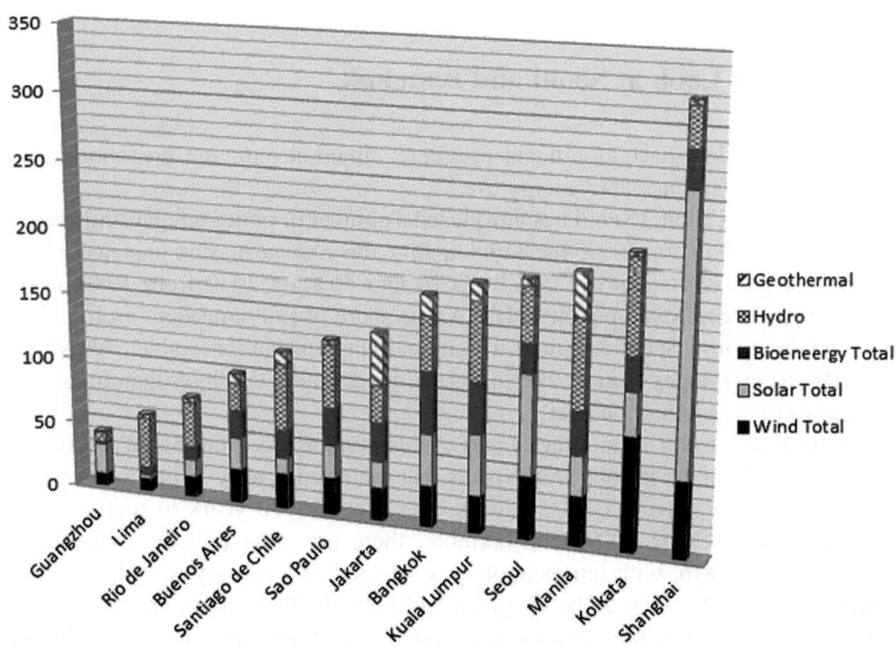

Fig. 4 Renewable energy business sub-sectors for 13 selected cities

are active in the solar sector, which matches the municipal focus on solar energy promotion (EIU 2010b). Lima's hydropower focus and Jakarta's pro-geothermal policies (Jakarta for example is the host of the 2012 World Geothermal Energy Summit) are also reflected in the local business profile (71 and 44% respectively). In contrast, Shanghai supports major initiatives in the wind sector, while 71% of the businesses are making money with solar photovoltaic. In Guangzhou, 64% of businesses are solar-related, while the local authorities focus on hydro energy (EIU 2010b). All other cities present a relatively balanced playing field among the various renewable energy sectors.

What do these findings tell us about the link between green city politics and the green urban economy? There are only a few examples in the sample where this type of synergy can be found. More than half (18 cities) have significantly less company offices (under 50). Among the 15 higher-ranking cities, Mumbai and also New Delhi stand out as having a disproportionally high amount of companies in the wind energy sector (84 and 61% respectively). The type of business however does not reflect the major municipal priorities regarding renewable energy; they focus on energy efficiency (Mumbai) and solar energy (New Delhi) respectively (EIU 2010b). In the mid-range, there are three cities where we can observe a degree of correlation between both business and politics scores: Seoul (5; 106), Bangkok (4; 91) and São Paulo (5; 89). A further assessment of the first two cities reveals that the green urban economies might indeed benefit from a growing synergy between green business and green politics.

4 A Closer Look at Seoul and Bangkok

Seoul and Bangkok have significant programs aimed at reducing greenhouse gas emissions through renewable energy use.

South Korea's capital Seoul has introduced a number of policies regarding energy and CO_2 reduction. In 2009, the municipality unveiled a master plan to reduce greenhouse gas emissions by 25% by 2020 and 40% by 2030 from 1990 levels (Connected Urban Development 2010). As the share of CO_2 emissions from industry is relatively low, Seoul envisions a significant increase in renewable energy use. By 2030 the city hopes to meet 20% of its energy demand with hydrogen fuel cells, solar power and geothermal heat.

The business sector is an integral part of this municipal vision: Seoul's aim is to create one million green jobs and expects that the municipality will, together with the private sector, invest US$45 billion over the next 20 years to realize these ambitions (C40 Cities 2009). Furthermore, there are clear policies to facilitate green-driven growth, with ten technologies selected as a priority for city support: hydrogen fuel cells, solar cells, IT, green buildings, LED lighting, green IT, green cars, urban environment recovery, recovering waste into resource and climate change adaptation technology. Seoul also plans to invest around US$ two billion (an average of US$100) in R&D by 2030 in order to provide systematic financial support (The Climate Group 2010).

There are a total of 106 renewable energy companies with an office presence in Seoul. More than two-third of these companies (75 companies) have some sort of solar energy product in their broader renewable energy portfolio (including solar thermal and solar PV). Sixty-five of these 75 companies offer solar PV products (no solar thermal) and 33 companies have an exclusive focus on solar PV (e.g. no other renewable energy products).[8] Seventeen of the 65 solar PV companies have their headquarters in Seoul and 19 are office branches (locations) from companies with their headquarters elsewhere. A total of 28 companies fall in the category 'subsidiary'. The company size varies across the spectrum. Out of the eight company headquarters that are specialized in solar PV for example, some are major industrial players (such as LG and OCI who added solar products to their existing portfolio). Others are solar-only companies (such as Nexolon and S-Energy).

Given the particular strength of the solar industry in the city, the envisioned R&D support for this technology is an interesting example of a synergy between green energy politics and green energy business.

In Thailand, the Bangkok Metropolitan Administration (BMA) has introduced policies and plans promoting renewable energy and climate change mitigation. The municipal government drafted its own '5 Year Action Pan for Global Warming Alleviation 2007–2012', which includes the promotion of renewable energy as one of its five main strategies. In 2009 the mayor of Bangkok spearheaded a regional initiative called "Cool ASEAN, Green Capitals" to mitigate climate change (ASEAN Secretariat 2009).

To achieve these renewable energy goals, the city has introduced several initiatives. One example is a campaign implemented by the BMA in cooperation with the Bangchak Petroleum Public Co. to collect used cooking oil and turn it into bio-diesel (Bangkok Metropolitan Administration 2009). The city also promotes the use of biofuels by the general public, setting an example by utilizing gasohol and biodiesel for BMA infrastructure and city vehicles (Jungrungruen 2011).

In a related development, there are a number of municipal waste-to-energy projects in Bangkok, and the city plans to upgrade more of its landfills in the future to generate electricity and to qualify as CDM projects (Jungrungruen 2011). Of the 723 MW of renewable power already selling to the grid, the vast majority (637) were generated from biomass, which is reflected in the large number of biotech companies in Bangkok. Moreover, many existing companies and industries have branched out into renewable energy. The sugar industry is a good example. Of the 47 sugar mills in Thailand, 35 now sell electricity generated from biofuels to the grid as a secondary source of income.[9]

---

[8] This specialization only refers to the renewable energy sector. For example, many companies produce a range of products, including renewable energy products: i.e. Avago Technologies in Seoul produce fiber optics, optical sensors, etc. but also solar PV products. This company does not produce any other renewable energy appliances and therefore, for the purpose of this chapter, is considered to be specialized in solar energy.

[9] According to the Chairman of the Bio-Electricity Working Group of Thai Sugar Millers Corporation, Interview, September 2010.

In total there are 91 renewable energy businesses listed in Bangkok specialized in a variety of energy sources, but in particular solar PV, hydropower, wind, and biomass. While many companies choose to locate their offices in the country's business center, most of the existing renewable energy projects are located outside of Bangkok. To demonstrate, there are 11 small renewable power plant projects operating in the Bangkok Metropolitan Electricity Authority area, as opposed to 166 in the Provincial Electricity Authority's region (EPPO 2011).

5 The Way Forward

The uptake of renewable energy in cities is only beginning, particularly in the developing world. Many of today's technologies are still too expensive to be widely adopted in non-OECD countries, and often require a favorable regulatory environment on a national level in order to become a feasible alternative for fossil fuel-based (or nuclear) urban energy. Nonetheless, an increasing number of municipalities have begun implementing a green agenda, including the scale-up of renewable energy.

The role of private actors can be a significant catalyst for supporting such a transition towards an alternative energy supply. The authors argue that the booming green energy business sector in particular can help green energy politics succeed. Cities and their (gradually) increasing leverage vis-à-vis other government levels offer new opportunities in this regard. However, when looking at recent developments, there are very few examples of how green costs can be offset by adding green revenues. Green-driven growth is still in its very beginnings. The presented cases across Latin America and Asia show that a mutually reinforcing relationship between green energy politics and business is not easy to pin down.

The presented findings demonstrate that where municipal authorities are committed to promoting renewable energy, such political ambition is seldom mirrored in the city's green energy business profile. Major business hubs such as Kolkata, New Delhi and Beijing host the largest number of renewable energy companies, mainly because they, as a city, present an interesting office location. In the case of Beijing and New Delhi, the proximity to political decision-makers can be considered an additional pull-factor. These municipalities do promote renewable energy solutions in general, but these decisions are made independently from the city's business profile in this sector.

In contrast, the cities Seoul and Bangkok reveal a more specific agenda for promoting renewable energy as well as related businesses. Bangkok's bioenergy initiatives and Seoul's solar energy targets are evidence of more concrete municipal action towards green-driven growth, and demonstrate the will to try out innovative technologies and experiment with new governance tools. In fact, Seoul provides an interesting example of how municipal renewable energy policies can create synergies with green energy business. The existing industries in the city were included in the priority subsidy program, including the city's highly visible solar PV sector.

The idea that Seoul plans to use know-how from city-based businesses to develop city-specific energy solutions is an interesting development in this regard.

The biggest problem is that even for these promising examples of green-driven growth, there is little we can say about what works best. In the case of Bangkok, the existing green energy companies might not make a lot of a difference in terms of concrete benefits for the city's revenues; and for Seoul it could be argued that the national (not city-level) policies are in fact the real driving force behind city-level developments. Yet, these two cities currently have a clear opportunity to build on this momentum. If they were to find a way to create benefits from having green energy business match the city's green energy policy goals other city leaders will certainly be interested to learn from them.

Despite this cautionary note, the authors emphasize that in the coming years new sources of revenue are likely to become an increasingly important enabler for municipal initiatives in support of a green urban economy. If the global green economy keeps growing and the current trend of municipalities going 'green' continues, the examples discussed in this chapter might offer some initial lessons learned for other cities that are willing to pursue renewable energy solutions within their jurisdictions. At the same time, much more trial and error is needed to determine the most effective strategies for municipal governance towards a synergy between green-conscious growth and green-driven growth.

References

ASEAN Secretariat (2009) Cool ASEAN, Green capitals – region to tackle climate change. Available via: http://www.asean.org/23783.htm. Cited 3 Aug 2011

Bangkok Metropolitan Administration (2009) Bangkok Assessment Report on Climate Change 2009. Available via: http://www.unep.org/dewa/pdf/BKK_assessment_report2009.pdf. Cited on 3 Aug 2011

C40 Cities (2009) CCAP Seoul. Available via: www.c40cities.org/docs/ccap-seoul-131109.pdf

Connected Urban Development (2010) Seoul, a clean and attractive global city. Available via: http://www.connectedurbandevelopment.org/cities/seoul. Cited on 3 Aug 2011

Economist Intelligence Unit (2010a) Green city index Latin America. Siemens Publication, Copenhagen

Economist Intelligence Unit (2010b) Green city index Asia. Siemens Publication, Copenhagen

Energy Planning and Policy Office (EPPO) Thailand (2011) Power data. Available via: http://www.eppo.go.th/power/data/index.html. Cited 3 Aug 2011

Euromonitor International (2010) World's fastest growing cities are in Asia and Africa, 2 March. Available via: www.blog.euromonitor.com/2010/03/special-report-worlds-fastest-growing-cities-are-in-asia-and-africa.html. Cited on 3 Aug 2011

IEA (2009) Cities, towns and renewable energy. IEA Publication, Paris

Jie L (2010) European green capitals: looking to cities of the future, executive summary and conference report. 2010 EXPO Shanghai chapters, Shanghai

Jungrungruen S (2011) Low carbon target and activity in Bangkok. Presentation by the Director of Air Quality and Noise Management Division, Department of Environment, Bangkok Metropolitan Administration, Bangkok. Available via: http://lcc.ait.asia/upload/activities/BMA_20May2011.pdf Accessed 3 Aug 2011

Kamal-Chaoui L, Alexis R (eds) (2009) Competitive cities and climate change. OECD regional development working chapters n° 2, Paris

Mans U (2010) Renewable energy in African cities: the case of Cape Town, South Africa. Chapter presented at ISOCARP Congress 2010, Nairobi
Mans U (2012) Linking resilience and green growth, how green business can contribute to more resilient cities in India. In: Otto-Zimmermann K (ed) Resilient cities: cities and adaptation to climate change. Proceedings of the Global Forum 2011. Local sustainability, vol 2. Springer, Dordrecht, forthcoming
New York Times (2009) The 2010 New York City executive budget, 1 May
REN21 (2011) Renewables 2011: global status report
Roland B (2011) Green growth, green profit, how green transformation boosts business. Palgrave Macmillan Publishers, Hampshire
The Climate Group (2010) Low-carbon cities – an international perspective. Available via: http://www.theclimategroup.org.cn/publications/2010-08-Low_Carbon_Cities-An_International_Perspective-en.pdf. Cited on 3 Aug 2011
UNEP SEFI and New Energy Finance (2009) Global trends in sustainable energy investment 2009. UNEP, Paris

Policy Instruments for Promoting a Green Urban Economy: The Changing Role of the State

Golam Rasul

Abstract Urban areas consume more than two-thirds of the world's primary energy and contribute nearly four-fifths to global greenhouse gas emissions. The promotion of a green economy and enhancement of eco-efficiency of urban economies are closely linked. Although the concept of a green economy has raised hopes among urban stakeholders to enhance eco-efficiency, there is little understanding on the available policy and economic instruments to realize this, as well as on the new role to be played by the state in promoting the green economy. This chapter briefly examines the concept and principles of a green economy, types of available policy and economic instruments and new governance mechanisms, particularly the evolving role of the state to foster a green urban economy. The analysis reveals that a transition to a green economy would require a fundamental shift in the design of policy and institutions and in the organization of economic activities so as to enhance efficiency and reduce unsustainable practices. Financial, economic, and market instruments need to be aligned to provide incentives to enhance energy efficiency, encourage recycling and reuse, and greening the urban economy. This necessitates a new role for the state, moving from its traditional command and control approach towards a market-based economic approach by facilitating and creating a conductive environment for the proper functioning of markets, the private sector and other actors. The findings of this chapter are expected to be useful in designing a green economy approach for developing green cities.

Keywords Policy instruments • Economic instruments • Regulatory instruments • Governance • Role of the state

---

G. Rasul (✉)
Economic Analysis Division, International Centre for Integrated Mountain Development (ICIMOD), Kathmandu, Nepal
e-mail: grasul@icimod.org

R. Simpson and M. Zimmermann (eds.), *The Economy of Green Cities: A World Compendium on the Green Urban Economy*, Local Sustainability 3,
DOI 10.1007/978-94-007-1969-9_15, © Springer Science+Business Media Dordrecht 2013

1 Introduction

Urban areas are the engine of economic growth in the world (Karlenzig 2011). Urban areas are not only a primary engine of economic growth but also providers of employment, shelter and services as well as centers of culture, learning and technological progress. However urban areas also consume more than two-thirds of the world's primary energy and contribute nearly four-fifths of global greenhouse gas emissions. Different projections indicate that by 2050, 70% of the world population will live in cities and urban population will grow from 3.5 billion in 2010 to 6.3 billion. Likewise in Asia by around 2,035 the urban population will have ballooned to more than 2.6 billion people (ADB 2006). Although cities occupy a relatively small land area their carbon footprint is large and cities have had a huge impact on the environment. Cities consume disproportionate amounts of resources and urban population cause much greater pollution per capita than those in rural areas. In their present condition, cities are not sustainable. They are destroying local environments and cultures, as well as the ability of local communities to solve their problems in a sustainable manner. One of the greatest challenges in reducing the atmospheric levels of greenhouse gases lies with buildings and housing (Levine et al. 2007; Anonymous 2009).

The concept of green economy has emerged in response to concerns about the adverse environmental, economic and social impacts of conventional growth oriented economic model (ICIMOD 2011). Although the existing growth-oriented approach increases gross domestic product (GDP) in many parts of the world, it has intensified resource extraction, accelerated depletion of natural resources, and degraded environmental quality and ecosystems' health upon which economic activities and human well-being fundamentally depend. The excessive and unsustainable use of resources has put serious pressure on land, water, environment and atmosphere, which has triggered multiple crises including global warming and climate change, energy crisis, food insecurity, economic decline and natural disaster.

The conventional economic perspective – grow first and clean up later – has proved problematic. The current climate and economic crisis clearly demonstrates that growth through overexploitation is unsustainable; in fact destroying natural capital hampers present and future livelihoods. The cost of cleanup could be very high and even impossible to repay when facing irreparable damage. For instance in China annual investment in pollution control has reached 1.44 billion RMB equivalent to 1.16% of GDP (Kunmin et al. 2007). There is now a growing realization that it is necessary to decouple economic growth from environmental degradation so that the path and pace of economic growth does not affect the environment, its ecosystems and natural capital.

Environmental challenges that urban areas face today such as deteriorating environment, energy crisis, waste management traffic congestion and reduction of green house emission have raised serious concerns for the welfare of current and future generations. Thus greening the urban economy has become the need of the hour (ICLEI 2011; UN-Habitat 2011). The promotion of green economy and enhancement of eco-efficiency of urban economies are closely related. While the concept of green economy has raised hopes for enhanced eco-efficiency among urban stakeholders, the mechanisms of this process – translation of green economy instruments into action and the role of the state both at central and local government levels – is poorly understood and rarely discussed.

The success of the green economy will depend on creating enabling and appropriate institutional framework conditions to translate green economy approaches and instruments into action (Barbier 2011; ICIMOD 2011). This requires a fundamental shift in the design of policy and institutions and in the organization of economic activities to enhance efficiency and reduce unsustainable practices. Although traditionally the command-and-control approach has been widely used in environmental management, it alone will not be effective in the pursuit of the green economy. In order to create a win-win synergy between economic growth, social development and environmental sustainability the role of market instruments is critical. Financial, economic, and market instruments need to be aligned to provide incentives to enhance energy efficiency, encourage recycling, reuse, and ultimately green the urban economy.

While market instruments are necessary they will not be sufficient to create the right conditions for the success of a green economy alone. The market often fails to deal with environmental externalities and cannot ensure public goods such as clean air, water, biodiversity or environmental quality at the levels necessary for wider society. As such the state plays a critical role in correcting the market, directing policy, and balancing the divergence between private and social interests. The state also plays a critical role in coordinating and regulating the market and in creating a space for the private sector, civil society, research organizations, development agencies and other relevant actors to participate in the green economy at the national, regional, and global levels. The task of dealing with known and emerging environmental challenges and tapping opportunities which emerge from the green economy requires: effective use of financial and economic instruments, greater involvement of private sector, alignment of institutional incentives, and the promotion of sustainable production and consumption practices. This necessitates a new role of the state, moving from its traditional command and control approach to a market-based economic approach which facilitates and creates a conducive environment for the proper functioning of markets, the private sector and other actors.

Despite the growing interest in promoting green economy, little effort has been made to understand the link between the state and the effective implementation of green economy instruments in urban areas. This chapter attempts to fill this gap by examining economic and public administration literature and drawing lessons from policy and institutional innovations in promoting a green economy in different countries.

The structure of the chapter is as follows. The second section briefly examines the concept and principles of green economy and its relationship with the green urban economy. The third section looks at the types of policy and economic instruments available in promoting a green urban economy followed by the evolving role of the state in designing, organizing, and coordinating financial and economic instruments to foster a green urban economy. The findings of this chapter are expected to be useful in designing a green economy approach for developing green cities.

2 Green Economy: Definition and Principles

Although the concept of green economy has been gaining currency no systematic clarification of its meaning or articulation of its principles has been agreed by all stakeholders. As a result different scholars and stakeholders see the green economy from different perspectives based on their disciplinary background and/or professional interests.

Scholars optimistic about the potential role of green technologies see green economy as a means of developing and using new products, services, innovation and techniques that promote environmental protection and enhance energy security (Henton et al. 2008; CEDEFOP 2009; ECO CANADA 2010). This perspective sees green economy just as a technological fix and innovation of clean technology and their wider adoption as a way to reduce environmental hazards and generate employment in new sectors. The role of the state is here one that encourages the adoption of new technology by providing financial incentives and reducing their associated entry costs.

In contrast to the narrow techno-centric perspective other scholars consider green economy as a process and mechanism of greening the existing 'brown economy' by reducing the ecological footprint (OECD 2009, 2010, 2011; DANIDA 2010). This perspective sees the green economy as a wider technical, economic and policy issue, which can enhance efficiency of resource use while reducing the ecological footprint. The solution here lies with correcting policy failures to get the price right for a smooth functioning of market forces, removing subsidies from fossil fuels and other environmentally harmful inputs, introducing corrective taxation as well as encouraging investments in green technologies. According to this approach social welfare will come through improved environmental quality and better resource management.

Another important perspective on green economy is put forward by the United Nations Environment Programme (UNEP). For UNEP a green economy is "one that results in improved human well-being and social equity, while significantly reducing environmental risks and ecological scarcities", it is "low carbon, resource efficient and socially inclusive" (UNEP 2011). UNEP also place strong emphasis on appropriate market and price mechanisms as a prerequisite to a green economy (Bina and Camera 2011).

The green economy also has a particular relevance for scholars concerned with social inequality. The benefits of a green economy are seen to range from reducing poverty and inequality to enhancing social harmony through improved distributional and social justice (Kennet 2007). According to such scholars the implications of the green economy are broad: it should lead to a fair distribution of resources, provide livelihood security, and a more egalitarian distribution of wealth, whilst also shifting public values, norms and behavior patterns to ensure human existence within the carrying capacity of the Earth, eradicate poverty, and improve quality of life (UNESCO 2011).

Although as yet there is no formal institutional framework or definition of a green economy, certain core principles can be discerned. The green economy:

- Recognizes the role of nature in sustaining economic growth as well as the limits to which the Earth can support economic activities and capacity to absorb wastes
- Takes action to preserve natural capital and its sustainable use
- Manages and safeguards environmental assets so that future generations can meet their own needs and are able to flourish
- Seeks management approaches and technologies that increase resource efficiency and reduce waste
- Internalizes externalities so that market prices reflect real social and environmental costs and benefits and the polluter bears the cost of pollution
- Promotes sustainable production and consumption, either reducing or eliminating unsustainable patterns, i.e., reduce, reuse, and recycle the materials
- Promotes economic, social and environmental justice

3 Policy Options and Instruments

A key challenge in transitioning to a green economy is around the question of 'how to' influence the behavior of myriad economic and social actors to move from unsustainable to sustainable practices. Key questions to be answered are: What policies and instruments are available at the disposal of policy makers? Which approaches and instruments are suitable and effective in what contexts? Which particular types of instruments work best under what conditions? This section provides a brief overview of the key options and instruments available for policy makers to promote a green economy.

The purpose of green economy policy instruments is to influence the behavior of economic agents so that they gradually move towards an environmentally friendly development path. Thus all green policies must focus on fuel efficiency, renewable energy, increased energy efficiency, material efficiency, substitution of intermediate and final goods and change in consumption pattern. Economic policies should be crafted in such a way that economic agents are encouraged to improve techniques, develop new options, find substitutes for environmentally damaging techniques, and follow low-carbon development path.

A variety of policy measures and economic instruments can be used to promote a green economy, which can be categorized into two broad groups: regulatory instruments and economic instruments. The regulatory instruments are based on state rules and regulations, which is commonly known as the command and control approach. The economic instruments, which rely on markets and the price mechanism, are used to internalize environmental externalities. The underlying premise for economic instruments is to correct the market failure by placing a cost on the release of pollutants. This will internalize the 'externalities' into the decision making process. For example placing a charge or a fee on every unit of effluent released into a river transforms the manufacturer's decisions regarding how much he will produce and how he will produce it. These policy instruments have different features and mechanisms of influencing the behavior of economic actors and might have different impacts on the transition to a green economy (Table 1).

Regulatory approaches are direct measures aimed at directly influencing the behavior of economic agents by setting environmental standards and enforcing them through command-and-control. Common regulatory instruments include standards, licensing, mandatory management plans and covenants. Regulation is an effective tool to facilitate the green transition by prescribing and controlling the behavior of the economic actors. Environmental regulations can contribute to reducing emission and improving urban environmental quality. The main strength of regulatory instruments is that regulation is generally binding: it includes all actors who undertake an activity described in the regulation, treating each in the same way. Regulatory instruments are however rigid and difficult to change. Because of their rigidity and difficulties in implementation and in precisely defining environmental quality, regulations often slow down innovation and they tend to benefit known incumbent practices and technologies (and thus their owners) over new and unknown ones. The effectiveness of regulation depends upon the knowledge, skills, or financial ability of regulated actors to comply as well as the regulators' ability to monitor and enforce regulation. Moreover, monitoring and enforcement costs are often very high in regulatory measures due to complexity in defining environmental standards and ambient quality (Cohen 1999). Besides, regulatory measures do not always provide an incentive for exceeding targets even if it might be cost-effective, as different economic agents may face different levels of compliance cost due to variations in technology, location or production processes.

Economic approaches are market-based instruments that indirectly influence the behavior of economic agents through the pricing mechanism. Such incentives or disincentives change the basis upon which producers and consumers make economic decisions. Governments can use a range of economic instruments such as charges, subsidies, grants, taxes, and tax concessions, deposit-refund systems, and market creation to stimulate behavioral change. Financial incentives can play an important role in enhancing innovation and its successful adoption and diffusion within a green economy. For instance a large scale adoption of renewable energy technologies in Germany, particularly of photovoltaic panels was primarily triggered by widespread subsidy between 2000 and 2008 in the form of low interest loans to investors (Jacobsson and Lauber 2006, also see Frondel et al. 2008). Similarly in China finance in energy efficiency and environmental protection has increased from 341 billion RMB in 2007 to 1,011 billion RMB in 2010 through a green finance program (Box 1). Similarly ecological fiscal transfers in Brazil have increased green areas in municipalities and their quality (Box 2).

Table 1 Characteristics of regulatory and economic policy instruments

Characteristics	Regulatory	Economic
Approaches	Command and control	Market-based
Key instruments	Standards, bans, permits, laws, licenses, registrations, administrative guidelines, directives, codes of practice, etc.	Grants, subsidies (grants, soft loans, green finance, green credit, and tax allowances); taxes, charges (effluent/emission charges, user charges, product charges and administrative charges), emissions trading, price intervention, liability insurance, creation, tradable emission, quotas, resource taxes, consumption taxes, congestion charges, non-compliance fees performance bonds, etc.
Operating mechanism	Reliance on laws and regulations	Reliance on markets and the price mechanism
	Legal penalties in judicial and administrative procedures	Operate by changing the costs and benefits incentives that people face when making decisions
Choice of instruments	No freedom of choice	Gives greater freedom to choice and compliance
	Need to comply the same standard, irrespective of marginal compliance costs faced by different economic actors	Economic actors are encouraged to make better choices and to share the burden of low-carbon development paths in socially beneficial ways
		Firms and industries can respond to low carbon development paths in a variety of ways, including process change, technology development and product modification
Environmental effectiveness	Direct	Relatively low
	Relatively high	
Efficiency	Low, little incentives for technical innovation as charges or fees are fixed irrespective of marginal compliance costs	High. Encourages innovation as well as allow buying and selling quota which reduces total compliance cost to the society while achieving the desired goal
		Can generate revenue, which can be used financing green activities
Administration and compliance costs	High transaction cost, high rent seeking	Relatively low transaction cost
Political feasibility	Low	High
Strength	Directness and certainty of outcomes	Provides more choices to comply with minimum social costs
Weakness	Difficult and costly to administer and enforce	Uncertainty of their impacts
	Economically inefficient as no economic mechanism operates to enable standards to be attained at least cost	Too indirect and relatively slow to influence change
	Offering scope for corruption as incorporating fines or penalties, which are usually set at levels too low to deter violators	

Box 1 Green Finance and Green Procurement Policy in China

In 2005, China State Council endorsed promotion of the economic policies conducive to environmental protection and in 2007, urged for strengthening financial services in the field of energy conservation and environmental protection. Since 2007, three green finance policies have been produced by the Ministry of Environmental Protection, the peoples' Bank of China, the China Banking Regulatory Commission, the China Security Regulatory Commission, and the China Insurance Regulatory Commission. Green finance policies in China include green credit, green insurance, and green security. In 2007, 106.3 billion RMB was provided as loans to companies to reduce emissions and conserve energy and Industrial and the Commercial Bank of China decreased its loans to polluting industries by 24%. As a result, bank loans for energy efficiency and environmental protection increased from 341 billion RMB in 2007 to 1,011 billion RMB in 2010 (Climate Group 2011). Similarly, green procurement is also gaining acceptance in China. For instance the Beijing Olympic Committee set environmental requirements for each category of procurement and gave priority to the purchase of products with environment friendly labels (MEP 2006). Similarly some local governments enact green procurement policies (Aizawa and Yang 2010).

Box 2 Ecological Fiscal Transfers: The Case of Brazil

In Brazil, some states have introduced ecological fiscal transfers (Imposto sobre Circulação de Mercadorias e Serviços [ICMS] Ecológico) to compensate municipalities for the opportunity costs of coverage of green areas and for protecting watersheds that benefit other municipalities. So far 16 of 26 Brazilian states have introduced the ICMS Ecológico in their states' constitutions and 13 have implemented ecological fiscal transfers. In 1991, the state of Paraná implemented a law that awarded 5% of ICMS revenue to municipalities in proportion to their protection of watersheds and coverage of green areas that benefited other municipalities, to compensate municipalities for their opportunity costs. The fiscal transfers to municipalities are determined by indices. In the case of protected areas for biodiversity conservation, these indices consider the size of the protected area, the size of the municipality, and the protected area's management category. Paraná has also decided to add a protected area 'quality index' to the calculation. This has created an incentive for increasing green area, and the quality of the green area has increased.

Source: Ring et al. (2011) and Farley et al. (2010).

Economic instruments can also be used to stimulate professional actors to develop innovative ideas and technologies for green growth For example subsidizing environmental innovation by small and medium-sized enterprises has led to numerous innovations in this sector. An important shortcoming of economic instruments is that subsidies may distort the market when used for a long period. They are expensive instruments and not sustainable in long-run.

3.1 Balancing the Instruments

A careful choice of policy instruments is critical for achieving the goal of a green economy. Such a choice necessarily depends on the characteristics of the problem to be addressed and the institutional environment in which they are implemented as well as economic, political, and social particulars. To evaluate the suitability of different policy instruments a number of criteria such as environmental effectiveness, economic efficiency, equity, and social and political acceptance can be used. The effectiveness of each of these instruments varies according to different sets of criteria which are explored in more detail in the following sub-section. To determine what policy would be appropriate for particular sectors or environmental issues a comparative evaluation of alternative options is necessary.

Environmental effectiveness can be measured by establishing the effectiveness of an instrument in reducing for example the carbon intensity or increasing energy efficiency. Economic efficiency can be measured in terms of cost-effectiveness and Pareto efficiency – where the gain of one individual is measured against the potential/resultant forfeit elsewhere. Equity is the distributional consequences of economic instruments on different socio-economic groups. The use of economic instruments may influence the price of a particular product relative to its substitutes and complements, and may affect different socio-economic groups differently. Thus when the feasibility of an economic instrument is assessed such distribution and equity issues must be considered. The effectiveness of policy instruments is largely dependent upon the involvement of diverse stakeholders and actors. The higher the acceptance of a policy instrument is the larger the chances of its effectiveness. Administration and compliance costs of implementation of a policy instrument is also important. It is necessary to strike a balance among environmental effectiveness, economic efficiency, equity, as well as administrative costs of implementation.

Determining the best policy for a particular environmental problem requires a comparative evaluation of each policy option. Following the successful experience of Singapore such as its strict emission standards in air, water, solid waste and noise quality, its tight land-use planning controls, and its innovative traffic congestion tax scheme for promoting environmental quality (Seik 1996), combining market-based instruments with regulatory instruments would be advisable. It may be noted that markets alone have been inadequate in addressing the conservation of environmental resources as environmental goods and services are not properly valued in markets. Building eco-efficient economies and promoting synergies between environmental protection and economic growth requires a targeted intervention.

It is therefore pivotal that economic and regulatory instruments are combined and balanced to achieve greater acceptance and effectiveness. A combination of regulatory measures and economic instruments may help to balance policy objectives more effectively than having either alone. Moreover, mixing instruments helps policy makers to exploit the advantages of each instrument and minimize the problems of implementation of regulatory measures. Economic instruments are not however a panacea for a green economy in themselves and require tailoring to specific contexts and situations. In order to realize the multiple objectives of efficiency, equity and environmental quality, a mixture of regulatory instruments, economic instruments and other measures such as persuasive instruments to stimulate green urban growth such as awareness raising, ethical and moral persuasion may be required. The use of economic instruments requires a strong institutional framework and capable civil servants. These are important so that the policy instruments can be appropriately designed, politically endorsed and effectively administrated.

4 The Evolving Role of the State

Opinion on the role of the state in economic affairs is widely contested. New liberal economic thought maintains that for market mechanisms to be efficient the state should play a minimum role in economic affairs. The market is trusted to allocate resources efficiently and generate desirable outcomes when market forces are uninterrupted. Rather than encroaching on these processes, state intervention should concentrate on maintenance of macro-economic stability, enforcing laws and contracts, and securing private property rights so that the market can work properly.

Many development economists on the other hand argue that the market does not automatically lead to efficient outcomes (Stiglitz 2005). For efficient outcomes the market has to be regulated to correct market failures, mitigate negative effects of natural economic cycles, ensure stability, and reduce uncertainties for private actors through economic policy and fiscal or budgetary provisions. There is a need for mobilizing diverse stakeholders and pulling their creativity, energy, and diversity to create enabling conditions, including establishing an appropriate institutional framework to translate market approaches and instruments into action. According to this perspective, the state has a specific role in economic affairs to regulate economic activities and particularly to maintain trust and legitimacy in the economic system.

Although the debate on the role of the state in economic affairs continues, the green economy approach requires a system-wide transition from unsustainable to sustainable practices and necessitates a comprehensive approach of economic and social change. This demands a fresh approach to the role of the state in economic affairs and goes beyond enforcing contracts or securing private property rights. The success of transition to green economy to a large extent therefore depends on the ability to create enabling conditions including an appropriate institutional framework to translate its approaches and instruments into action, to strengthening economic foundations and to shift into a new period of low carbon green development

(Hezri and Ghazalib 2011). Growing evidence suggests that without appropriate policies, strategies and incentives, certain countries are likely to be unable to decouple environmental degradation from economic growth. Local innovation will be critical for the development of green urban growth, particularly innovative policies and strategies in city planning, transport systems, waste management, service delivery and work practices. The following are some of the ways that the state can promote urban green economy.

- **Setting the right policy framework**. The transition to green economy requires long-term and coherent policy framework with right incentives and sanctions to ensure that both public and private sectors move in the direction of sustainable production and consumption. It is the state's responsibility to frame appropriate development strategies and policies to maintain macro-economic stability as well as creating micro-economic conditions for investment and good governance to enhance efficiency and reduce unsustainable practices. Public policies not only shape the incentives and disincentives of adopting particular methods of production but also influence the demand and supply of environment-friendly goods and services by influencing consumer behavior. Well-designed policies are therefore necessary for facilitating the transition to green economy so that environmental quality in urban areas can be improved without sacrificing economic growth.
- **Strengthening governing institutions.** While well-designed policies are necessary, they are not sufficient for facilitating the transition to green economy. For translating policies and strategies into action, appropriate governance mechanisms and instruments are required. For promoting good governance and ensuring smooth functioning of markets the right institutions at central and local governments are required.
- **Sound management of economic affairs**. Transition to a green economy also requires sound management of economic affairs, macro-economic stability, efficient functioning of markets, and predictable and credible economic policies, which create an enabling environment for business and investment in green sectors. Efficient economic institutions and policies that entail improving the efficiency and effectiveness of governing institutions; developing sound financial markets with appropriate regulations and supervision; enhancing the legal and regulatory environments; and improving the quality of the public sector policy, legal, and regulatory framework are vital.
- **Coordination of decentralized economic planning.** For a transition to a green economy macro-economic and sectorial policies need to be well coordinated. All actors, producers and consumers including private sector, NGO and civil society need to be involved in green economy activities. Although economic activities are organized by private entrepreneurs, efficient coordination among different sectors of economy and diverse stakeholders is vital to promote green economy. The state's actions are needed for coordination among diverse actors for long-run economic stability as markets alone cannot coordinate all economic activities and generate desired economic and environmental outcome, due to externalities and coordination failure among different sectors. As such there is a need of right mixture of state-market interactions.

- **Structuring market forces**. **Markets** do not operate in a vacuum. Smooth functioning of markets requires certain institutional setting such as well defined property rights, free functioning of market forces, and maintaining trust and legitimacy in the economic system. Effective role of the state is thus critical for smooth functioning and tapping the power of the market in facilitating innovation and guiding economic activities towards sustainable development path.
- **Developing a shared vision and mobilizing diverse actors**. Transition to green economy is a complex process, which interacts with social, political and cultural factors among various forces. It entails promoting different philosophical orientation; a new attitude towards economy which recognizes that economic growth is not an end itself but a means to achieve human well-being. As such, it is important to develop a shared vision and mobilize collective actions for a sustainable future – moving from quantity to quality of economic growth, from GDP to human well-being and quality of life, from economic growth to economic security, from completion to cooperation and from individual to collective action. This demands an effective participative system of governance, where diverse actors such as the private sector, business community, civil society, market forces and the government act together to foster necessary conditions for innovation, investment and competition that can build a strong, stable and sustainable economy.
- **Creating enabling environment for private sector**. Private business and investors respond to incentives. Development accelerates when a society adopts institutions and policies that create incentives for its citizens to save, invest and innovate. Aligning financial, economic and market instruments to incentivize sustainable production and consumption practices and dis-incentivize unsustainable practices and lifestyles encourages investment in the green sectors. To foster a favorable environment for the private sector to embrace the green economy, it is necessary to set the right policy incentives for enterprises and consumers to invest in these green technologies.

These evolving tasks and responsibilities demand a new role of the state, moving from a command and control approach towards more transparent and accountable governance. Where the state and bureaucracy are capable of developing innovative policy options, the benefits of a green economy can be maximized and negative impacts minimized, reconciling developmental and environmental goals.

5 Conclusion

The foregoing discussion suggests that a transition to a green economy will require a fundamental shift in the way we design and implement economic policies so that we can bring forward changes in current practices of production and consumption in the urban economy. In the process of greening the urban economy, market instruments will become more important to stimulate innovation of new technologies, attracting investments in urban green sectors and enhancing eco-efficiency to promote urban economic growth without sacrificing environmental quality. Financial and economic instruments need to be applied judiciously in order to alter incentive structures and to induce necessary changes in production and consumption practices in the choices of different economic actors – including the private sector – thereby encouraging the development of green markets. Fiscal, monetary and trade policy as well as policies governing the private sector need to be aligned to incentivize sustainable production and consumption practices.

While economic instruments play a central role in influencing the market, the role of state is also critical in order to guide transformation and facilitate green investment, green business practices and green consumption. This demands new methods of engagement with the private sector, business community and citizens. The state needs to create a space where markets and citizens can stimulate technological and institutional innovations for going green. This requires judicious use of regulatory instruments and establishing sound frameworks for making environmental considerations central to economic decision making, prioritizing government investment, and spending in areas that stimulate the greening of urban sectors and also to influence the market. This entails making existing institutions more effective and functional at national and local levels in urban planning, developing transport systems, and promoting innovation and efficiency in providing urban services.

References

ADB (2006) Urbanization and sustainability in Asia: case studies of good practice. Asian Development Bank, Manila

Aizawa M, Yang C (2010) Green credit, green stimulus, green revolution? China's mobilization of banks for environmental cleanup. J Environ Dev 19(2):119–144

Anonymous (2009) Buildings and their impact on the environment: a statistical summary. Available via: http://www.epa.gov/greenbuilding/pubs/gbstats.pdf

Barbier EB (2011) The policy challenges for green economy and sustainable economic development. Nat Resour Forum 35:233–245

Bina O, Camera FL (2011) Promise and shortcomings of a green turn in recent policy responses to the 'double crisis'. Ecol Econ 70:2308–2316

CEDEFOP (European Centre for the Development of Vocational Training) (2009) Future skill needs for the green economy. The European Centre for the Development of Vocational Training (CEDEFOP), Publications Office of the European Union, Luxembourg

Climate Group (2011) Financing China's clean revolution report preview, China charts 'clean industrial revolution' to power its new economy. http://www.theclimategroup.org/what-we-do/news-and-blogs/New-report-China-charts-clean-industrial-revolution-to-power-its-new-economy/. Accessed on 25 Feb 2012

Cohen MA (1999) Monitoring and enforcement of environmental policy. In: Tietenberg T, Folmer H (eds) International yearbook of environmental and resource economics, Vol 3. Edward Elgar publishers, Cheltenham

DANIDA (2010) Discussion paper on green economy for Danida strategy process 2009. Available via: http://www.povertyenvironment.net/files/PEP15-Green%20Economy.pdf

ECO CANADA (2010) Defining the green economy. ECO CANADA, Alberta, Canada

Farley J, Aquino A, Daniels A, Moulaert A, Lee D, Krause A (2010) Global mechanisms for sustaining and enhancing PES schemes. Ecol Econ 69:2075–2084

Frondel M, Ritter N, Schmidt CM (2008) Germany's solar cell promotion: dark clouds on the horizon Ruhr Economic papers. Ruhr-Universität Bochum (RUB), Department of Economics, Bochum
Henton D, Melville J, Grose T, Maro G (2008) Clean technology and the green economy: growing products, services businesses and jobs in California value network. The California Economic Strategy Panel, Sacramento
Hezri A, Ghazalib R (2011) Social aspects of the green economy goal in Malaysia: studies of agriculture, renewable energy and waste initiatives. Paper presented at the UNRISD conference on green economy and sustainable development: bringing back the social dimension conference, 10–11 Oct 2011, Geneva
ICIMOD (2011) Green economy and sustainable mountain development: taking the Hindu Kush-Himalayan Region as a reference, background paper. International Conference on Green Economy and Sustainable Mountain Development, ICIMOD, Kathmandu. www.icimod.org/resource/3992
ICLEI (2011) Briefing sheets, green urban economy. Available via: http://local2012.iclei.org/fileadmin/files/ICLEI_Green_Urban_Economy_Briefing_Sheet_20110215.pdf
Jacobsson S, Lauber V (2006) The politics and policy of energy system transformation-explaining the German diffusion of renewable energy technology. Energ Pol 34(3):256–276
Karlenzig W (2011) Transforming the urban economy in Shanghai manual: a guide for sustainable urban development in the 21st century. United Nations Bureau International des Exposition, Seattle
Kennet M (2007) Editorial: progress in green economics: ontology, concepts and philosophy. Civilisation and the lost factor of reality in social and environmental justice. Int J Green Econ 1(3/4):225–249
Kunmin Z, Zongguo W, Liying P (2007) Environmental policies in China: evolvement, features and evaluation. China Popul Resour Environ 17(2):1–7
Levine MD, Ürge-Vorsatz K, Blok L, Geng D, Harvey S, Lang G, Levermore A, Mongameli Mehlwana S, Mirasgedis A, Novikova J, Rilling H, Yoshino (2007) Residential and commercial buildings. In: Metz B, Davidson OR, Bosch PR, Dave R, Meyer LA (eds) Climate change 2007: mitigation. Contribution of working group III to the fourth assessment report of the intergovernmental panel on climate change. Cambridge University Press, Cambridge/New York
MEP (Ministry of Environmental Protection) (2006) Green procurement in green Olympic. Retrieved 31 Mar 2010, from MEP website: http://kjs.mep.gov.cn/zghjbz/xgbd/200612/t20061204_96965.htm
OECD (2009) Declaration on green growth. Adopted at the council meeting at ministerial level on 24 June 2009, C/MIN(2009)5/ADD1/FINAL. Organisation for Economic Co-operation and Development, Paris
OECD (2010) Interim report of the green growth strategy: implementing our commitment for a sustainable future. C/MIN(2010)5, 10 May 2010, Organisation for Economic Co-operation and Development, Paris
OECD (2011) Towards green growth: a summary for policy makers. Organisation for Economic Co-operation and Development, Paris
Ring I, May P, Loureiro W, Santos R, Antunes P, Clemente P (2011) Assessing fiscal transfers for conservation policies and their role in a policy mix. Paper presented at the 9th international conference of the European society for ecological economics, Istanbul, 14–17 June 2011
Seik FT (1996) Urban environmental policy. The use of regulatory and economic instruments in Singapore. Habitat INTL 20(1):5–22
Stiglitz JE (2005) The ethical economist. Foreign Aff 84(6):128–134
UNEP (United Nations Environment Programme) (2011) Towards a green economy: pathways to sustainable development and poverty eradication. A synthesis for policy makers. Available via: www.unep.org/greeneconomy
UNESCO (2011) From green economies to green societies. Available via: http://unesdoc.unesco.org/images/0021/002133/213311e.pdf
UN-Habitat (2011) Urban patterns for sustainable development: towards a green economy. UN-Habitat working paper January 2011, UN-Habitat, Nairobi, Kenya

Local Sustainability: Driving Green Urban Economies Through Public Engagement

Rosalie Callway

Abstract Across the globe we find examples of local initiatives that stimulate green economies either led by communities themselves or through the formal engagement and mobilization of the public by local government and other local institutions. There are a huge range of sustainability projects and green businesses that benefit from local skills, innovation and community support. This chapter seeks to highlight how adopting a publicly 'embedded' approach, one that is culturally and socially-tied to local communities, will produce more lasting and equitable outcomes.

Keywords Inclusive green economy • Community engagement • Wellbeing • Behavior change

1 Introduction: Overview – Aims and Context

This chapter reviews public engagement as an embedded part of the move towards a green local economy. The term 'public engagement' refers to the involvement of people within a process – whether a commercial enterprise, community venture, the delivery of policy or otherwise. Engagement can occur at three levels:

- Cognitive – how people *think* about things
- Emotional – how people *feel* about things
- Behavioral – how people *act* (Bobek et al. 2009)

Local government has numerous tools to support community participation in green economies. The rationale for increased public engagement is that it will

R. Callway (✉)
Sustainability Consultant, London, UK
e-mail: rfcallway@gmail.com

R. Simpson and M. Zimmermann (eds.), *The Economy of Green Cities: A World Compendium on the Green Urban Economy*, Local Sustainability 3,
DOI 10.1007/978-94-007-1969-9_16, © Springer Science+Business Media Dordrecht 2013

increase trust, buy-in and, if the process is sufficiently open, can bring new ideas and innovation, improve the equality of impacts, promote longer term outcomes and wider uptake.

Furthermore, public engagement may also help address two challenges faced by local government and other actors in moving towards a 'green economy'. Firstly, clarifying the direction and focus of green economic policy, and secondly, building local resilience in the face of increasing global connectivity.

- **Dialogue on future economic models:** The transition to a sustainable economy will require an open dialogue about economic policy in the context of promoting social and environmental wellbeing. According to the UN Environment Programme (UNEP), a green economy is "*an economy that results in improved human well-being and reduced inequalities over the long term, while not exposing future generations to significant environmental risks and ecological scarcities*" (Scrieciu 2010). People's well-being is a central concern for local government and therefore, public inclusion in the development, delivery and refinement of a green economy is vital.
- **Interconnectedness and resilience:** A recent review of the 1972 '*Limits to growth*' book indicates that the original forecast of correlated population, consumption and pollution growth still hold largely true today (see Fig. 1) (Hall and Day 2009). These interrelated trends imply a green economy must address human interactions across a broad range of goods and services. Furthermore, these linkages are interacting on a global stage. As such cities cannot ignore the impacts of global markets to their local communities. For example, global food prices changes resulting in local food insecurity. In the face of such global socio-economic ties, a central element of promoting local green economies must be to build resilience within communities.

A summary of consultations running up to the Rio+20 Summit in 2012 in Brazil, the Caribbean, India and Mali, concluded that any new approach to economic development would only be meaningful and legitimate if it is built on a set of broad principles: sustainable development, equity, resilience, accountability and citizen empowerment (IIED 2011). In turn, the 'green' in green economy should not only be technocratic but tied to social objectives.

According to research in Bangladesh and Indonesia the characteristics of a sustainable and socially inclusive business should comprise of:

- Provision of products, services or decent work to lower-income communities in ways that stimulate more sustainable production and consumption patterns;
- If outcompeting locally produced goods and services, then offer superior eco-social qualities to existing options, and provide local employment;
- A mixed-ownership economy;
- Provision of new sources of capital to community members;
- Access to new markets for communities on stable and transparent terms;
- Transfer of appropriate technology and skills to community members;

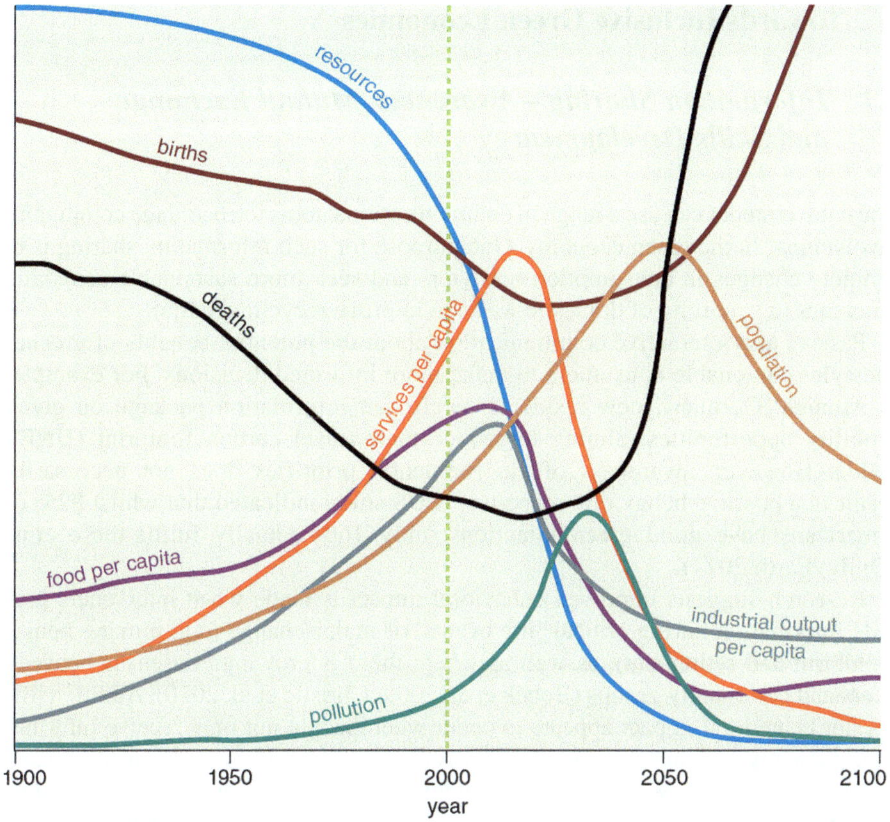

Fig. 1 'Limits to Growth' model projections relating a growing population, resources and pollution, with an additional timescale between 1900 and 2100 (Hall and Day 2009)

- Generate a Return on Investment as part of a scalable business without future reliance on external subsidies;
- Support Good Governance and enabling conditions locally and nationally;
- Involve participatory monitoring, evaluation and learning that address each of the preceding characteristics to inform future strategy and operations (Bendel 2011).

This chapter reviews four broad approaches to developing a more inclusive green local economy with case studies covering:

1. Information sharing
2. Active involvement
3. Policy and financial support
4. Design, monitoring and evaluation

2 Towards Inclusive Green Economies

2.1 *Information Sharing – Awareness, Mutual Exchange and Skills Development*

Local government can use a range of communicational tools to encourage community involvement in the green economy. One purpose for such information sharing is to promote changes in consumption behaviors and seek more sustainable economic outcomes (e.g. sorting of domestic waste to improve recycling value).

Passive and interactive communication about the potential benefits of greener lifestyles can enable consumers to make more informed decisions. For example, in Munich, Germany, new residents receive an information package on green mobility opportunities, aiming to reduce their travel carbon footprint (UNEP 2011). However, awareness of environmental priorities does not necessarily result in a positive behavioral response. A US study indicated that whilst 82% of Americans have good green intentions, only 16% actually fulfill these aims (OgilvyEarth 2011).

Research suggests improved behavioral impact is made when interveners provide information during critical 'life events' of major change (e.g. moving house, childbirth and retirement), as well as where there is growing consensus between peers and community groups (Bobek et al. 2009; Christie et al. 2010). Additionally, greater behavioral impact appears to occur when people not only receive information passively but are also directly involved in green initiatives from the outset. A recent review of UK community engagement in the food sector found the greatest sustainability and behavioral impacts when such programs were inclusive at an early stage (CAG Consultants 2011).

Listening to and understanding local producers and consumers are key steps to moving towards green economies.

One of local government's objectives in promoting green economies should be to empower citizens, giving them the tools and information they need to make sound economic decisions of their own and contribute to larger economic debates. Information about policy decisions on the green economy and corporate decisions and practices should be freely accessible to the public. Legal frameworks also need to be in place to guarantee accountability and transparency, and build public confidence in formal institutions. Along with educational institutions and mainstream media, communication tools such as community radio, local newspapers and youth campaigns can be used to reach a more targeted audience (IIED 2011).

An International Finance Corporation study of water services for poor communities in Manila, Philippines illustrates the value (commercial, behavioral and sustainability impacts) of early involvement by community groups, in terms of access to information and decision-making (Case Study 1).

Case Study 1 Inclusive Information and Decision-Making in Manila Water

The 'Tubig Para Sa Barangay', or 'Water for Poor Communities' program, is based on a clear business case: underserved, low-income households are willing to pay for safe, reliable water; connecting them creates new markets and cuts costs arising from inefficiencies and illegal connections. Partnerships between local government, community-based organizations (CBOs) and the private utility, Manila Water, included communities in the design and implementation of new water supply systems.

Communities are central to Manila Water's inclusive business model. By visibly placing water meters in side-by-side arrangements in public areas, meter monitoring becomes easier and the community can regulate itself as water use and fees become more transparent. In informal settlements or very low-income areas where land ownership is a problem, bulk metering and cost-sharing programs support self-monitoring through collective responsibility. The community also assigns or elects individuals to administer collections, monitoring and maintenance, directly supporting local employment. These methods help build a sense of local ownership and responsibility that enhances the system's good repair, promotes timely payment, and discourages water pilfering. This results in superior service and water quality for the community and lower costs for Manila Water.

By 2008, revenue generation had increased nearly 24 times, from ₱277 million to ₱6.803 million since the start of the project in 1999. Water services reached 1.6 million people, 99% gained 24-h water availability, as compared to 26% in 1999. Customers pay 20 times less than previously charged by water vendors (per cubic meter rates). New work opportunities from fee collection, monitoring meters and pipe servicing have resulted in jobs benefitting over 850 families.

Source: IFC (2010).

2.2 Active Engagement – Community-Led Eco-Enterprise

Community groups globally are already stepping-up to deliver the green economy by themselves. For example, in the city of Naples, Italy, various social enterprises are working to turn their city's overflowing waste into community resources (Case Study 2).

Case Study 2 Naples – Community Groups Tackle Naples' Rubbish Problem

Since the mid-1990s, the southern Italian city of Naples and surrounding provinces have struggled with the proper disposal of trash. Tensions peaked in 2008, after municipal workers refused to pick up any more garbage, partly due to overflowing landfills. Accusations of mafia involvement and government incompetence left citizens appalled over the piles of rubbish on the streets.

Neapolitans are now taking on their city's trash problems. *Ambiente Solidale*, a local civil society organization and others, are distributing recycling bins to homes and businesses throughout the region. Marginalized citizens, including local Roma, are employed as sorters to pick through glass, metal, textiles, and plastics. This work has inspired some towns, such as Portici, to achieve higher recycling rates than cities in wealthier Northern Italy.

According to Marco Traversi, Ambiente Solidale director, one trick to reduce municipal waste is to embed recycling as a habit with the young so it becomes second nature in adulthood. When Ambiente Solidale started in 2007, one of its first projects targeted elementary schoolchildren, teaching how to reuse and recycle waste.

Other tactics like flash mobs and guerilla gardening are behind Naples' grassroots cleanup. *CleaNap* uses social media tools to gather people quickly and clean up the city's stunning piazzas, as well as separating heaps of rubbish.

Another group, *Friarielli Ribelli* (Rebel Broccoli), supports community gardening, as another step towards revitalizing Naples. Their self-financed activities have created flowerbeds throughout the city and taught residents how to convert food waste into compost.

A new collection service, *Reteca*, is collecting unwanted computers and other electronic gadgets; and another Ambiente Solidale initiative will streamline processing of used printer cartridges. Naples is searching for more ideas. The Euclid Network, a London-based NGO, has organized a competition to help fund social entrepreneurship ventures so Naples will not only clean up, but thrive.

Sources: Kaye (2011) and Halbert (2011).

Links: *Ambiente Solidale*: http://www.ambientesolidale.it

Friarielli Ribelli: http://friarielliribelli.blogspot.com/p/chi-siamo.html

Euclid Network: http://www.euclidnetwork.eu

Municipal authorities should see such community actors as partners in building the green economy. Councils should work with them to stimulate activities throughout their cities. For example, the London Borough of Lambeth, UK, has embraced the 'Transition Town Brixton' movement with financial, information and research support, supporting the creation of a local currency, recycling, and renewable energy generation (Case Study 3).

Case Study 3 Transition Brixton

Transition Town Brixton (TTB), following the example of other transition towns across the globe, is seeking to respond to the pressure of Peak Oil and climate change. As a community-led organization it's activities include: setting up a local currency, the 'Brixton Pound'; Community renewable energy project 'Brixton energy' and a zero waste initiative 'Remade in Brixton'. The London Borough of Lambeth has supported these initiatives that together are promoting local businesses, employment, low carbon development and community resilience. Lambeth has a population of 270,000 with Brixton its administrative and commercial center.

'Brixton Pound' – In 2009, Brixton launched its own currency. There are now more than 130 business accepting and trading in '*B£s*'. More than B£35,000 is 'sticking to Brixton'. The currency stimulates an independent local economy and conversations about localization. More of the money spent in Brixton stays there, helping local businesses thrive in the face of recession and increased competition. It also helps reduce Brixton's carbon footprint by supplying more of its needs locally. Lambeth Council supports the B£ with funding and in-kind support, as well as an online B£e-currency pilot.

'Brixton Energy' – aims to put solar panels on the Loughborough housing Estate. It involves Lambeth Council, the Low Carbon Zone, United Residents Housing and the Loughborough Estate Management Board. Brixton residents can invest in a Community Share Offer. The project aims to:

- Generate local energy, as 'Brixton Energy'
- Develop a community investment vehicle
- Increase resilience by reducing dependence on big utilities
- Use profits to educate about energy efficiency, tackle fuel poverty and develop partnerships to provide energy efficiency services
- Provide training and employment for local people

continued

Case Study 3 (continued)

'*Remade in Brixton*' – A community-led zero-waste initiative, focusing on waste prevention, repair, reuse and recycling. The TTB team engages with residents, businesses and schools to promote local waste reduction, reuse and recycling. They also develop skills of remaking and repair that are vital to the creation of local green enterprises and sustainable employment.

Sources: LGID (2011).

Links: Brixton Pound: http://brixtonpound.org

Transition Town Brixton: http://www.transitiontownbrixton.org

Another UK example of a regional green economy partnership is Sustainability West Midlands, which brings together environmental leaders from local government, voluntary groups, housing associations, private sector and academia. Importantly, the program invests directly in providing guidance and skills to community groups. They have trained and now support 40 community mentors to help convene new community eco-enterprises in councils across the region (Ross 2011).

2.3 *Incentives – Policy and Financial Support*

Local government can help to set the context within which to inspire and guide new inclusive green businesses. Funding and supporting such work during a time of economic downturn is tough but, if well designed, can bring savings to service delivery, cut clean-up costs, stimulate income generation, skills development and other environmental benefits.

In the run up to the UN climate negotiations, the city of Durban innovatively addressed the opportunity of the thousands of delegates attending the meeting through a new carbon credit initiative (Case Study 4).

Case Study 4 Innovative Funding for Climate Adaptation Enterprises

In an effort to offset emissions from the UN Climate negotiations the government of Durban set up a carbon credit scheme. The scheme enabled delegates to invest in local reforestation, employing local people to replant native trees. The initiative implicitly ties-in several sustainability strands together – biodiversity rehabilitation, income generation, skills development, and offsetting carbon emissions through a market-based mechanism, the '*Community Ecosystem Based Adaptation*' (CEBA) initiative.

continued

Case Study 4 (continued)

eThekwini's deputy head of environmental planning and climate protection, Debra Roberts, said; "*The Durban Ceba Initiative is one of the most exciting elements of the city's broader COP17 greening program. It has been adopted as the official voluntary offset mechanism for COP17. Delegates, corporates and residents of Durban will be able to contribute towards the project by buying 'Ceba credits' to play their part in helping offset the environmental impact associated with hosting COP17*'.

She explained: "*Each Ceba credit will cost about $10 (R80) and the money raised will be used for the official Ceba greening site along the Umbilo River catchment area. Unemployed people from the neighboring communities will then be employed as 'green collar' workers to first remove alien plants and trees.*"

Part of the money will be used to plant indigenous trees along the Umbilo River, with the pilot reforestation site located at Paradise Valley, near Pinetown. The project is a community-centered climate adaptation model. "*We want this project to not only benefit the ecological sustainability of the river, but the social sustainability of poor communities along it.*"

Source: ICLEI news, accessed online 3 Dec 2011 http://www.iclei.org/index.php?id=12500

Link: Durban CEBA http://durbanceba.org

2.3.1 Inclusive Markets for Natural Capital

Building market economies around biodiversity and ecosystems has a strong base. Economists estimate that losses in global biodiversity cost trillions of dollars annually and directly affect people's quality of life and future well-being.

Policy makers have proposed various market tools to better secure natural capital. One such tool is called 'Payments for Ecosystem Services (PES)'. PES recognizes that local communities are dependent on ecosystems for their livelihoods and cultural traditions. For example, over 90% of the world's poorest people are estimated to depend on forests. PES can help the better valuation and sustainable management of ecosystems through supporting these communities (Forest Trends 2011a).

However, this type of financial tool has not always been successful in delivering equitable benefits to communities and requires careful implementation. Some risks include:

- It only recognizes the human value of ecological systems, and does not help protect other services. Bundling of services might go some way to overcoming this problem.
- Participation can be difficult for poor communities. They may lack land titles or the necessary knowledge to manage administrative tasks required by PES schemes;

- Further impoverishment of the poor might occur (e.g. if they are required to pay for services that were previously provided for free);
- There are difficulties in attributing an economic value to environments of high cultural value (Wertz-Kanounnikoff 2006).

PES can exacerbate problems where land tenure is insecure. Case studies show that there is the potential for marginalization, repression, exploitation and the internal division of communities (ELDIS 2011). It's estimated that indigenous peoples and local communities hold defined land tenure rights in only 2% of forests throughout Africa .Yet conservationists have long pointed out that "*the recognition of indigenous rights to forest-resources management leads to successful management practices*" (Ecosystem Market Place Community Portal 2011).

Recognizing these challenges, 'Forest Trends', a global forest coalition, has set up a 'Communities & Markets' program to promote community participation in market-based conservation mechanisms. The program aims to promote awareness about the value of community engagement for effective conservation and seeks to strengthen community capacity to participate in and benefit from PES schemes (Forest Trends 2011b). Projects in Uganda and elsewhere highlight the importance of local government investing in good governance, transparency and community engagement if they are to be trusted by international ecosystem and climate adaptation investors. Local governments need to increase the level of community-owned and administered land to bring wider community benefits from PES and similar mechanisms.

2.3.2 Sustainable Procurement

Sustainable procurement is another important tool to shift businesses towards more inclusive and green outcomes. Local government can encourage business to adopt clear environmental and social objectives through applying clear sustainable procurement criteria (ICLEI 2007). For example, Queensland State Government in Australia has incentivized Biodiversity Conservation enterprises through their competitive tendering-criteria. They invite organizations to develop innovative, semi-permanent and low impact sustainable nature-based tourism accommodation throughout Queensland. Their assessment rating tool at the 'Expressions of Interest' stage seeks proposals that demonstrate harmony and integration with the natural and cultural values of a site through: appropriate facility layout and design; construction materials and methodologies; energy, water and waste management; and operational and visitor protocols and activities. Community inclusion and benefits are a central part of their assessment criteria (Queensland State Government 2011).

2.3.3 Local Governance for Inclusive Environmental Mainstreaming

The International Institute for Environment and Development (IIED) reviewed the local governance conditions required for successful environmental mainstreaming.

This is important in setting the context for green economies. Conditions that better integrate environmental and social objectives into policy include:

1. Legislative system: Supporting environmental protection and social justice.
2. Institutional mandate: All sectorial and decentralized institutions tackle environment as a cross-cutting issue within their work;
3. Public concern: Public demands to tackle environmental degradation and care for environmental assets are significant and well-expressed;
4. Public and media advocacy: Mass organizations and NGOs are able to raise difficult policy issues in relation to the green economy.
5. Leadership: Government and local leaders prepared to listen, to change policy, to act and to be accountable;
6. Communication and transparency: Offer many ways to access, share and feedback information about environment-society-economy links;
7. Cooperation: Shared initiatives and processes allowing actors to collaborate (adapted from IIED 2010).

With the right policy frameworks in place, local government can create the space for like-minded businesses, entrepreneurs, investors, academics, and community groups to cooperate over the future development of a local green economy (Monaghan 2011). This convening and facilitating role appears to be an important component to mobilizing inclusive green economies.

2.4 Design, Monitoring and Evaluation

Before embarking on any participative approach local government needs to be clear about the design of that process. A group of UK charities undertook a two-and-a-half year research project exploring how and why individuals participate in their communities, producing the '*Pathways through participation*' report. They made three key findings for those who wish to foster community participation, in this instance, towards building a green economy:

1. Participation is personal – it must be viewed first and foremost from the perspective of the individual/s taking part;
2. Participation can be encouraged and made more attractive – opportunities need to respond to people's needs, aspirations and expectations;
3. Significant barriers to participation can be entrenched – issues of power and inequality in society are critical to understanding how and why people get involved and stay involved (Involve 2010).

The report recommends being realistic about what participation can and cannot achieve – "*policymakers [need] to be clear about the purpose of the participation they want to see happening…It also requires institutions, organizations and groups to recognize that participation is dynamic and that opportunities need to be flexible*".

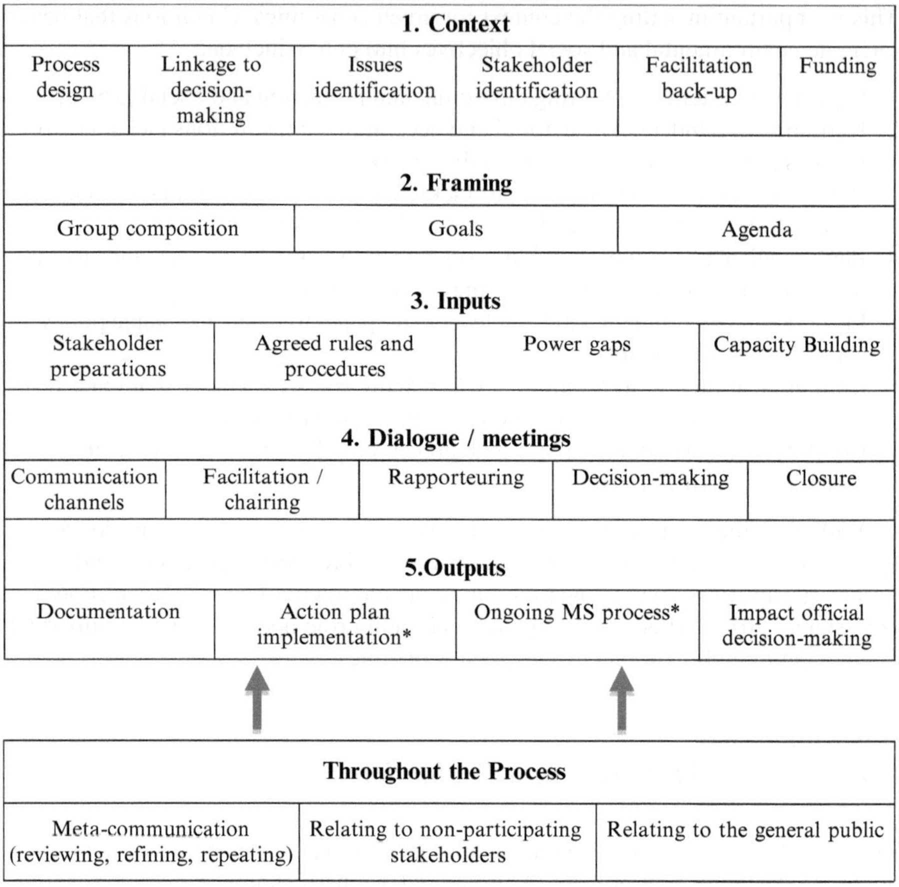

Adapted from Hemmati, 2002
* Optional

Fig. 2 Key steps in participative processes – overview

In 2002, Dr. Minu Hemmati undertook a global review of leading participative examples that had produced positive sustainability outcomes and good governance. Based on these findings, Hemmati recommended steps for designing effective participative processes. Whilst the steps refer to multi-stakeholder contexts, they are still relevant for local governments when seeking a participative model to delivering a green economy (Fig. 2).

A variety of participative tools can be used to promote engagement. Table 1 lists examples of the types of tools commonly applied. Some participative resources are also available online. For example, 'Involve' developed a free tool providing support for planning participative processes, 'peopleandpariticipation.net' (Involve and Headshift 2011).

Assessing the progress towards delivering a green economy, particularly the sustainability impacts and degree of engagement of local communities, depends

Table 1 Participative tools for an 'inclusive' green economy

Type	Methods
Information sharing	Community fairs
	Workshops
	Conferences and seminars
	Open days and drop-in sessions
	Discussion packs
	Outreach processes
	Road shows and exhibitions
	Web-based, new media
	Marketing campaigns
	Advise centers
	Networks
	Observational experiences e.g. site visits, smart metering
Direct involvement	Community 'lifestyle' projects
	Energy Service Companies (ESCOs)
	Micro-renewables
	Housing development associations
	Food cooperatives
	Green apprenticeships e.g. groundwork UK
Incentives – policy and finance	Grants
	Tax breaks
	Payments for Ecosystem Services (PES)
	Bonds/guarantees
	Policy guidance
	Funding skills development e.g. partnering with local universities and colleges
Design, monitoring, evaluation	Focus groups
	Opinion polls
	Surveys and questionnaires
	Consultation documents
	Interviews
	Carbon footprinting

Adapted from: Dialogue by Design (2011), Whitmarsh et al. (2011), and Monaghan (2011)

upon effective design and monitoring throughout the process. A planned participative process and monitoring system helps identify whether the tools being used are successful, and highlights whether policies need changing to improve impacts (Cranidge 2011).

Indicators for assessment can be formed through a participatory process allowing individuals to better use and learn from them. Social, ecological and economic indicators need to be aligned with local government policy areas and accountability frameworks (Berlin and Hunter 2011). For example, the 'development dividend' developed by David Disch is a simple set of 15 indicators, assessing local and community benefits of projects which has been applied to assess various bio-energy projects globally (see Table 2) (SEI 2011).

Another free online resource developed by WWF-UK and CAG consultants, called '*The Community Engagement and Sustainable Development Tool*', specifically

Table 2 'Development Dividend' sustainability indicators

Environmental	Economic	Social
• Does the project reduce local air pollution? • Does it reduce local water or soil pollution? • Does it reduce natural resource degradation? • Does it improve waste management? • Does it 'green' the energy production?	• Does the project create new employment? • Does it source material or inputs from local supplies? • How does it initiate technology transfer? • Does it extend public infrastructure? • Does it contribute to the energy security of the country?	• How did the project involve local stakeholders? • Does it have a clear rural and/or pro-poor focus? • Does it contribute to health and safety standards? • Does it share some of the profits? • Does it provide training and education?

measures the impact community engagement has on changing and maintaining behaviors towards sustainable lifestyles (Cranidge 2011)

3 Future Opportunities?

Integrated approaches, such as the One Planet development model developed by WWF and later adopted in the UK (Sutton and Brighton), USA (Sonoma Mountain Village), and Portugal, with others planned in South Africa, China, Australia, and Canada, offers a good model for inclusive green economies (Bioregional 2011). This is because they set combined well-being and environmental objectives as their starting point, rather than focusing on economic growth objectives in isolation.

The discussion and case studies included here highlight that local government need to undertake a package of measures to ensure communities are engaged and able to benefit from green economies. Such measures need to target all three areas of public engagement: cognitive understanding (via information sharing), emotional involvement (information and direct participation) and behavioral engagement (resources, training, information, active involvement). Successful engagement in the future will benefit from an 'active research' approach, applying learning (positive and negative) from collaborative initiatives to find effective ways of joint-working. Whilst each local government and community is unique, there are many common experiences to community engagement. Sharing on regional and international scales will also help to identify alternative approaches and solutions.

Many recent examples of the shift to local green economies are coming from communities themselves, as exemplified by the Transition Town movement and others. Such locally grown movements can often hit road blocks created by local government themselves, from legal barriers, bureaucracy in planning or shifting priorities of political parties. Local government needs to be a facilitator and enabler, rather than an obstacle in this area. This will be vital to support resilient, innovative, and progressive local economies in the long term.

Acknowledgements Thanks to CAG, Involve, Dr Minu Hemati and Clive Mitchell for their comments and inputs.

References

Bendel J (2011) Evolving partnerships: a guide to working with business for greater social change. Greenleaf Publishing, Sheffield. Available via: http://www.greenleaf-publishing.com/product-detail.kmod?productid=3351

Berlin R, Hunter A (2011) Co-constructing the sustainable city: how indicators help us 'grow' more than just food in community gardens. Available via: http://en.wikipedia.org/wiki/Urban_agriculture

Bioregional (2011) One planet living case studies. Available via: http://www.bioregional.com/oneplanetliving/one-planet-living-case-studies

Bobek D, Zaff J, Li Y, Lerner RM (2009) Cognitive, emotional, and behavioral components of civic action: towards an integrated measure of civic engagement. J Dev Psychol 30:615–662. Available via: http://www.national4-hheadquarters.gov/comm/Bobek_FINAL.pdf

Christie I, Jackson T, Rawles K (2010) International dimensions of climate change. Report 1.3: ethical, social and behavioral impacts of climate change. Foresight report. UK Government Office for Science, p 33. Available via: http://www.bis.gov.uk/assets/bispartners/foresight/docs/international-dimensions/11-1018-ethical-social-behavioural-impacts-of-climate-change

Cranidge E (2011) Informal interview with Emma Cranidge from CAG Consultants about 'The Change Tool' – a tool for evaluating sustainability impacts of community engagement. Available via: http://community-engagement.org.uk

Dialogue by Design (2011) Available via: http://designer.dialoguebydesign.net/Docs/Instructions.htm

Ecosystem Market Place Community Portal (2011) Ugandan tribe struggles to maintain forests and access benefits. Available via: http://community.ecosystemmarketplace.com/pages/dynamic/opinion.news.page.php?page_id=6160&eod=1

ELDIS (2011) Concerns regarding payments for ecosystem services. Available via: http://www.eldis.org/go/topics/resource-guides/environment/key-issues/payments-for-ecosystem-services/concerns-regarding-payments-for-ecosystem-services. Cited 4 Dec 2011

Forest Trends (2011a) Ugandan tribe struggles to maintain forests and access benefits. Blog Date Published: 07/05/2011. Available via: http://community.ecosystemmarketplace.com/pages/dynamic/opinion.news.page.php?page_id=6160&eod=1

Forest Trends (2011b) Communities & markets. Available via: http://www.forest-trends.org/program.php?id=58

Halbert W (2011) An invitation for innovation: Naples opens its doors to cross-border co-operation. Blog from The Guardian's Social Enterprise Network. Available via: http://www.guardian.co.uk/social-enterprise-network/2011/aug/04/cross-border-social-innovation

Hall C, Day J (2009) Revisiting limits to growth after peak oil. Am Sci 9:230. Available via: http://www.esf.edu/efb/hall/2009-05Hall0327.pdf

Hemmati M (2002) Multistakeholder processes for governance and sustainability – beyond deadlock and conflict. Earthscan, London. Section II. Designing an multi-stakeholder process – a detailed guide, p 211. Available via: http://www.earthscan.co.uk/?tabid=37&st=basic&se=hemmati

ICLEI (2007) Procura+ Manual. A guide to cost-effective sustainable public procurement, 2nd edn. Preface. Available via: http://www.nuigalway.ie/buildings/documents/procura__manual.pdf

IFC (2010) Scaling up inclusive business: advancing the knowledge and action agenda. IFC, Washington, DC

IIED (2010) Integrating environment and development in Viet Nam. Achievements, challenges and next steps. Paper resulting from the Viet Nam Environmental Mainstreaming 'Lessons Learned Review' of March 2009 organised by IIED in association with the Viet Nam/UNDP Poverty Environment Programme (2010). Available via: http://www.environmental-mainstreaming.org/documents/Vietnam%20Learning%20Group%20report.pdf

IIED (2011) Green economy: developing country stakeholders have their say. Green Economy Coalition. Available via: http://www.greeneconomycoalition.org/sites/greeneconomycoalition.org/files/GEC%20dialogue%20synthesis_FINAL.pdf

Involve (2010) Pathways through participation: what creates and sustains active citizenship? Summary report. Available via: http://pathwaysthroughparticipation.org.uk/wp-content/uploads/2011/09/Pathways-through-Participation-summary-report_Final_20110913.pdf

Involve and Headshift (2011) People and participation website. Available via: http://www.peopleandparticipation.net

Kaye L (2011) Community groups tackle Naples' rubbish problem. Blog from The Guardian's Sustainable Business and Waste Recycling Hub, 18 August 2011. Available via: http://www.guardian.co.uk/sustainable-business/naples-waste-management-problem-community-groups

LGID (2011) Local government improvement and development 'Low Carbon Futures'. Available via: http://www.idea.gov.uk/idk/core/page.do?pageId=19882346

Monaghan P (2011) Sustainability in Austerity. How Local Government Can Deliver During Times of Crisis. Greenleaf Publishing

OgilvyEarth (2011) Mainstream green: moving sustainability from niche to normal. Available via: http://www.ogilvyearth.com/wp-content/uploads/2011/05/Mainstream_Green_Exec_Summary.pdf

Queensland State Government (2011) Tourism project website. Available via: http://derm.qld.gov.au/parks_and_forests/activities_in_parks_and_forests/nature-tourism/pdf/sustainability-guidelines-eoi.pdf

Ross B (2011) Sustainability West Midlands 'Community Engagement' web pages. Available via: http: www.Sustainabilitywestmidlands.org.uk/projects/?Voluntary+sector+Commu nty+Strand+of+the+West+Midland+Low+Carbon+Economy+Programme/1649

Scrieciu S (2010) Greening the economy: real climate economics and sustainable development. Available via: http://realclimateeconomics.org/wp/archives/617

SEI (2011) Bioenergy projects and sustainable development. Which types of bioenergy projects offer the greatest development benefits? SEI policy brief. Working paper no. 2011-04, by Carrie Lee and Michael Lazarus

UNEP (2011) Towards a green economy: pathways to sustainable development and poverty eradication. Part II. Investing in energy and resource efficiency. Cities chapter, p 133

Wertz-Kanounnikoff S (2006) Payments for environmental services: a solution for biodiversity conservation? Institute for Sustainable Development and International Relations, Winnipeg, Canada. Available via: http://www.eldis.org/go/topics/resource-guides/environment/key-issues/payments-for-ecosystem-services&id=24425&type=Document

Whitmarsh L, O'Neil S, Lorenzoni I (2011) Engaging the public with climate change – behaviour change and communication. Earthscan, London

We Know Enough: Achieving Action Through the Convergence of Sustainable Community Development and the Social Economy

Sean Connelly, Sean Markey, and Mark Roseland

Abstract Ten years ago in Johannesburg, there were over 6,000 communities across the world that had taken tangible steps towards implementing sustainability. However, while many have conducted visioning exercises and hired consultants to draw-up sustainability plans, far too often those plans remain on the shelf. In short, we face an implementation gap. Barriers to implementation are less about our technical capacity – we know enough about viable alternatives and solutions – and more about the mobilization of citizens and their governments to enact structural change. In addition, communities are struggling to deliver on the holistic promise of sustainability. Sustainability suffers from policy inflation of increased expectations to deliver development that is economically, socially, and environmentally sound, yet has failed to acknowledge the increasing capacity gap for implementation. We have made great progress on both the economic and environmental dimensions, well encapsulated by the burgeoning green economy. However, Agyeman et al. (2003) and others remind us that the social aspects of sustainability are lagging. The purpose of this chapter is to address these two implementation gaps: mobilization and socializing sustainability. Our approach to these challenges is framed within the context of two concepts: sustainable community development and the social economy.

S. Connelly (✉)
Centre for Sustainable Community Development, Simon Fraser University, Burnaby, BC, Canada
e-mail: swconnel@sfu.ca

S. Markey
School of Resource and Environmental Management, Simon Fraser University, Burnaby, BC, Canada
e-mail: spmarkey@sfu.ca

M. Roseland
School of Resource and Environmental Management, Centre for Sustainable Community Development, Simon Fraser University, Burnaby, BC, Canada
e-mail: roseland@sfu.ca

R. Simpson and M. Zimmermann (eds.), *The Economy of Green Cities: A World Compendium on the Green Urban Economy*, Local Sustainability 3,
DOI 10.1007/978-94-007-1969-9_17,

It is our hope to contribute to the discourse surrounding sustainable development and to offer insights, drawn from our research, into how to mobilize sustainable forms of development that offer a truly balanced and holistic interpretation of the sustainability ideal.

Keywords Sustainable community development • Social economy • Implementation gap • Local government

1 Introduction

The Rio+20 event provides a timely opportunity for reflection and reconsideration of how to continue to advance the cause of sustainable development. At the time of the original summit in 1992, we were witnessing the dawn of more mainstream public awareness about environmental issues. This was marked in Canada with massive environmental protests (e.g. Clayoquot Sound), federal government round-table initiatives, and the mainstream appeal and purchase of green products (real or not). This awareness, however was not ably matched by a capacity to implement and offer tangible examples of how to live, do business, travel, and eat in sustainable ways. The machine of protest and awareness-raising had not had the time or resources to adequately invest in solutions. The end result was an implementation gap that explains, in part, a waning of public interest and commitment to meaningful change for sustainability.

In 1992, the ingredients for change included awareness and some level of political engagement, but not the technical capacity for meaningful, structural change. Addressing this gap has been perhaps the most exciting aspect of the past 20 years – in Canada and internationally – as there has been an explosion of innovative solutions in planning, design, transportation, infrastructure, education, food production and distribution, and so on. It is no longer possible to be a generalist in the sustainability sector due to the myriad of advanced specialties.

By 2002, there was an increasing sense of crisis, as knowledge about the state of a wide range of global environmental systems – from climate, water, biodiversity, desertification – showed a continued trend in negative directions. Kofi Annan, the then secretary-general of the UN, laid out the goals of the Johannesburg summit quite clearly. He stated that humanity was at a critical juncture and that we had a choice: we could choose despair, an uncertain future, and conflict over resource use, or we could choose hope, a future that is better for people and better for the environment (Annan 2002). Unfortunately, for much of the past decade, we have chosen the status quo and postponed making that choice.

Now in 2012, despite all of our technical capacity – known and proven strategies for how to live and do business in more sustainable ways – we find ourselves facing another implementation gap. Communities across Canada have conducted visioning exercises and hired consultants to draw-up sustainability plans only to have too many of those plans remain on the shelf. In short, we face another implementation gap.

This time, however, the barriers are less about our technical capacity – we know enough about viable alternatives, solutions – and more about the mobilization of citizens and their governments to enact structural change. A second barrier concerns our inability to deliver on the holistic promise of sustainability. Sustainability suffers from policy inflation of increased expectations to deliver development that is economically, socially, and environmentally sound, yet has failed to acknowledge the increasing capacity gap for implementation. We have made great progress on both the economic and environmental dimensions, well encapsulated by the burgeoning green economy. As Agyeman (2005) and others remind us however, the social aspects of sustainability are lagging.

The purpose of this paper is to address these two implementation gaps: mobilization and socializing sustainability. Our approach to these challenges is framed within the context of two concepts: sustainable community development and the social economy. It is our hope to contribute to the discourse surrounding sustainable development and to offer insights, drawn from our research, into how to mobilize sustainable forms of development that offer a truly balanced and holistic interpretation of the sustainability ideal.

2 Sustainable Development and Sustainable Community Development

Sustainable development is a concept that has achieved widespread recognition in the public, private and NGO sectors, yet at the same time it has been interpreted in different and often competing ways. Despite the diverse and contested meanings attached to concepts of sustainability, they all fundamentally begin by recognizing the mismatch between increasing human demands on the Earth and the ability of finite natural systems to cope with those demands (Williams and Millington 2004).

Sustainability is used at a variety of spatial scales, from the global to the local and there have been different approaches proposed for implementation. At global and national levels, the Rio commitments to Agenda 21 and the Johannesburg plan of implementation demonstrate the political commitment to sustainable development, although concrete actions are more difficult to identify. The private sector has also begun to make credible commitments towards sustainability, despite the fact that for many businesses, the term has been interpreted to mean staying in business. However, it is at the local level where significant action towards implementation can be observed (Beatley 2000; Roseland 2012).

Over 6,000 local governments from across the globe have committed to planning processes to make their communities more sustainable (ICLEI 2002). Many communities have incorporated sustainable development principles into their community vision statements and decision-making processes.

A focus of sustainable development at the local level explicitly acknowledges that we all live in local communities where our everyday actions contribute to environmental, social and economic problems, and therefore it is at the local level that solutions need to be developed through a culture of community involvement, multi-stakeholder participation and consensus-building that our cities and communities will be made more sustainable (Otto-Zimmermann 2002). Local governments also play a crucial role in sustainable community development as it is in local communities where the tangible impacts of global social, environmental and economic trends play out, as well as being the locally elected, representative and accountable bodies responsible for local decision-making (Brugmann 1994; Parkinson and Roseland 2002; Roseland 2012).

If local governments are to serve as solutions to local and global environmental problems, they need to build capacity to address those problems in a more integrated manner. Sustainable Community Development (SCD) provides a "systems approach" that recognizes that local communities are complex systems with multiple inter-relationships at different scales (rural/urban, local/national/global) and among different issue-based sectors (e.g. the linkages between local air quality, transportation planning and land use densities) (Connelly et al. 2009). SCD involves widespread community awareness-raising and integrated municipal involvement aimed at bringing about a shared understanding of what sustainability means and how to achieve it throughout all sectors of municipal government and the wider community (James and Lahti 2004; Gruder et al. 2007).

Sustainable Community Development (SCD) has the potential to address these issues due to the focus on integration at the local level and the connections that can be made by mobilizing citizens and governments to take action on critical issues specific for their communities. SCD applies the concept of sustainable development to the local or community level by integrating sustainable development principles, long-term planning processes and specific community priorities. The community capital framework (see Fig. 1) is a useful way of framing the complexity of integrated development at the community scale and to describe the meaning and objectives of SCD. The key to integration is capacity-building and mobilization of citizens and their governments in the coordination, balancing and catalyzing community capital (Roseland 2012).

However, in both the sustainability literature and in practice, there has been a lack of emphasis placed on the relationships between social, cultural, economic and equity issues and the environment – issues that in most communities remain politically sensitive (Pugh 1996). Sustainability practice suggest that much of the focus has been on reducing the environmental impacts of development (Kenworthy 2006; Mazza and Rydin 1997), particularly in areas that lend themselves more readily to quantification, measurement and win-win solutions such as air quality, transportation, and waste management. These types of initiatives tend to avoid political conflicts by focusing on technological solutions and efficiency arguments to promote change, rather than addressing more conflictual social issues around equity and justice. The recycling movement that began in the 1980s under the blue box program is a good example. While promoted under the motto of "reduce-reuse-recycle", the social and political consequences of addressing issues around over-consumption were dominated more by the technological fix of sorting recyclables. One promising approach to address this gap is the social economy.

Fig. 1 Community capital framework

3 Social Economy/Community Economic Development

The social economy has emerged as a community response to negative impacts of social and economic restructuring. The social economy suffers from definitional confusion (see McMurtry 2009), however, it has been loosely defined to include activities by democratically controlled organizations that integrate a social and economic mission (Lionais and Johnstone 2009; Neamtan 2009; Amin et al. 2002). The social economy provides an inspirational alternative of how communities and organizations are addressing new and complex social, economic and environmental challenges. Despite many similarities in terms of core values and objectives, there has been little dialogue in theory or in practice between social economy and sustainability researchers and practitioners.

The social economy can be defined to include those organizations and associations that use the market to pursue explicit social objectives. The size, scope and impact of the social economy vary greatly from region to region across Canada. In Quebec, the social economy emerged as a form of economic activity that reflected distinct social and cultural values. The Chantier de l'economie sociale defined the social economy as an overarching concept that unified a variety of pre-existing non-profit, mutuals, cooperative and community economic development (CED) initiatives under a common framework that was able to gain legitimacy and support from the province. Elsewhere in Canada, the social economy is less advanced and less organized, despite the long history of non-profit societies, cooperatives, mutual associations and foundations that pursue economic activities for social gain. Despite the lack of a national policy framework for the social economy and definitional variability, researchers have estimated the scale of social economy be $79.1 billion in Canada (7.8% of GDP), employing over two million people annually and is growing rapidly (Amyot et al. 2010).

In the US, the Obama administration has signaled commitment to setting up an Office of Social Entrepreneurship that would promote social entrepreneurship and non-profit action to address community problems by creating funding programs that reward innovation, tax incentives for partnerships between non-profits and businesses and new development aid that better mirrors the existing flexible supports for the private sector. Social economy organizations are also mobilizing and developing partnerships internationally through the International Centre of Research and Information on the Public, Social and Cooperative Economy (CIRIEC) that ensures that the social economy is part of discussion of global problems, such as through the World Social Forum.

The social economy field is evolving rapidly. Faced with growing demands for social service provision and declining government grants, many non-profit societies, cooperatives and mutual associations are turning to social enterprise ventures in which they sell or provide goods and services to the market to serve social and economic purposes, while returning profits to the organization rather than maximizing profits to shareholders. Recent crises of rising energy costs, rising food costs and global financial system collapse and the resulting economic bail out have only increased interest in the potential of alternative economic models such as the social and green social enterprises.

For decades, the primacy of economic development through the freeing of markets from government intervention dominated economic, social and environmental policy. However, the optics of the global economic recession that emerged out of the sub-prime lending bubble and the resulting billion dollar bail outs for large corporations that were "too big to fail" while individuals and communities experienced the brunt of the fallout has led many to question the focus of our social, economic and environmental policies. As a result, there is increasing interest in local and regional initiatives and organizations that are able to use the economy to achieve explicit social and environmental outcomes.

Despite the promise of the SE to deliver an economy that is more responsive to local needs, that is more equitable and more just, SE approaches have been criticized for not addressing societal transformation, for paying limited attention to environmental concerns and often working to maintain existence on the margins of the larger capitalist system, and in fact embrace the mainstream system in its approach (Amin et al. 2002). For example some food banks working under the charity model can be considered in a similar manner to the blue-box recycling program discussed above. Food banks play a critical and much needed role in providing food for people facing hunger, yet have also been criticized for depoliticizing hunger and poverty and contributing to resolving agribusiness's waste problem through federal tax incentives for corporate food donations of inadequate, inappropriate and unstable food (Poppendieck 1999). In essence, food banks working under the charity model have provided the impression that corporations (or individuals) clearing their shelves of spoilt and nearly spoilt food are addressing issues of poverty and hunger.

4 The Green Economy: Convergence of Social Economy/CED and SCD

In the post-Rio+20 decade, how will we ensure that we take action and make a conscious choice for a hopeful future, rather than one of despair at global and local scales? The Rio+20 process has identified the potential of a new "Green Economy" as a means of shifting us back from the brink and placing us on the path for hope. However, focusing on the green economy without attention to how we mobilize citizens, build capacity and provide supportive social infrastructure for innovation is likely to lead us down the path of the status quo. There exists considerable potential for bridging sustainable community development and social economy approaches – bringing environmental considerations into the social economy and using the social economy to advance equity concerns within sustainability. One way to think about this potential convergence in theory and in practice is through recognition of the shared values, underlying concepts and worldviews that exist for strong sustainable community development and strong social economy (Connelly et al. 2011).

Williams and Millington (2004) have characterized the diversity of definitions for sustainability along a spectrum from weak to strong sustainability based on underlying concepts and worldviews of the relationships between the environment, economy and society. At a conceptual level, weak SCD approaches recognize that economic growth needs to do a better job of accommodating environmental issues, but fundamental changes to the concept of growth are unnecessary. Weak SCD approaches can be characterized as taking a human-centered approach to human-environment relations that assumes that environmental (and social) problems will be resolved through technological changes, more economically efficient use of resources, and substituting manufactured capital for depleted natural capital.

In contrast, strong SCD approaches begin with an explicit recognition of the finite nature of the Earth and the need to reduce demands placed on it. Substitutability of manufactured capital for natural capital is limited by environmental characteristics of uncertainty and irreversibility. From a strong SCD perspective, well-being can be enhanced through development of different forms of community capital (social, human, cultural, physical, economic and natural) (Roseland 2012) and not just through quantitative measures of growth, wealth and consumption. Collective action, finite limits to growth and social innovation are at the core of the strong sustainability perspective and solutions to local and global environmental problems are rooted in social rather than technological change (Rees 1995).

Similar to SCD, conceptual debates about social economy activities represent a spectrum; from weak to strong, from high road to low road (see Lewis and Swinney 2007), or from pragmatic to utopian (Fontan and Shragge 2000). Weak SE approaches are criticized for not addressing societal transformation, providing minimal attention to environmental concerns and often working to maintain existence on the margins of the larger capitalist system, and in fact embrace the mainstream system in its approach (Amin et al. 2002). Strong social economy initiatives on the other hand are focused on community-based actions that incorporate the principles of equity, redistribution, solidarity, mutuality and meeting social needs rather than maximizing profit (Pearce 2003). A critical point of differentiation is whether social economy/ enterprise activities are able to generate their own capital, rather than relying on an ongoing subsidy from the derivatives of the mainstream economy and the politics of redistribution.

5 Examples from Canadian Cities

Canadian cities of all sizes have made commitments to move their cities towards sustainability. There are numerous concrete examples of cities that have been successful in implementing their sustainability projects or initiatives. Each year, the Federation of Canadian Municipalities (FCM) recognizes the efforts of municipalities that have implemented innovative sustainability projects in a range of categories with their Sustainable Community Awards (see Table 1 for select summaries of 2011 award winners).

These examples offer proven solutions and tangible alternatives to business as usual and more of them are needed. They provide tangible improvements to the natural environment and in most cases represent significant cost savings to the municipality. Projects such as these will not however transform our cities into more sustainable places on their own. These examples demonstrate that while Canadian cities have been effective at making the connections between environment and economic concerns by relying on an ecological modernization framework of technological innovation and financial incentives to drive efficiency (Fisher and Freudenburg 2001), they continue to struggle with incorporating social justice concerns that other researchers claim are at the heart of sustainability (Agyeman and Evans 2004).

While these are noteworthy and innovative examples at addressing environmental concerns in cities, they represent the "low-hanging fruit" – those changes that are easiest to make because they do not challenge the status quo. They make financial sense and are focused on the relationships between the environment and the economy however they do not address the political conflicts between environment, economy and equity goals that are at the heart of sustainable cities (Campbell 1996).

In addition, these types of environmental issues are also commonly thought of as the "business" of local government (i.e. municipal infrastructure) (Evans et al. 2005), areas where local government has jurisdiction and control, further reinforcing the bias towards environmental issues in local government sustainability initiatives at the exclusion of social justice and equity issues (Saha 2009).

Many Canadian cities have initiated sustainable development projects in areas such as green building programs, affordable housing, open space preservation, recycling, climate change, and smart growth, but they are largely occurring on a project-by-project or issue-oriented basis rather than being connected and integrated throughout municipal governments and their larger communities.

Table 1 Selection of FCM sustainable community award winners 2011 (Adapted from FCM 2012)

Category	Winner
Brownfields	The Town of Blue Mountains (ON) new municipal building was developed on a brownfield site to spur revitalization of a historic part of the community and to promote additional infill development. The LEED Gold green building certification also serves as a model of how energy efficient design can reduce operating costs for future private sector developments. It is expected that the project will result in improved property assessments and sales tax revenues from new surrounding businesses
Energy	The City of Medicine Hat (AB) set goals of community-wide residential energy consumption reductions and targets of 25% renewable energy sources by 2025. A conservation charge was added to utility bills to raise awareness among city residents about the cost of energy use and to establish local renewable energy projects
Neighborhood development	The City of Burnaby (BC) and the SFU Community Trust engaged in a cooperative planning process to develop UniverCity, a mixed-use and transit-oriented neighborhood based on sustainability principles. UniverCity is currently home to 3,000 residents within walking distance to the University and live in homes that use 13% less energy than the Metro Vancouver average and plans are underway for the development of a neighborhood energy utility that would provide for all space and hot water for the campus, reducing GHG emissions by over 10,000 t per year. Revenues from the development have allowed for the creation of a $25 million endowment to support sustainability teaching and research at SFU
Waste	The Regional District of Nanaimo (RDN) (BC) began their Zero Waste Program in 2002 with a goal of reaching a 75% diversion rate by 2010. While not reaching that interim target, they have achieved a diversion rate of 70% by implementing a waste licensing system, expanded disposal bans, single-family organics collection and education and promotion. Per capita waste disposal has decreased, the RDN has saved $16 million in capital costs by extending the life of the landfill and the project has increased employment in recycling and composting fields
Transportation	The City of Toronto (ON) is focused on becoming a City where people walk more and drive less. The City's walking strategy is designed to transform and integrate its streets, public spaces and neighborhoods with public transit, cycling and other sustainable modes of travel. It is anticipated that the Walking Strategy will promote a healthier population, improve access to shops and services and revitalize neighborhoods. As of 2008, 58% of downtown residents walked to work (up from 44% in 2001), cycling had doubled and only 6% of residents drove to work (down from 16%).

Linking to the social economy movement provides interesting avenues for expansion of these types of initiatives and to engage a broader cross-section of community members in making tangible changes in their daily lives. For example, the Ottawa Renewable Energy Cooperative allows any Ottawa resident to jointly own and invest in renewable energy projects in their community while earning a reasonable return on their investment (OREC 2012). Small-scale renewable energy projects are often capital intensive and are beyond the means of a single household. The cooperative model allows members to jointly own and invest in renewable energy projects in their city larger than what they might individually own and to provide their property and buildings as sites for installation. As a result of social economy and SCD convergence, up-front cost for individuals to invest in neighborhood scale renewable energy projects is overcome. This project has been facilitated by a provincial policy that has encouraged the installation or approval of over 3,500 MW of new solar, wind, hydro and biogas projects in Ontario since 2009.

Car sharing initiatives are another example of social economy and SCD convergence. There are over 16 car-sharing organizations across Canada, representing over 2,000 cars and over 54,000 drivers, with the majority being run as cooperatives (Car Sharing Canada 2012). Car sharing initiatives typically emerge out of a desire for residents in a particular place to reduce automobile dependency and associated environmental impacts. Using social economy models such as cooperatives provides an opportunity for individuals to collectively have access to an automobile when needed, to participate in decision-making about the organization and to generate revenues to expand the fleet of available cars, thereby further decreasing automobile usage. Municipalities are recognizing the benefits of supporting car sharing organizations, as each car-share car removes 6–8 cars on local roads and reduces individual members driving by 50%. For example, the City of Vancouver reduces by five the minimum parking requirements in a new development for every Car Share car provided by the developer, saving developers hundreds of thousands of dollars (Modo 2012).

Finally, local food initiatives and farmer's markets offer a promising opportunity to connect efforts by non-profits, cooperatives and local food organizations focused on building a just and sustainable local food system. In Vancouver (BC) for example, local organizations from all aspects of the food value chain (production, retail, processing, distribution, consumption, etc.) have come together to re-build the local food system infrastructure based on the creation of the New City Market Local Food Hub (New City Market 2012). It is envisioned as a physical space that, in addition to making local food easier to get, will strengthen connections between consumers and producers and provide functions that bring all aspects of a local food system (from farm to plate) under one roof. It will also improve the viability of local farming in terms of fair wages and working conditions by increasing options for local sales and value-added production. The project is currently in the pre-development phase, but its success will ultimately depend on how local food system infrastructure is used differently to foster relationships between food system actors and to promote greater consumer awareness of local food. It also provides a tangible opportunity for increased community mobilization for citizens and the City in support of Vancouver's "Greenest City" goals (Vancouver 2012).

6 Conclusion

The framework of sustainable community development (SCD) illustrates a "systems approach" that recognizes that cities are complex systems with multiple inter-relationships at different scales (rural/urban, local/national/global) and among different issue-based sectors (e.g. the linkages between local air quality, transportation planning and land use densities). Linking SCD with the social economy provides for widespread community awareness-raising and integrated municipal involvement aimed at bringing about a shared understanding of what sustainability means and how to achieve it throughout all sectors of municipal government and the wider community (James and Lahti 2004; Gruder et al. 2007).

The scale and scope of sustainability problems and solutions require solutions that move beyond simply greening the economy. As the FCM case studies above demonstrate, there are plenty of innovative opportunities that cities across the country can and should be doing to "green" their economies. Opportunities exist for municipal governments to support and facilitate organizations that are engaged in work in their communities that reflects the convergence of strong sustainability and strong social economy approach.

We need to move beyond "green economies" and work towards communities that are environmentally, socially, and economically sustainable. Pieces of the puzzle exist amongst diverse communities and in the minds of those who have dedicated time and energy to confronting this challenge. However, despite the hard work of people and communities around the world in pioneering solutions to the sustainability challenge this work often exists in a vacuum, with both success stories and failures remaining only with fading local knowledge. The challenge of sustainability is simply too big for the hard lessons learned to remain local secrets. The convergence of SCD and social economy activities provides opportunities to build broader networks, to scale-up impacts and to access increased resources, knowledge and experience about implementing sustainability initiatives in local communities and to mobilize citizens as active participants in those processes. Supporting convergence of SCD and social economy creates unique opportunities for collaboration that can lead to innovative approaches to implementing sustainability and result in tangible successes. Learning and sharing experiences of these innovations will be critical for advancing the sustainability agenda beyond the green economy. Pando (www.pando.sc), a new dynamic network of collaboration among sustainability researchers, academics, and practitioners with a shared goal of overcoming challenges related to community sustainability provides an opportunity to share knowledge and experience and assess the potential of increased convergence.

Policy makers and practitioners at the local level must work to ensure that initiatives that link sustainability and social economy are provided with the appropriate operational and regulatory settings to realize their transformative influence. Measures for success of transformative sustainability initiatives that incorporate multiple-bottom line outcomes need to incorporate the fact that these initiatives are often competing with mainstream economic activities that are heavily subsidized and do not account for negative social, economic and environmental externalities. Placing emphasis solely on the "green" economy risks reinforcing the perception that sustainability and social economy initiatives are only worthwhile if they make sense in a narrow economic sense.

References

Agyeman J (2005) Sustainable communities and the challenge of environmental justice. NYU Press, New York

Agyeman J, Evans B (2004) 'Just sustainability': the emerging discourse of environmental justice in Britain? Geogr J 170(2):155–164

Agyeman J, Bullard R, Evans B (2003) Just sustainabilities: development in an unequal world. MIT Press, Cambridge, MA

Amin A, Cameron A, Hudson R (2002) Placing the social economy. Routledge, London

Amyot S, Downing R, Tremblay C (2010) Public policy for the social economy: building a people-centred economy in Canada. Public policy paper series number 3. Canadian Social Economy Research Partnerships. Available via: http://www.socialeconomyhub.ca/sites/default/files/PublicPolicyPaper3_0.pdf. Cited 2 Feb 2012

Annan K (2002) Beyond the horizon: will the future be bountiful? Time Magazine 26:22–23

Beatley T (2000) Green urbanism: learning from European cities. Island Press, Washington, DC

Brugmann J (1994) Who can deliver sustainability? Municipal reform and the sustainable development mandate. Third World Plan Rev 16(2):129

Campbell S (1996) Green cities, growing cities, just cities? Urban planning and the contradictions of sustainable development. Am Plan Assoc 62(3):296–312

Car Share Canada (2012) About car sharing. Available via: http://autoshare.com/ca/ca.html. Cited 2 Feb 2012

Connelly S, Markey S, Roseland M (2009) Strategic sustainability and community infrastructure. Can J Urban Res Can Plan Policy Suppl 18(1):82–104

Connelly S, Markey S, Roseland M (2011) Bridging sustainability and the social economy: achieving community transformation through local food initiatives. Crit Soc Policy 31(2):308–324

Evans B, Joas M, Sundback S, Theobald K (2005) Governing sustainable cities. Earthscan, London

Federation of Canadian Municipalities (2012) Sustainable community award winners 2011. Accessed 1 Feb 2012 at: http://fcm.ca/home/awards/fcm-sustainable-communities-awards/2011-winners.htm

Fisher DR, Freudenburg WR (2001) Ecological modernization and its critics: assessing the past and looking toward the future. Soc Nat Resour 14(8):701–709

Fontan J-M, Shragge E (eds) (2000) Social economy: international perspectives and debates. Black Rose Books, Montreal

Gruder S, Haines A, Hembd J, MacKinnon L, Silberstein J (2007) Toward a sustainable community: a toolkit for local government. University of Wisconsin, Madison

ICLEI (2002) Local Agenda 21 Survey. UNDESA. Available via: http://www.iclei.org/documents/Global/final_document.pdf. Cited 28 Mar 2010

James S, Lahti T (2004) The natural step for communities: how cities and towns can change to sustainable practices. New Society Publishers, Gabriola

Kenworthy JR (2006) The eco-city: ten key transport and planning dimensions for sustainable city development. Environ Urban 18(1):67–85

Lewis M, Swinney D (2007) Social economy and solidarity economy. Making Waves 18(4):9–15

Lionais D, Johnstone H (2009) Building the social economy using the innovative potential of place. In: McMurtry J (ed) Living economics: Canadian perspectives on the social economy, co-operatives and community economic development. Emond Montgomery, Toronto, pp 105–128

Mazza L, Rydin Y (1997) Urban sustainability: discourses, networks and policy tools. Prog Plan 47(1):1–74

McMurtry JJ (ed) (2009) Living economics: Canadian perspectives on the social economy, co-operatives, and community economic development. Emond Montgomery, Toronto

Modo (2012) Developers for Modo. Available via: http://www.modo.coop/developers. Cited 1 Feb 2012

Neamtan N (2009) Social economy: concepts and challenges. Universitas Forum 1(3):1–5

New City Market (2012) Visioning workshop report. Available via: http://www.newcitymarket.org/about/visioning-workshop-report-2010/. Cited 1 Feb 2012

Ottawa Renewable Energy Cooperative (OREC) (2012) What's new with OREC? Available via: http://ottawarenewableenergycoop.ca/. Cited 1 Feb 2012

Otto-Zimmermann K (2002) Local action 21: motto-mandate-movement in the post-Johannesburg decade. Local Environ 7(4):465

Parkinson S, Roseland M (2002) Leaders of the pack: an analysis of the Canadian 'Sustainable Communities' 2000 municipal competition. Local Environ 7(4):411–429

Pearce J (2003) Social enterprise in anytown. Calouste Gulbenkian Foundation, London

Poppendieck J (1999) Sweet charity? Emergency food and the end of entitlement. Penguin, New York

Pugh CDJ (1996) Sustainability, the environment and urbanization. Earthscan, London

Rees WE (1995) Achieving sustainability: reform or transformation? J Plan Lit 9(4):343–361

Roseland M (2012) Toward sustainable communities: resources for citizens and their governments, 4th edn. New Society Publishers, Gabriola Island

Saha D (2009) Empirical research on local government sustainability efforts in the USA: gaps in the current literature. Local Environ 14(1):17–30

Vancouver, City of (2012) Greenest City 2020. Available via: http://vancouver.ca/greenestcity/. Cited 2 Feb 2012

Williams CC, Millington AC (2004) The diverse and contested meanings of sustainable development. Geogr J 170(2):99–104

Seven Conditions for Effective Green Governance

Rula Taher Qalyoubi

Abstract The development and deployment of renewable energy could be perceived as a green energy revolution that may be instrumental in jump-starting economies out of recession. This chapter argues that out of environmental urgency we need to collectively act and systematically reorganize for a top-down approach to facilitating the introduction of renewable energy. This argument can be validated through historical lessons such as public funding for the establishment of the backbone infrastructure of the Internet. The outcome of this public funding was declining marginal costs of digital enterprises. Hereto institutional reforms were necessary, which were restricted to supply-side management. To expedite the introduction and integration of renewables, governmental intervention needs to cover research and development in augmenting human capital, science, and law, as these are at the same time the main resource capabilities and constraining factors. These factors can be renegotiated through a serious examination of what this chapter proposes to be the seven necessary and sufficient conditions to facilitate green economy outcomes.

Keywords Renewables • Sustainable economic growth • Government intervention • Governance

1 Background

Only few deny that global climate change presents us with a major test that confronts humanity, and poses challenges on many levels. How we pass this test depends on the availability and the viability of options that policy makers, locally and nationally, have at their disposal. According to the Patz et al. (2007), as climate change intensifies

R.T. Qalyoubi, Ph.D. (✉)
3rd Group Consultancy Engineering (3G) . M1 Pearl Plaza, 32nd St., Khalidiya
P.O. Box 94197, United Arab Emirates
e-mail: Rula.Qalyoubi@3rdgroup.ae; rt.qalyoubi@gmail.com

R. Simpson and M. Zimmermann (eds.), *The Economy of Green Cities: A World Compendium on the Green Urban Economy*, Local Sustainability 3,
DOI 10.1007/978-94-007-1969-9_18,

we will begin to experience an increasing rate of catastrophic events. Problems of the global South will also have impacts on the global North such as environmental refugees, political instability, increases in the cost of doing business-as-usual, social unrest and economic stagnations, while the global North will also have to deal with their own environmental problems.

To lessen the effects of global warming and to remain below a 2°C increase in this century, national and local government bodies need to proactively coordinate efforts and facilitate development paths that are in line with growth without adding pressures on the environment (IPCC 2007). According to UNEP 2011, it has been recognized that green growth paths produce reliable sustainable economic growth when compared to the 2030 carbon-based growth projections. The economic green growth projections will surpass business-as-usual by 2050 as modeled by UNEP (2011). The UNEP 2011 reports clarify that green growth is the outcome of synergized action plans aimed at all agents in the market: government bodies, households, financial investors and businesses. A green economy will be the outcome of systematically changing the way we do business and titling the balance against carbon-based industries. For some government bodies, it involves refining standing procedures and amending their environmental laws and processes. Yet others need to revisit their priorities and strategies to create a sense of necessity and mandate stakeholders to conform to the new paradigm.

Both the rebound effect of the energy sector, which offsets the decline in industrial energy intensities, and the volatile nature of energy prices have delayed investments in renewable energy technologies. Long term government planning should not be tied to short-term market variations, since decoupling offers opportunities to gain a competitive advantage in eco-efficiency economies. As such, governments need to avoid tying long-term strategies to the short run fluctuation of energy markets. Government bodies ought instead to build long-term renewable agendas that guarantee economic prosperity, environmental equity and social inclusion. This involves short-term green planning with long-term renewable energy targets.

There are currently three key factors that can enable green economies by accelerating introduction of renewable energies: financial resources, physical infrastructures, and maturity of renewable energy markets. When favorable these conditions enable government, NGO and private entities to duplicate successful experiences around the world and to realize the financial and environmental improvements of sectors' performances. These three factors have the potential to result in systematic adjustments to existing institutions, but can also have shortcomings:

- Abundance of financial resource does not in itself eliminate scarcity problems. Areas with abundant energy resources (such as oil, forests, minerals, water, sun or wind) may develop relative scarcity as opposed to absolute scarcity. Scarcity ought to be the driver behind investment for the development of renewable technologies and financial capital ought to be tied to renewable technologies that have the greatest impact on reducing reliance on oil based products and practices.
- In the absence of legal frameworks, physical infrastructures can be used to facilitate transactions and exchanges that compete with green goals. A suitable regulatory system ought to facilitate green economic activities and at the same time create

the proper incentives in order to drive green transactions and remove barriers to private investments in eco-efficiency.
- Maturity of renewable energy markets is important. If however the alternatives to fossil fuel practices are not economically feasible for mass dissemination, this can slow introduction of renewable energy technologies. Such technologies ought to be available to all market agents and be economically viable when compared to current practices.

These factors are critical preconditions of a green economy although not sufficient in themselves to guarantee the environmental, economic and social sustainability criteria required for a fully functioning green economy. Without a comprehensive assessment of the renewable energy strategies currently in place, required structural changes will go unnoticed. Using a bottom-up approach to the creation and implementation of renewable technologies, financial resource, physical infrastructure and mature energy markets may prove more successful at the local level. While this approach remains important, grassroots movements and small-scale renewable initiatives will not be enough to counter the centuries of damage. In addition there may be teething problems when small-scale initiatives are implemented on a larger scale.

2 Conditions for an Effective Intervention

Government bodies have to examine their own complexity in the face of environmental challenges, the interdependencies of local frameworks, and the maturity of their institutions in order to identify the opportunities at their disposal. It is only when governments answer the "what, how and why" questions that cities can begin to formalize their priorities and conceptualize dis-incentives to conventional carbon-based thinking. A smooth transition to a green economy depends upon seven related conditions which are both necessary *and* sufficient to produce substantial changes, and which will enable governments to take on a more proactive role in reversing the course of carbon-based activities. These include: political commitment, a legal and regulatory foundation, implementation of financial instruments, technological viability, human capital formation, appropriate institutional setup and a common language. Each is addressed in detail as follows.

2.1 Political Commitment

Examining the success of enterprises around the world requires a review of all relevant enabling conditions. The fast paced fiber-optic world we live in today would not be possible without the backbone of the Internet and the NASA program. The side-effect of governments' commitment to satisfy their own goals has given rise to an IT revolution – such societal leapfrogging would not have been possible in the absence of Internet infrastructure. The positive externalities associated with telecommunication accounts to more than 50% of the world's productivity as direct benefits and even more as indirect benefits (Prescott 1997). This kind of government intervention will be essential to the green growth paradigm.

Political commitment to the green growth path presupposes that fiscal and monetary tools are used to steer economies away from business-as-usual. This would include investing in research and development (R&D), demonstration, deployment and the commercialization of different renewable technologies in production and consumption activities. It was estimated that an annual global expenditure of US$50–100 billion on R&D would be needed to achieve a green outcome scenario by 2050 (IEA 2010). This includes R&D expenditure shortfall on advanced vehicles, bioenergy, CCS, nuclear fission, smart grid, solar energy, wind energy, higher-efficiency coal and industrial energy efficiency.

Despite this financial gap and the current global economic contraction, governments could solve these challenges through strategic coordination aimed at sustainability goals. In the wake of the 1990s financial crises both Finland and Korea made strategic moves to increase R&D funding for renewables while scaling back on other public expenditures coupled with phasing out fuel subsidies. Currently, Finland and Korea have the competitive advantage in innovative technologies in renewable energies. Korea's New Green Deal established 17 new growth ventures and supported funds to research potential alternative energies.

2.2 *Legal Foundation*

With political commitment, regulatory reforms can enable policies to take shape which sponsor the greening of economies on both local and national levels. If decisions and policies are not anchored in a binding and enforceable framework, they become merely ornamental in nature. While we do not need to green all regulatory systems concurrently in order to respond to climate change challenges, legal standards must be assessed in a systematic approach.

Legal reforms require revisiting existing laws and tuning their jurisdictions through amendments and directives. In some areas, this may require energizing the legal stand by conducting gap analysis, which entails comparing current environmental laws and procedures with international best practices in order to assess the current incentive system. With proper incentive systems reinforced with green mandates and directives, agents are eased in the new environment either through voluntary or coercive actions. For example by examining existing zoning laws along with planned projects policy makers can restructure the incentive system in order to pave the way for green growth. Advanced law amendments could be aimed towards closed-cycle manufacturing. The model of Kalundborg in Denmark demonstrates that wise planning can generate income and economic growth without adding environmental pressures (Yu et al. 2009). This distinctive example of using wastes from certain manufacturing operations as raw materials for industry was the result of coordinated efforts and a sustainability mindset in action.

Honest efforts focused on increasing the efficiencies of the current setup can amount to millions of dollars' in benefits. This can be in line with mandating new building retrofits which can be phased in within 5 years (McKinsey & Company 2009). Moreover, it was estimated that the net present value (direct energy benefits) was US$210 million and US$1.26 billion for targets of 10 and 40% energy savings targets respectively. When indirect productivity is included, the net present value increased to US$2.81 billion and US$1.34 billion with 10 and 40% energy savings targets respectively (WBCSD 2010).

2.3 *Financial Instruments*

Policy makers backed by a regulatory body have the power to initiate a paradigm shift using three simultaneous measures. Firstly, putting proper mechanisms in place to prevent natural resource use and abuse. This entails reevaluation of the two types of resource usage, one being resource-use intensity and the other the depletion of natural stock. To internalize the cost of resource-use intensity, a "just price" doctrine will be required which will readdress market prices in order to properly reflect the energy used for production. Funds procured in this way can be utilized to finance green initiatives. Correspondingly city balance sheets should include "ecosystem payments" providing liability for the maintenance of natural stock. Adding this missing element into the accounting procedures would need to be mandated across all municipalities in order to prevent budget variations within resource maintenance. Adding payments and expenses made in order to maintain natural capital is the first step towards accurate accounting models.

Secondly, government bodies need to reevaluate subsidy provisions and redirect funds towards ventures that provide permanent solutions to energy security issues. Currently the global subsidy system that oil as well as automotive and petrochemical industries enjoy has proven a handicap to actions related to the procurement and implementation of renewable energy. According to UNEP (2011) estimated fossil-fuel subsidies amount to US$500 billion–700 billion. If a green paradigm is to become an implementable strategy then policy makers will have to exercise preferential treatment to non-carbon based activities, sending a distinct message to industry, businesses and commerce that private firms need to reorganize their priorities. This could push business R&D expenditure towards green business dealings and procurement practices and should invigorate private renewable energies development.

Finally, to reduce the financial risks associated with private green investment the government should boost their R&D support, providing stimulus packages aimed at increasing the efficiencies of existing systems and supporting the development and demonstration stages of the renewable energy agenda. This could take the shape of targeted or time-limited financial mechanisms such as loan guarantees and insurance mechanisms (OECD 2011). When risks (and costs) are reduced, business ventures are far more likely to adopt a variety of ways to green their production and to become innovative in finding more environmental efficient solutions.

If the above three measures are strategically tailored this could lead to a scenario in which green practices are more cost effective and could even outstrip business-as-usual. The financial sector should be ready to finance the deployment of renewable technologies as the risk of doing business is lowered due to governmental R&D sponsorship. The market incentive approach leads to economies of scale, reduces costs and increases profits for renewables enterprises that will eventually push against all carbon-based practices. This will pave the way for efficient renewable technologies to be available at low marginal cost – a pillar for the commercialization of renewable energies.

2.4 Technological Viability

Even with financial instruments in place, the choice of renewable technologies may prove to be a significant challenge. There are three steps by which choices in technology can be tested, the first of which is to use public funds to support working groups to debate the type of renewable technology most suited to current geopolitical and climatic conditions. In order to have a proactive debate academics and practitioners should be part of the dialogue as both parties represent different solutions to the energy challenge: one more theoretical involving research and the other developed from practical experience. The criterion of technology debates must be in line with technology suitability in order to meet the designated goal and to avoid duplications of efforts.

Secondly, government funding for frequent sensitivity analysis will enable a clear vision of all options regarding renewable technologies. Because technological innovation is a dynamic process, frequent upgrades to existing technologies from new and innovative forms of renewable energy must be anticipated. New technology must be analyzed in order to establish whether they can add real value or properly address existing energy problems. This requires the establishment of R&D systems that test the viability of all new renewable technology.

Thirdly, governments should champion resource-gap assessments. Such dialogues promote stakeholders to participate in finding solutions by acknowledging resource-gap difficulties. This gap analysis should be sponsored by different government bodies in coordination with practitioners, research groups and venture capitalists at the national and local levels. Historical experience tells us that governments unilaterally committed to nuclear programs to solve gap-resource problems experienced short-term successes with incurred longer-term and immeasurable environmental damage. From 1974 to 2009, governments' expenditures on R&D compared to alternative (nuclear) energy programs amounted to more than double when compared to renewable technologies (IEA 2010).

Governments can trigger private partnership investments in renewable energy investigations. Successful R&D public policies can target particular forms of renewable technology or can adopt a broad market approach ranging from fiscal incentives, capital grants, feed-in tariffs or direct subsidies. In addition non-market R&D public strategies can be used to test the feasibility of biofuel requirements, quota-based schemes and tradable quotas among different sectors. Both market and non-market mechanisms ought to have multiple goals such as making the introduction of renewables faster, easier and more efficient. Moreover when private sector bodies champion renewable energy ventures they become part of the process and more engaged with solutions towards renewable energy strategies.

Sound choices on behalf of governments should pave the way for positive governance network which should involve private stakeholders and through which alterations to policy and existing practices will enhance rather than disrupt the efficiency of existing systems. China, India and Brazil have all adopted "open" renewable innovation strategies and their established collaborative partnerships ensure that costs are shared and risks spread, allowing for a wider array of contributors (OECD 2009). Concurrently, establishing, intellectual property rights and harmonization of technical standards should pave the way for the commercialization of renewable technologies within different parts of the market.

2.5 Human Capital Formation

Investments into human capital is a crucial expenditure that should not be compromised and as part of public funding ought to enhance the know-what, know-how and the know-why (David 2001). Education, vocational training and research are extremely vulnerable areas during economic downturns. During recessions, governmental bodies (at both local and national levels) exercise greater budgetary discretion and often downsize publically funded activities such as education and training. Stripping educational budgets reduces the value of human capital which could otherwise help to stimulate the economy out of recession. It is during economic downturns that education and other human capital building vehicles need extra support to fund training and facilitate the formation of a cadre that fits the new green ideal.

It has been empirically validated that investments into human capital tend to promote faster recovery from economic stagnation (Nelson and Phelps 1966). It is therefore a necessity that during recessions educational budgets are boosted rather than welfare budgets. Financial assistance including grants should be allocated to educational institutions that will increase green skills among unemployed and displaced individuals. Furthermore educational organizations that have community outreach programs should receive preferential funding, especially if their focus is to solve local energy problems. In this manner green education investments become implicated in a network of financial instruments aimed towards solving environmental challenges, making human capital a vital part of the green movement. Conditional fund allocations should be used to build a network of cooperation between educational entities, businesses and the community.

Lessons can be learned from Finland where during the economic crisis in the early 1990s, the government made a commitment to avoid cuts on essential services favoring R&D and educational institutions. Finland experienced a quick recovery from recession with a world-class educational system and highly skilled workforce. Financial support to programs like TEKES (the Finnish Funding Agency for Technology and Innovation) boosted the contribution of exports in renewable energy technologies by SMEs to 8%. In 2009, TEKES expanded its funding strategies aimed at cleantech industries in order to share risks and encourage cooperation between the public and private sector. The Finnish TEKES experience shows that the positive effects associated with public investment can be manifested in building societies' social capital and can increase productivity and innovative solutions. In 2010 the added contribution of cleantech amounted to EUR 18 billion with an annual growth rate of 10% (TEKES 2011).

2.6 *Dynamic Adjustments*

Institutional adjustments are required which will adopt and adapt to innovative solutions, otherwise institutions will face "creative destruction" through inaction. Despite the presence of regulatory rigidity U.S. utility companies in Vermont adopted the "Cow Power" program, circumventing local government stalemate and providing energy directly to customers (Wang et al. 2011). Farmers found it more efficient and environmentally sound to invest in manure digesters and selling the surplus to customers but at higher prices for guaranteed services. Agents on both sides of the exchange eventually incurred higher transaction costs for energy provision in order to correct for the "non-market" failure brought on by the systems' inflexibility.

The evolution of institutional processes has to be dynamic in nature and changes should be built on existing and successful rules. The Village Council of Wildpoldsried in Bavaria, Germany, passed a local green initiative in 1997 with modest goals to attract new industries and to bring in new revenue (Singh 2011). The council equipped new installations with solar panels, build biogas digesters and installed seven windmills. Today the village sells power back to the national grid generating 321% more energy than it requires. Thus far, returns on this investment have amounted to US$5.7 million for 2,600 villagers (though the proportion of initial public investment is hard to tell). The insight of the village council lies with simultaneously seizing opportunities to create a cleaner environment using renewable sources and at the same time generated income through consultation and tourism.

The above two experiences present opposite storylines: flexible government institutions are more likely to accommodate green technological change compared to government bodies that are locked into their institutional framework. There is no scale to determine institutional flexibility other than their apparent responses to accredited studies on green practices.

2.7 *Common Language*

Common language entails the standardizations of targets, benchmarks, indicators and measurement units and methods. Moreover, common language is about a unified code of practices, streamlined green accreditation of products and services, and consistent decision making processes on green issues across government. This is an important criterion as it eliminates the ambiguities and doubts associated with an emerging breed of buzz words which quickly become redundant and which government officials frequently need to relearn. Choice of target thresholds and indices may be important in order to establish a benchmark comparison but ought to be used with caution. By and large, instruments ought to be modified by region and reflect local environmental circumstances. This may require some trial and error but more importantly requires having the necessary human capital base to provide knowledge both of local conditions and on the availability and viability of tools.

Government sponsorship of initiatives (and their associated indicators) ought to provide for reliable institutional growth and development without restructuring existing rules which are successful. In this way governments are engaged in meliorating rather than satisficing. When President Obama signed on Executive Order 13514 in 2009, the administration mandated all federally funded entities to report on Strategic Sustainability Performance Plans (The White House 2009). This new measure is a sharp double edge sword; for one it forces federal entities to comply with sustainability principles with aims to modify existing practices. But the new measure is also used as a purposeful scale to help determine increases in the efficiency for all existing systems. Although, this requires the generation of additional reports on these savings this can be done under the current accountability structure which is superior to reinventing the wheel.

A further measure of collaboration is that it builds on information sharing between different government bodies and civil society. Internet-based information hubs ought to provide citizens with quick and easy access to background checks regarding the production of goods and services. In this manner consumers can vote for green practices using their purchasing power, and can influence producers to accommodate green preferences. Such information hubs would need sponsorship from local and national governments to expedite the transfer of green-based knowledge into the public domain. In this manner the new green approach becomes part of civic life and embedded in the fabric of societies. Such internet-based green information hubs should facilitate regional collaboration and allow national partners to share information, methods and success stories that are suitable to regional circumstances.

3 Conclusions

These seven conditions are not only necessary but will be sufficient to move economies towards convergence on green principles. With political commitments in place and incremental reform focused on managerial change, policy makers can draw processes away from a carbon-based paradigm. Through green innovation we have been given a chance to get our priorities right and this chance can facilitate more profitable investment. If used intelligently, environmental legal reforms together with green financial instruments can even the playing field for green practices, goods and services, without economic tradeoffs. Empowering civil society with green know-how and skills will enable communities to enjoy a smooth green transition with positive employment effects in the long-term. Governments that aim at a higher diffusion rate of renewable energy ought to have the institutional flexibility to accommodate the transition and the sponsorship of R&D to jump start green growth paths.

References

David PA (2001) Knowledge, capabilities and human capital formation in economic growth. New Zealand Treasury, Working paper 01/13, Wellington

IEA (2010) Global gaps in clean energy RD&D: update and recommendations for international collaboration. IEA report for the Clean Energy Ministerial, Paris

IPCC (2007) The long term perspective. In: Core Writing Team, Pachauri RK, Reisinger A (eds) Summary for policy makers. Climate change 2007: synthesis report. Contribution of working groups I, II, and III to the fourth assessment report of the intergovernmental panel on climate change. IPCC, Geneva. This version: IPCC website. Cited 2 Feb 2012

MaKinsey & Company (2009) Pathways to a low-carbon economy. Version 2 of the global greenhouse gas abetment cost curve. Available via: http://solutions.mckinsey.com/climatedesk/getfile.aspx?uid=a7310e0e-ba5b-4217-aaee-773d68744338&fp=design%2fClimate+Change+Center%2fPathwayToLowCarbonEconomy_FullReportA.pdf&ru=default%2fen-us%2fhidden%2fduplicatedownload.aspx

Nelson RR, Phelps ES (1966) Investment in humans, technological diffusion, and economic growth. Am Rev 51(2):69–75

OECD (2009) Policy responses to the economic crisis: investing in innovation for long-term growth. OECD, Paris

OECD (2011) Towards green growth – a summary for policy makers. OECD, Paris

Patz JA et al (2007) Heat advisory – how global warming causes more bad air days. Natural Resource Defense Council (NRDC) The Earth's best defense. Online access: http://www.nrdc.org/globalwarming/heatadvisory/heatadvisory.pdf

Prescott EC (1997) Needed: a theory of total factor productivity. Research Department Staff Report 242. Federal Reserve Bank of Minneapolis and University of Minnesota, Minneapolis

Singh T (2011) German village produces 321% more energy than it needs! Inhabitat: design will save the world. Available via: http://inhabitat.com/german-village-produces-321-more-energy-than-it-needs/comment-page-1/#comment-350561. Cited 13 Dec 2011

TEKES (2011) World class sustainable solutions from Finland – sustainable urban development, sustainable buildings, sustainable solutions. Available via: http://www.tekes.fi/u/sustainable_solutions.pdf. Cited 25 Dec 2011

The White House (2009) Federal leadership in environmental, energy, and economic performance. Office of the Press Secretary. Available via: http://www.whitehouse.gov/assets/documents/2009fedleader_eo_rel.pdf. Cited 20 Dec 2011

UNEP (2011) Towards green economy – pathways to sustainable development and poverty eradication: a synthesis for policy makers. UNEP, Nairobi

Wang Q, Thompson E, Parson R, Rogers G, Dunn D (2011) Economic feasibility of converting cow manure to electricity: a case study of the CVPS Cow Power program in Vermont. J Dairy Sci 94(10):4937–4949

World Business Council for Sustainable Development (2010) Vision 2050: the new agenda for business. Available via: http://www.wbcsd.org/vision2050.aspx. Cited 20 Dec 2011

Yu Q, Li H, Wang J (2009) Promoting industrial symbiosis network through public-private partnership: a case study of TEDA. IEEE 978-1-4244-2902-9. Available via: http://www.symbiosis.dk/en/bibliotek. Cited 10 Jan 2012

Part IV
Strategies and Approaches for Greening Urban Economies

Introduction: Strategies and Approaches to Greening Urban Sectors

Shay Kelleher

Keywords Approaches • Sectors • Introduction • Green urban economy

Although cities only cover between 1 and 2% of the Earth's land surface,[1] a relatively small amount, they account for between 70 and 80% of total greenhouse gas emissions as well as the majority of global GDP.[2] The carbon/ecological footprint of cities combined with their economic importance, within the context of an increasing urbanization trend, make urban areas the most important arena for a transition to a Green Economy. In relation to greenhouse gas emissions, Gouldson et al. in the first chapter of Part IV highlight a key question by asking how to decarbonize a city. Cities are not just regarded as source of environmental problems but they also offer many of the solutions, particularly as a result of the various opportunities that exist in urban areas which have been detailed in Part II.

Urban areas have particular structural and technological potential as well as agglomeration economies which provide numerous opportunities for a Green Economy (see UNEP 2011). Urban systems offer opportunities particularly in terms of energy and resource efficiency (see UNEP 2011). The concentration of finance, innovation, knowledge, economic activity and people in cities provides a dense and nucleated manifestation of human activity in a number of urban sectors. By greening these urban sectors, and therefore this concentration of human activity, a Green Urban Economy can then be wielded as a source of global sustainability.

[1] Schneider A, Friedl MA, Potere D (2009) A new map of global urban extent from MODIS satellite data. Environ Res Lett 4(4):044003. doi:10.1088/1748-9326/4/4/044003

[2] Half of global GDP (US$ 30,000 trillion) in 2007 came from 380 cities in developed-regions, 220 largest cities in developing-regions contributed another 10% (McKinsey Global Institute 2011).

S. Kelleher (✉)
ICLEI – Local Governments for Sustainability, Bonn, Germany, Bonn, Germany
e-mail: urban.research@iclei

R. Simpson and M. Zimmermann (eds.), *The Economy of Green Cities: A World Compendium on the Green Urban Economy*, Local Sustainability 3,
DOI 10.1007/978-94-007-1969-9_19, © Springer Science+Business Media Dordrecht 2013

However for urban policymakers and local government decision makers the issue is what is required to green this activity and therefore to green the urban economy?

Part IV aims to answer this question by analyzing the application of the Green Economy to urban strategies at city level in sector specific areas. It also reflects upon methods and approaches that assess and realize the economic opportunities in cities for the development of a Green Urban Economy. Gouldson et al. open Part IV by highlighting that higher levels of risk and uncertainty can be major barriers to action which "make it hard to develop a political, a business or a social case for investment in low carbon options". Their chapter illustrates ways to overcome these barriers while simultaneously analyzing the impact of low carbon growth on the urban economy. Philip Monaghan illustrates the example of Low Carbon Enterprise Zones as an increasingly used policy innovation for a fossil fuel free urban economy.

Le Yin Zhang's contribution on City Development Strategies is a city-level approach detailing how urban governments can accelerate the transition to the Green Economy. It details how urban governance can use policy and institutional frameworks to prioritize the growth of green economic activities to realize the potential associated with urban form and energy use, service and infrastructure provision. Shay Kelleher and Art von Lehe's chapter illustrates the application of a comprehensive city strategy by using Portland's Economic Development Strategy as an example for green cluster development.

Matthew French and Christophe Lalande highlight another urgent agenda: slum upgrading and sustainable housing in the context of the Green Economy and poverty eradication. They argue that the building sector has great potential, particularly in relation to energy efficiency and greenhouse gas reductions, as well as being a vital sector for constructing and building housing for the urban poor. The chapter underscores the opportunities that exist, specifically in the developing world, where rapid urbanization instigates an increasing demand for housing.

Another urban sector is addressed with Santhosh Kodukula's chapter on EcoMobility. It addresses the benefits of EcoMobility for urban areas and its importance for greening the transport sector. There exists great potential to green cities by promoting an integrated urban transport system. By exploring the various approaches to urban planning and transport this contribution discusses how walking and cycling friendly cities complimented by sustainable public transport can bring significant social, economic and environment benefits.

Water is a fundamental element for livelihoods and economies, which Poyyamoli et al. and Caroline King and Boshra Salem's chapters explore from different angles. The former illustrates the importance of treating domestic grey water in a rapidly urbanizing India. This contribution discusses the use of constructed wetland systems as an example for domestic waste water treatment, while conserving biodiversity. Through a cost assessment method King and Salem highlight the cost of groundwater degradation and the associated potential economic loss from this degradation. This contribution illustrates how economic evaluation methods are and can be applied to environmental resources, which can provide useful tools for local decision makers when implementing policies.

Measures to reduce waste close to the source of generation and innovative approaches to waste disposal and water treatment are crucial, as discussed in Gamaralalage and Premakumara's chapter on decentralized composting in Asian cities. This contribution analyzes new innovative waste disposal techniques in the cities of Surabaya, India and Matale, Sri Lanka. It highlights the benefits of reducing waste close to the source of generation by adequately recycling organic waste through decentralized composting; a method for dealing with the growing problem of waste disposal. This research also shows how effective waste management solutions can bring strong environmental and health benefits, while realizing "economic benefits such as generating incomes, creating job opportunities for the urban poor, and minimizing costs for municipal governments".

Raquel Moreno-Penaranda's contribution highlights the importance of the local economy and the preservation of local resources for local production and consumption. This contribution also discusses the need for preserving biodiversity for cultural, economic and social benefits. In conjunction with Salem and King's contribution, this chapter highlights the importance of placing an economic value on ecosystem services. In addition, Jonathan Quarterly's application of the Hedonic Price Model emphasizes this point further by illustrating the economic loss as a result of the deforestation of the Kumasi Forest Reserve in Ghana. Furthermore, international climate financing mechanisms like the Clean Development Mechanism (CDM), which can help to establish such costs through carbon pricing, are further detailed by Sippel and Michaelowa. In their chapter they show the opportunities that exist for the application of the CDM to specific urban sectors, most notably energy supply, buildings and transport, as a means for urban mitigation measures against environmental degradation as well as a source to realize further financial capital.

Part IV details many of the principles explored in Part II with a greater emphasis on approaches and individual urban sectors – sectors that when combined in the complex web of urban synergetic relations constitute the urban economy. It also analyzes which strategies and methods can be used to accelerate and realize the opportunities of greening the urban economy. Cities worldwide are diverse, and vary in terms of their geographical and political attributes. Local circumstances and unique characteristics must therefore be taken into account when applying the Green Economy to urban sectors. It needs to be recalled that various other approaches, including various financial mechanisms and institutional instruments aimed at internalizing externalities, pricing pollution, amongst others, also exist. In conjunction with the discussed strategies and approaches, further tools and research need to be pursued and utilized to inform policy direction, encourage private actors to be involved and to mobilize public support in a collaborative manner which has been suggested in Part III. Such support and involvement, in collaboration with local government action can allow for a successful application of Green Economy principles to urban sectors, and therefore the urban economic system.

References

McKinsey Global Institute (2011) Urban world mapping the economic power of cities. McKinsey & Company. www.mckinsey.com/mgi. Cited 21 Mar 2012

UNEP (2011) Towards a green economy: pathways to sustainable development poverty eradication. UNEP, Nairobi

The Economics of Low Carbon Cities: Approaches to a City-Scale Mini-Stern Review

Andy Gouldson, Niall Kerr, Corrado Topi, Johan Kuylenstierna, and Richard Pearce

Abstract What is the most effective and efficient way to decarbonize a city? There are hundreds of low carbon options available and although they present a significant opportunity to reduce energy bills and carbon footprints there is often a lack of reliable information on their performance. The higher levels of risk and uncertainty that emerge as a result can be a major barrier to action, making it hard to develop a political, a business or a social case for investment in low carbon options. This in turn makes it harder to set targets, stimulate investment, target interventions, develop new delivery mechanisms and so on. To address this problem, we describe the approach adopted in a city-scale mini-Stern review that reviews the performance and assesses the scope for the deployment of hundreds of low carbon options at the city scale. The review identifies the business case for major scale investments in low carbon options at the city scale, examining the investment needs and payback periods associated with different levels of decarbonization. The results of a case study from the UK are presented. These highlight the presence of very significant and commercially viable opportunities for decarbonization at the city scale – and the potential economic benefits associated with exploiting these. However, it also highlights the scale of the challenge

A. Gouldson (✉) • N. Kerr
School of Earth and Environment, University of Leeds, Leeds, UK
e-mail: a.gouldson@see.leeds.ac.uk; N.Kerr@leeds.ac.uk

C. Topi • J. Kuylenstierna
Stockholm Environment Institute, University of York, York, UK

R. Pearce
Quantum Strategy and Technology, Warrington, UK

R. Simpson and M. Zimmermann (eds.), *The Economy of Green Cities: A World Compendium on the Green Urban Economy*, Local Sustainability 3,
DOI 10.1007/978-94-007-1969-9_20,

and the need for innovations in a number of areas if major scale investments are to be secured and delivered. To push the debate forward, the chapter concludes by proposing ten key steps to deliver major scale investments in low carbon cities.

Keywords Economics • Low-carbon • Cities • Mini-Stern

1 Introduction

What is the most effective and efficient way to decarbonize a city? There are hundreds of low carbon options available and although they present a significant opportunity to reduce energy bills and carbon footprints there is often a lack of reliable information on their performance. The higher levels of risk and uncertainty that emerge as a result can be a major barrier to action, making it hard to develop a political, a business or a social case for investment in low carbon options. This in turn makes it harder to set targets, stimulate investment, target interventions, develop new delivery mechanisms and so on.

In an attempt to address this problem, we report the results of a city-scale mini-Stern review. Applying a methodology developed by the authors and tested in the Leeds City Region in the UK, this review starts by examining the cost and carbon effectiveness of a wide range of the low carbon options that could be applied in households, industry, commerce and transport. It then explores the scope for the deployment of each measure at the city scale. This involves having some understanding of the size and the composition of the domestic, commercial (public and private), industrial and transport sectors, and their potential to absorb investment in a wide range of low carbon measures. The review then identifies the business case for major scale investments in low carbon options at the city scale, examining the investment needs and payback periods associated with different levels of decarbonization. It also considers the wider economic implications of such transitions – with a particular emphasis on the opportunities for job creation in the low carbon and environmental goods and services sector.

The study highlights the presence of very significant and commercially viable opportunities for decarbonization at the city scale – and the potential economic benefits associated with exploiting these. However, it also highlights the scale of the challenge and the need for innovations in a number of areas if major scale investments are to be secured and delivered. To push the debate forward, the chapter concludes by proposing ten key steps to deliver major scale investments in low carbon cities.

2 Conducting a City-Scale Mini-Stern Review

Our approach is based on a robust evaluation of the costs and benefits of different levels of decarbonization at the city-scale. We use UK Committee on Climate Change data on the potential energy, cost and carbon savings from thousands of low carbon measures. We take into account changes in the economy and the wider

Table 1

Stages in the Analysis
• A review of thousands of energy efficient and low carbon options– focusing on energy efficiency and small scale renewables.
• A realistic assessment of the costs and the energy (and hence cost and carbon) savings associated with each measure.
• A review of the scope for the deployment of these measures in households, commerce, transport and industry at the local level.
• Identification of the most cost and carbon effective options for different sectors.
• Mapping the capacity of the low carbon goods and services sector to absorb investment and create employment.
• The development of baseline forecasts taking into account background trends, price increases, electricity decarbonisation and up-take of energy efficiency and small scale renewable options.

energy infrastructure, but focus primarily on the potential for the wider deployment of energy efficiency measures and small-scale renewables.

Throughout the analysis we use realistic projections of the energy, cost and carbon savings emerging from different measures. Typical interest rates (8%) and energy prices are used and ambitious but realistic scenarios for the rate at which different technological and behavioral options are adopted. Projected savings are adjusted to take into account implementation gaps and rebound effects, and the scope for the adoption of different measures is adjusted to take into account hard to reach households and businesses.

The stages in the approach are set out in Table 1.

The key variables included in the analysis are presented in Table 2.

Based on a methodology that is explored in more detail in Gouldson et al. (2012), we then identify the different levels of decarbonization that could be achieved with different levels of investment, with a distinction drawn between three levels of investment:

- The cost effective level – this includes all of the measures that would more than pay for themselves over their lifetime.
- The cost neutral level – this includes all of the measures that could be afforded if the benefits from the cost effective measures were captured and reinvested in further low carbon options.

Table 2

List of variables
• Baseline trends
• Range of applicable low carbon measures
• Capital cost of each measure
• Operational costs of each measure
• Hidden and missing costs of each measure
• Energy savings per measure
• Financial savings per measure
• Carbon savings per measure
• Scope for deployment at the city scale
• Rate of deployment at the city scale
• Total costs and carbon savings
• Cost and carbon savings for different levels of investment, decarbonisation
• Employment and wider economic effects of different levels of investment, decarbonisation

- The realistic technical potential level – this includes all of the measures that could realistically be adopted, regardless of their cost effectiveness.

As a first case study to test the method, we apply it to the Leeds City Region (LCR). The LCR has a population of three million people, an economy worth £52 billion a year and an annual energy bill of £5.4 billion a year. The LCR is a fairly typical northern European city region in terms of its size, the composition of its building stock, the make-up of its economy and so on.

3 Case Study Results

3.1 The Potential for Carbon Reduction – Investments and Returns

We find that – compared to 1990 levels – the Leeds City Region could reduce its carbon emissions by 2022 via:

- 12.9% through cost effective investments that would pay for themselves (on commercial terms) over their lifetime. This would require an investment of £4.9 billion, generating annual savings of £1.2 billion, paying back the investment in 4.1 years but generating annual savings for the lifetime of the measures.

- 18% through cost neutral investments that could be paid for at no net cost to the Leeds City Region economy if the benefits from cost effective measures were captured and re-invested in further low carbon measures. This would require an investment of £11.6 billion, generating annual savings of £1.6 billion, paying back the investment in 7.3 years but generating annual savings for the lifetime of the measures.
- 18.8% with the exploitation of all of the realistic potential of the different measures. This would require an investment of £13 billion, generating annual savings of £1.7 billion, paying back the investment in 7.6 years but generating annual savings for the lifetime of the measures.

3.2 Impacts on Future Energy Bills

These figures are particularly significant in the context of projected energy price increases. We calculate that the 2011 Leeds City Region energy bill is £5.4 billion per year, but we forecast that this will grow to £7.2 billion by 2022 – a 33% increase in the LCR annual energy bill. With exploitation of the effective measures, this £1.8 billion increase in the annual energy bill could be cut by £1.2 billion. With exploitation of the cost neutral measures it could be cut by £1.6 billion, and if all of the realistic potential was exploited it could be cut by £1.7 billion. The Leeds City Region could therefore insulate itself against projected energy price increases to a very large extent through investments in energy efficiency and low carbon options.

3.3 The Wider Context – Other Influences on LCR Carbon Emissions

To put these energy savings and carbon reduction figures into a wider context, we find that:

- With other things constant, background trends in economic growth combined with changes in the energy and carbon intensity of GDP will lead to a 2% increase in LCR carbon emissions between 1990 and 2022.
- Higher energy prices will impact on demand, and this will lead to a 12% drop in LCR carbon emissions compared to the 1990 baseline by 2022. The total effect of the background trends plus the response to higher energy prices will be a 10% drop in LCR emissions between 1990 and 2022.
- The decarbonization of the national electricity system will lead to a 13% drop in LCR carbon emissions by 2022. The total effect of background trends, the impacts of price increases and the decarbonization of the national electricity supply system will be a 23% drop in LCR emissions between 1990 and 2022.
- The total effect of all of the above plus the exploitation of all of the cost effective low carbon options will be a 36% drop in LCR carbon emissions between 1990 and 2022.

- The total effect of all of the above plus the exploitation of the remaining cost neutral options will be a 41% drop in LCR emissions between 1990 and 2022.
- The total effect of all of the above plus the exploitation of all of the remaining realistic potential will be a 42% drop in LCR carbon emissions between 1990 and 2022.

The impacts of these price effects, grid decarbonization and cost effective, cost neutral and realistic potential are shown in Fig. 1.

3.4 Wider Impacts on Employment and Economic Growth

We also calculate that the levels of investment required to realize these reductions in energy bills and carbon footprints could have wider economic benefits within the Leeds City Region:

- Over the next 10 years, the levels of investment needed to exploit all cost effective measures with employment generating capacity would lead (directly and indirectly) to the generation of 4,443 jobs and to growth in GVA of £211 million per year.
- Over the next 10 years, the levels of investment needed to exploit all of the cost neutral measures with employment generating capacity would lead (directly and indirectly) to a further 5,226 jobs and to GVA growth of £230 million per year.
- In total therefore we predict that the levels of investment needed to exploit all of the cost effective and cost neutral measures with employment generating capacity would lead to the generation of 9,669 jobs over the next 10 years and to GVA growth of £442 million per year.

4 Discussion

The analysis highlights that within the Leeds City Region there is very considerable potential to reduce energy use and carbon footprints through cost effective and cost neutral investments on commercial terms. However, the fact that these opportunities exist on this scale is obviously not enough to ensure that they are actually exploited. Incentives – no matter how strong they are – have to be matched with appropriate capacities if progress is to be made.

These capacities relate both to the supply side (the capacity to secure and distribute major scale investments) and to the demand side (the willingness to absorb such investment and the ability to capture and recover the associated benefits). Considerable attention is currently being paid to these issues, but as a brief summary we identify ten key steps that my need to be taken if the considerable opportunities are to be exploited. These are set out in Table 3.

The mini-Stern review methodology set out above clearly helps to build understanding of the scale of the opportunity. This can be critical in developing political, business and public commitment to city-scale decarbonization.

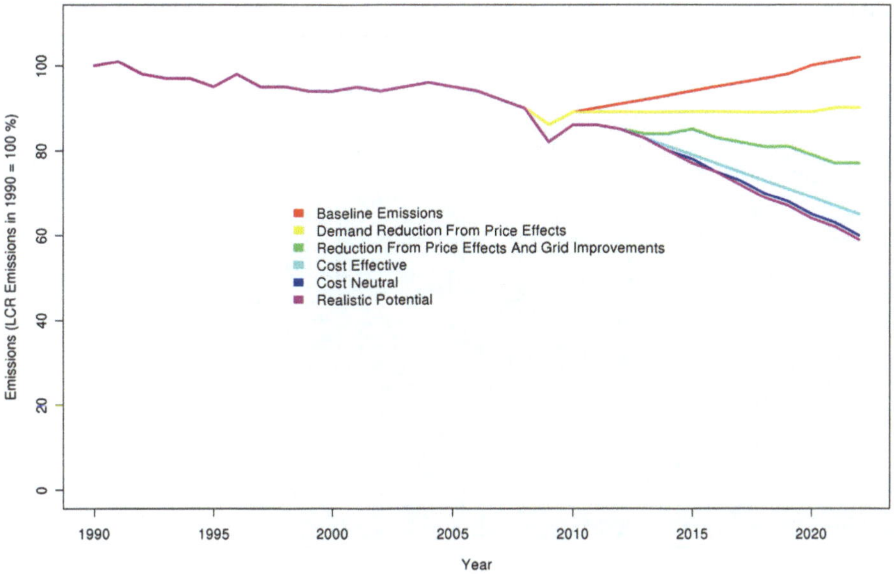

Fig. 1 Impacts on carbon footprint

However, the mini-Stern results are perhaps best seen as a prima facie case for major investments – more needs to be done to turn this prima facie case into action.

Clearly the results of a mini-Stern review can raise awareness, but they also help to reduce risk. By providing robust data on the performance of thousands of low carbon options, risks are reduced and action becomes increasingly possible. It is however important to note that the risks that could deter action come in a wider range of forms. There are political risks for key decision makers and financial risks for investors. There are risks that potentially significant start-up costs may not be recovered. There are risks that funds may be secured but that end-user demand is low, making it hard to distribute funds. There are risks that investments may not generate the returns predicted, or that recipients of investments may default on repayments. And there are legal risks for all involved. All of these risks need to be managed and mitigated if the potential is to be exploited.

One key way of doing this is to consolidate the opportunities for investment that are spread throughout the city into a single investable opportunity. Institutional investors prefer to invest at scale, and so opportunities in the tens or even hundreds of millions are much more likely to attract investment than smaller or more fragmented opportunities. Establishing a legal entity with the appropriate status and a capacity to receive, distribute, recover and pay-back major investments is one way of doing this.

To do this however, new business models may be needed. City-scale Energy Service Companies (ESCOs) could perhaps do this but based on a different business model than that applied in other ESCOs to date. For example, they could be

Table 3

Ten Steps to Deliver Major Change
• Understand the scale of the opportunity
• Increase awareness, reduce risk and uncertainty
• Consolidate the multiple options in to an opportunity at scale
• Design trusted and reliable delivery vehicles
• Develop new business models
• Draw in complementary finance
• Underwrite first phase investments
• Promote take - up
• Monitor, report, verify and learn
• Unlock second phase investments

established as a not-for-profit Community Interest Company with public and private sector support. Public-private partnerships seem to offer a more attractive combination of trust and capacity to deliver than either the public or private sectors could offer on their own. The not-for-profit element could also be key in ensuring that the most profitable opportunities are not `cherry picked', leaving the less attractive and harder to reach options behind for future entrants. Making a delivery vehicle not-for-profit and requiring it to invest in a blend of the most cost and the most carbon effective options may be one way of decarbonizing a city in the most effective and efficient way.

Complementary finance could also help to stimulate early stage investments from the private sector. Especially for the first-movers where the start-up costs and the risks are comparatively higher, government support for demonstration schemes could be key. Government could also stimulate complementary financing by introducing a legal obligation for energy companies to invest in energy efficiency measures, or by providing feed in tariffs for small scale renewables. In the UK for example, the introduction of the Energy Company Obligation (ECO) has led energy companies to provide money at the local level – if this is done at scale this can be an easier way of discharging a legal obligation than the energy company seeking to engage directly with thousands of households or businesses to promote reductions in energy demand.

Even with all of the above, it may still be necessary for a local authority to underwrite the first tranche of investment. In an age of austerity where local authorities are reluctant to take risks or to commit to major new initiatives, this requirement for underwriting can be a major barrier to change. If however a number of 'cornerstone clients' who would receive and then repay major investments in energy efficiency can be found, then the risks of under-writing can be drastically reduced. Such clients – who would be virtually guaranteed to repay any early investments made – could include the local authorities themselves, but also hospitals, universities and potentially also larger companies.

For a wider audience, it may still be necessary to promote take-up of the investments and of the various low carbon options that they might finance. There are numerous reasons why decision makers in households or businesses may be reluctant to participate in such schemes. Lack of awareness, lack of trust, short-termism, risk aversity and general inertia are amongst the key barriers that may have to be overcome if widespread take up is to be secured.

Where investments are secured and distributed there is a need to monitor the performance of each stage of the process. Although not without precedent, there are many new elements here and rapid learning from the first movers and the early identification and transfer of best practice is likely to be necessary. But – in economic speak – there are likely to be positive spill-overs from the first movers as they are likely to learn lessons that will be of great value to later movers. This is a classic market failure that provides a clear justification for some level of government support in pioneering demonstration schemes.

The major incentive for all of the above to happen is that private sector finance is likely to flow into city scale retrofits if the early stage investments prove to be successful. From a local authority perspective, under-writing early stage investments could easily lead to private sector flows of finance of ten times that amount if the first phase is successful. Stimulating this scale of investment at a time when inward investment is hard to secure can be a major incentive to city scale decision makers. Seeing those investments enhance competitiveness, create employment, reduce fuel poverty and cut carbon footprints are critically important added attractions.

5 Conclusions and Recommendations

The analysis presented here – based on the case study of the Leeds City Region – suggests that the economic returns on major scale investments in low carbon cities could be very significant indeed. Many of the measures would pay for themselves in a relatively short period of time, they would generate significant levels of employment and economic growth in the process, and if done well there may be a wider range of indirect benefits (not least from being a first mover in this field). The political and business case for very large investments in the low carbon economy is very strong indeed.

If these investments were made, the analysis suggests that cities such as the Leeds City Region could cut their carbon emissions by around 40% by 2022 at no net cost to their economies. Decarbonizing on this scale and at this rate should be possible. The technological and behavioral options are readily available, the energy and financial savings associated with these are clear (even based on conservative assessments), the investment criteria are commercially realistic if some key barriers can be overcome, and the deployment rates have been judged to be challenging but still realistic.

The transition depends however on political and social capital as well as financial capital. The levels of ambition, investment and activity needed to exploit the available potential are clearly very significant indeed. Enormous levels of investment are required, and major new initiatives are needed with widespread and sustained influence in the domestic, commercial and industrial sectors. And of course we need to think about some major innovations, particularly in stimulating the supply of and the demand for major investment resources. We need to think about innovative financing mechanisms, based on new forms of cost recovery and benefit sharing and new ways of managing risk. And we need to develop new delivery mechanisms that can stimulate and sustain demand for investment in low carbon options by overcoming the many potential barriers to change.

We should also recognize that economics is not the only discipline that has something useful to say on the transition to a low carbon economy/society. A wider analysis should also consider the social and political acceptability of the different options, as well as issues relating to the social equity and broader sustainability of the different pathways towards a low carbon economy and society. We also need to think about 'future proofing' investments to consider their compatibility with the more demanding targets for carbon reduction and with the different levels of climate change that are likely to come after 2022.

Reference

Gouldson A, Kerr N, Topi C, Dawkins E, Kuylenstierna J, Pearce R (2012) The economics of low carbon cities: a mini-Stern review for the Leeds City Region. Centre for Low Carbon Futures, Leeds. Available via: www.lowcarbonfutures.org

City Development Strategies and the Transition Towards a Green Urban Economy

Le-Yin Zhang

Abstract This chapter discusses the potential role of City Development Strategies (CDS) in accelerating the transition towards a green urban economy. It first examines different definitions, scopes and key issues of a green economy. It then addresses the question of how CDS could accelerate this transition by examining the challenges and potential of the green economy, the dynamics of the transition, and the nature of CDS. The chapter has three main arguments. First, although the green economy is expected to achieve a balance between economic growth, social equity and environmental sustainability, the driving force lies in the growth of green economic activities, which should therefore be prioritized. Second, the transition to the green urban economy is not a narrow technical change, but a change of a broad and all-encompassing socio-technical regime. Therefore, it requires a strong articulation of the vision, mobilization of resources, and coordinated actions. Third, the CDS has the potential to meet these requirements. However, to fulfill this role CDS's current practice has to become more effective in developing a shared vision and mobilizing support, and be more focused on strategies that develop the urban economy. Moreover, city managers have to become more entrepreneurial.

Keywords City development strategies • Transition • Green urban economy

1 Introduction

The global efforts to revive economic growth in the aftermath of the 2008 financial crisis has provided unprecedented impetus to the transition towards a green economy (Barbier 2010; Steiner 2010), a concept that was first popularized in the 1980s

L.-Y. Zhang (✉)
Development Planning Unit, University College London, London, UK
e-mail: le-yin.zhang@ucl.ac.uk

R. Simpson and M. Zimmermann (eds.), *The Economy of Green Cities: A World Compendium on the Green Urban Economy*, Local Sustainability 3,
DOI 10.1007/978-94-007-1969-9_21,

and early 1990s (Bahro 1984; Jacobs 1991; Pearce et al. 1989; Roberston 1983). While the UN Secretary General advocates it as an investment priority area for the fiscal stimulus packages implemented by several governments in 2009, the EU adopted a 10-year Green Growth Strategy in 2010. Furthermore, the green economy is slated to be one of the top two themes for the Rio+20 Forum in 2012. International organizations such as UNEP (2011), UNEP/ILO/IOE/ITUC (2008) and UN-Habitat (2011) have all added their voice to the promotion of this transition. The fact that cities are crucial domains for the green economy, accounting for up to 70% of global CO_2 emissions (UNEP 2011; UN-Habitat 2011) compels us to consider how such an economy can be developed in cities around the world. This chapter explores the potential role of city development strategies (CDSs) in facilitating this development.

This chapter proceeds as follows. It first examines different definitions of a green economy and their respective scope and key issues. The following part then addresses the question of how CDS can accelerate the transition by considering the natures of this transition and CDS. It concludes that the CDS approach as currently practiced needs to become more effective in mobilizing society-wide support and be more strategic in developing the relevant economic activities.

2 Defining the Green Urban Economy

A review of the literature shows that there exists no consensual definition of a green economy. Instead a range of definitions can be identified. The broadest identifies a green economy as essentially an antithesis of the conventional economic model with all its failings in economic, social, environmental terms (Bahro 1984; Jacobs 1991; Milani 2000; Roberston 1983, 1990). The following quotation from the early 1990s seems to be still relevant today:

> Much Green writing has been devoted to the goal of far-reaching social change. A Green society is envisaged as 'post-industrial'. In it people live in politically decentralized communities. Voluntary changes in values and lifestyles lead to a drastic reduction in material consumption, allowing a harmonious relationship to be developed between humankind and the natural world. Paid industrial employment correspondingly declines, as people find greater satisfaction in other forms of work, at home or in the community. A principal goal of economic policy is to increase self-reliance, particularly for local economies and 'bioregions' (Jacobs 1991, xvii).

Such a definition would seem to be impossibly broad, whose performance would be difficult to assess in the short or medium terms. More importantly, its feasibility is questionable, since industrialization remains to be an important strategy for development for many developing countries (Zhang 2011). Alternatively, one could look to the work that is commonly credited with having popularized the concept of a green economy (Pearce et al. 1989) for guidance. Unfortunately, however, that source offers no explicit definition of a green economy, but implicitly equates it with the one that delivers sustainable development, as defined by the World Commission on Environment and Development (WCED) (1987). This requires balancing the three pillars of long-term economic growth, social equity and environmental sustainability.

For a more analytic understanding, a recent conceptual note of Fulai Sheng (2008), a staff member with UNEP, provides helpful hints. He sees 'two faces of a green economy': one 'conventional' (equivalent to 'the greening of the economy') and the other 'modern'. In the conventional sense, "[a] green economy is typically understood as an economic system that is compatible with the natural environment, is environmentally friendly, is ecological, and for many groups, is also socially just." By contrast, in the modern sense, a green economy is "an economic system that is dominated by investing in, producing, trading, distributing, and consuming not only environmentally friendly but also environmentally enhancing products and services." The key difference here seems to be two-fold: the 'modern' definition is narrower and more focused on the underlying economic activities.

It is evident that the 'conventional' definition still dominates the broad policy agenda. In its *Towards a Green Economy* report, UNEP (2011) defines a green economy as one that results in "improved human well-being and social equity, while significantly reducing environmental risks and ecological scarcities. In its simplest expression, a green economy is low carbon, resource efficient, and socially inclusive." Indeed, the report argues quite persuasively that addressing social inequity and poverty reduction is an integral part of the green economy agenda and sustainable development agenda. The *Green Jobs* report, published jointly by UNEP/ILO/IOE/ITU (2008), states that "[a] green economy is an economy that values nature and people and creates decent, well-paying jobs." (p. 4).

However, in policy practices in the developed countries where there is greater awareness of the green economy, it appears to be principally associated with actions or policy measures addressing climate change. More specifically, the focus is on the development of clean energy and the improvement of energy efficiency, although wider concerns are also addressed. A study of 25 US regional and national reports on the green economy finds that, although all the reports place clean energy at the core of the green economy, there is significant variation in how much emphasis is placed on other environmental issues or/and job quality: 16 of the reports also covered transportation and infrastructure, and only three dealt with job quality (Chapple 2008).

To summarize, at least three quite distinct definitions of a green economy can be found. While the broadest encompasses the three pillars of sustainable development, the narrowest and the most instrumental is mainly concerned with green energy and energy efficiency. The intermediate definition covers economic activities that are environmentally friendly and enhancing. It is worth noting that the broader the definition, the harder it becomes to prioritize in policy action and in evaluating the outcome. One possible way of overcoming this problem is to start with a narrower definition and broaden it over time.

3 Operationalizing the Green Urban Economy

Broadly speaking, the fundamental issues of developing a green economy have not changed since the 1980s: they concern the tensions, and their resolution, between the three pillars of sustainable development:

1. Between the social and the economic: the pursuit of growth needs to be balanced with intra-generational equity to meet basic needs and strengthen livelihood;
2. Between the social and the environmental: while the preservation of the environment promotes inter-generational equity, values/culture (in the social domain) can influence how environmental issues are pursued;
3. Between the economic and the environmental: while internalizing environmental costs and benefits improves economic decisions and performance in the long-run, activities in the two domains affect each other, more often than not, in conflict (Munasinghe and Swart 2005).

It is self-evident that, to reduce or resolve these tensions, synergies or co-benefits must be sought and maximized. Under the Agenda 21, which has been widely adopted by local governments around the world, awareness of sustainable development in terms of the balance between the three pillars has been raised. However, actual results are highly uneven and in many cases rather limited. It seems that policies and programs of sustainable development have become politicized and have failed to move beyond a statement of commitment. The critical question of how to resolve the tensions and achieve the required balance has received insufficient attention. Moreover, there is a lack of understanding that this balance is likely to grow out of a process of unbalanced developments.

Regarding this issue, useful insights can be drawn from the literature on economic development. Contributing to the debate between the theories of balanced growth and unbalanced growth, Hirschman (1968) argued that balanced growth, especially in the earlier phase of development, is untenable because of the lack of the capacity to pursue all activities simultaneously; nor is it desirable because the cost of undertaking certain activities would decline once other investments have been made. In his view, the most important cause of underdevelopment is the lack of a 'binding agent', who is able to identify opportunities, combine necessary resources/conditions, make and carry out investment decisions. Such an agent is characterized by a 'growth perspective', comprising 'not only the desire for economic growth but also the perception of the essential nature of the road leading toward it.' (p. 10). He suggested that development theory and policy's task is to examine under what conditions development decisions can be called forth through 'pacing devices' or 'inducement mechanisms'. On the other hand, it has been argued (Chang 2003a) that 'good institutions' such as equity have historically been the consequence, rather than the cause, of economic development. Therefore it is necessary to prioritize economic development.

We believe that the green economy is a great development opportunity for many cities, with real potential to contribute to the resolution of the aforementioned tensions.

However, to achieve this potential, it is necessary to prioritize the growth of green economic activities, as such growth will be good to the poor (more on this below) and the environment. For this purpose, city development strategies (CDSs) should embody the spirit of the 'binding agent' and help to provide 'pacing devices' and 'inducement mechanisms' for this growth.

3.1 The Challenges and Potential of the Green Economy

In the transition to the green economy, the stake is high and the cost of inaction will be heavy. As the UNEP/ILO/IOE/ITUC (2008, p. 4) report observes, "[C]ompanies and regions that become leaders in green innovation, design, and technology development are more likely to retain and create new green jobs. But workers and communities dependent on mining, fossil fuels, and smokestack industries—or on companies that are slow to rise to the environmental challenge—will confront a substantial challenge to diversify their economies."

The green economy even in the limited sense has huge potential for two reasons. First, it is dynamic and fast growing so that it is capable of driving the economy forward. Estimates show that the global market volume for environmental products and services currently runs to about US$1,370 billion, with a projected US$2,740 billion by 2020. On the other hand, it's estimated that worldwide employment in alternative energy would soar—possibly as high as 2.1 million in wind energy and 6.3 million in solar photovoltaics (PVs) by 2030, and in the order of 12 million jobs in biofuels-related agriculture and industry. In particular, tremendous job growth is expected in installations and maintenance of solar PV and solar thermal systems (ibid, Executive Summary).

Second, the green economy covers many economic activities at different levels of labor and technology intensity so that it offers a variety of opportunities to people of different skill levels and economies at different stages of development. A review of 25 recent studies on the green economy (Chapple 2008) finds a huge range of relevant activities, covering both production and consumption (see Fig. 1). These range from energy and utilities, green building, transportation, clean tech manufacturing and other green manufacturing, clean tech R&D, waste management, chemistry and materials, environmental services to green financial services and eco-tourism.

Although the majority of current employment opportunities in renewables are in developed countries (more than 2.3 million according to incomplete counting), UNEP/ILO/IOE/ITUC (2008) estimate that 1.39 million are in three developing countries (China, India and Brazil). Replacing traditional cook stoves with recently developed biomass cooking technologies for nine million households in India alone would create 150,000 jobs, which requires only a few months' training (p. 10).

But how can cities make this transition and exploit these opportunities? There are many ways of doing so. For instance, out of the nine types of policy instruments identified by the *Green Jobs* report, ranging from subsidies to eco-labeling and R&D budgets, a majority of these are available to the city government. However, the

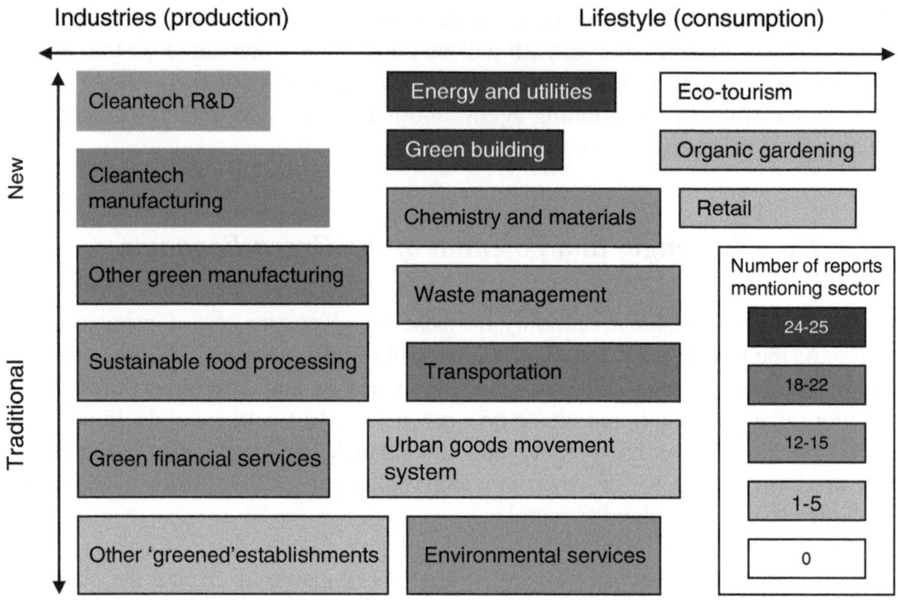

Fig. 1 Sectors of the green economy according to recent studies (Adapted from Chapple 2008)

crucial challenges are two-fold: to develop and articulate the right vision on the one hand; and to make real progress in developing selected green industries on the other hand. This combination would ensure that the city will not only have a stake in the future, but also receive immediate benefits from improved productivity (thus income) and job creation. This is supported by the literature on the transition.

3.2 *Dynamics of the Transition*

Past studies have significantly advanced our understanding of the dynamics of the transition. While earlier writings focused on the role of environmental economics and the role of technological innovation (the so-called "techno-economic paradigm"), research since the turn of the century has adopted a "socio-technical" perspective, which defines transitions as 'changes from one sociotechnical regime to another' (Geels and Schot 2007, p. 399). According to Smith et al. (2005, p. 1493), "[S] ocio-technical regimes are relatively stable configuration of institutions, techniques and artifacts, as well as rules, practices and networks that determine the 'normal' development and use of technologies". The techno-economic paradigm assumes that decarbonization of the energy system will be brought about by technological breakthroughs in other energy technology fields, thus focusing on the question of how technological novelties emerge and 'break through' (therefore giving emphasis to 'innovation' and 'diffusion'). By contrast, the 'socio-technical system' approach thinks more holistically about the interrelationships and interpenetration of the social, technical, political, economic and institutional spheres, symbolized by the 'seamless web' metaphor proposed by Hughes in 1986 (Maassen 2009).

Furthermore, it has been suggested that regime transformation is a function of three factors (Smith et al. 2005):

1. The degree to which the selection pressure is articulated towards a particular problem or direction of transformation on the part of regime members (prerequisite).
2. The degree to which the resources required are available either within or beyond regime members.
3. The extent to which responses to these pressures are coordinated across regime members.

The implication is that to effect the transition towards a green urban economy, it is essential to ensure that the selection pressure (for the green economy) is strongly articulated, resources mobilized and the responses coordinated. These accord well with the nature and scope of CDS formulation and implementation, at least in theory.

3.3 The Nature and Practice of CDS

CDS as a policy tool first emerged in the developed world in order to deal with the aftermath of the severe economic recessions in the 1970s and the deindustrialization following it. This history has also brought with it an emphasis on consensual and multi-stakeholder development of strategies (Harris 2002; Parnell and Robinson 2006). However, many CDSs have been undertaken in developing countries since the late 1990s under the tutelage of the World Bank and the Cities Alliance, a partnership involving a dozen of bilateral and multi-lateral development agencies. The Cities Alliance's 2010 *Annual Report* (Cities Alliance 2010) shows that during 2000–2010, a total of 247 projects were funded. This includes CDSs, which constitute one of the two main programs within the Alliance, as well as other kinds of projects.

According to the Cities Alliance website, "A city development strategy is defined as an action-oriented process, developed and sustained through participation, to promote equitable growth in cities and their surrounding regions to improve the quality of life for all citizens. A CDS helps cities integrate a strategic development approach and a long-term perspective into their urban planning. With a CDS, cities move beyond planning around the short-term political or donor-funding cycle to considering where they should be in 20 or 30 years, and the steps that need to be taken to achieve those goals." Similarly, the Asian Development Bank (ADB) (2004) writes: "A city development strategy combines strategic and corporate planning for a unit, that is, a city. It represents the statement of rules for making decisions on the development of a city and outlines what is expected, who is responsible for implementation, and what resources are needed to achieve the goals." (p. 6) As such, CDS appears to be well suited to play the triple roles required by the transition in terms of articulation of the vision, mobilization of resources and coordination of actions across regime members.

However, to realize this potential, the current CDS practice has to avoid two kinds of problems that have often thwarted its purpose. On the one hand, although overriding emphasis has been placed on participation, full and meaningful participation has proved to be difficult to achieve, even in the most conducive environment. For instance, in the post-apartheid Johannesburg, where the ruling ANC was strongly committed to the democratization of policy making and planning, both the labor unions and the business community dropped out of this process due to their dissatisfaction with the proposed changes (Parnell and Robinson 2006). The implication is that better mechanisms need to be found to encourage meaningful and broad-based participation on the one hand, and to compensate those who lose out from the changes. The latter is necessary in any process of major industrial restructuring (Chang 2003b), and can only be done with additional resources generated from significant growth.

Second, the CDS process has to revert to its former focus on long-term economic development strategies. Unfortunately, as Harris (2002) commented, "[i]n many, the aspiration for an economic development strategy for the city has been lost, so that the policy of helping the poor can mean little more than assessing what benefits the city can deliver to the poor, the passive recipients of municipal largesse, not how to lift the whole labor market and so the incomes of the poor, to empower them through work" (pp. 2–3). The CDS for the green economy must overcome this paternalistic streak and seek to help its population to exploit the real potential of green jobs.

4 Conclusion

To serve the purpose of building a green economy, the CDS needs to be explicit about the challenges of this new economy and fully recognize the global, regional and national contexts within which this challenge has arisen. It must be recognized that realizing this transition is not just about clean energy innovation or transfer of clean energy technologies. It is the development of a new sociotechnical regime, involving all spheres of the system. Therefore, the development and articulation of the vision must be embedded in wide societal support.

On the other hand, it should be recognized that it would be necessary to prioritize the growth of green economic activities, not the least because of resource limitation and the need to exploit optimal development sequences. Priorities must be given to the decarbonization of the energy system and improvement of energy efficiency, and to the exploitation of green job opportunities. The specific activities involved obviously would depend on the comparative and competitive advantages of the city.

Finally, there is the agency issue. It was more than 20 years ago when David Harvey (1989) wrote about the transformation from 'managerialism' to 'entrepreneurialism' in urban governance in the developed world. This was made necessary, he argued, by global and national contexts such as globalization, decentralization and deindustrialization. Today, the same (or similar) forces are increasingly affecting all cities. Therefore, to be a successful builder of a green economy, the city managers must become more entrepreneurial.

References

Asian Development Bank (ADB) (2004) City development strategies to reduce poverty. ADB, Manila
Bahro R (1984) From red to green. Verso, London
Barbier EB (2010) A global green new deal: rethinking the economic recovery. Cambridge University Press/UNEP, Cambridge
Chang HJ (2003a) Kicking away the ladder: development strategy in historical perspective. Anthem Press, London
Chang HJ (2003b) Globalization, globalization, economic development and the role of the state. Zed Books, London
Chapple K (2008) Defining the green economy: a primer on green economic development. The Center for Community Innovation, Berkeley
Cities Alliance (2010) 2010 Annual Report: an alliance in transition. Cities Alliance, Washington, DC
Geels FW, Schot J (2007) Typology of sociotechnical transition pathways. Res Policy 36:399–417
Harris N (2002) Cities as economic development tools. Urban Brief, Woodrow Wilson International Center for Scholars, December, 4 p
Harvey D (1989) From managerialism to entrepreneurialism: the transformation in urban governance in late capitalism. Geogr Ann 71B(1):3–17
Hirschman AO (1968) The strategy of economic development, 1958th edn. Yale University Press, New Haven/London
Hughes TP (1986) The seamless web-technology, science, etcetera, etcetera. Soc Stud Sci 16(2):281–292
Jacobs M (1991) The green economy. Pluto Press, London/Concord
Maassen A (2009) Social science perspectives on energy transitions. Proceedings of the ICE-Energy 162(4): 161–167
Milani B (2000) Designing the green economy. Rowman & Littlefield Publishers, Lanham/Oxford
Munasinghe M, Swart R (2005) Primer on climate change and sustainable development: facts, policy analysis and applications. Cambridge University Press, Cambridge
Parnell S, Robinson J (2006) Development and urban policy: Johannesburg's city development strategy. Urban Stud 43(2):337–355
Pearce D, Markandya A, Barbier EB (1989) Blueprint for a green economy. Earthscan, London
Roberston J (1983) The sane alternative. James Roberston, Ironbridge
Roberston J (1990) Future wealth. Cassell, London
Sheng F (2008) A green economy: conceptual issues. UNEP, Nairobi
Smith A, Stirling A, Berkhout F (2005) The governance of sustainable socio-technical transitions. Res Policy 34:1491–1510
Steiner A (2010) Policy statement at the opening of the Governing Council/Global Ministerial Environment Forum at its eleventh special session, Bali, Indonesia, 24 February 2010. UNEP, Nairobi
UNEP (2011) Towards a green economy: pathways to sustainable development and poverty eradication. UNEP, Nairobi
UN-Habitat (2011) Global report on human settlements 2011: cities and climate change. Earthscan, London/Washington, DC

UNEP/ILO/IOE/ITUC (2008) Green jobs: towards decent work in a sustainable, low-carbon world. UNEP, Nairobi

World Commission on Environment and Development (WCED) (1987) Our common future. Oxford University Press, Oxford

Zhang L-Y (2011) Is industrialization still a viable development strategy for developing countries under climate change? Climate Policy 11(4):1159–1176

Low Carbon Enterprise Zones: Towards a Fossil Fuel Free City Economy

Philip Monaghan

Abstract City-led 'low carbon enterprise zones' are rapidly on the rise across the globe – from Canada (energy efficient lighting in Toronto) and China (solar energy in Baoding) through to Spain (water efficiency in Zaragoza) and the UK (marine energy in Liverpool). For political leaders these designated areas of green growth are a silver bullet solution to the twin perils of unsuccessful climate deal negotiations and surviving an economic depression that lingers in many parts of the world. Whilst such zoning should 'in principle' be welcomed, it is vital that local leaders devise a suitable framework to govern such schemes. This chapter offers a critical analysis of the latest insights from this fast emerging practice and puts forward new thinking to assist cities in their decision-making process, and ultimately to make our societies more resilient. For instance, who owns the intellectual property rights for co-produced schemes that go to commercial success? Are the technologies involved in a genuine move toward renewable energy? Does it alleviate poverty and local unemployment? Is this low carbon trade displacing high carbon trade? Are their innovative ways to finance the setup of these zones in the current economic climate?

Keywords Carbon • Economic competitiveness • Governance • Jobs • Land use planning

1 Introduction

According to the US Pew Environment Group, 2010 was a landmark year for global clean technology investment reaching a record US$246 billion. This was also the first time that investment in renewable energy overtook nuclear power (Harvey 2011).

P. Monaghan (✉)
Infrangilis, NS 406 Royal Mills, 2 Cotton Street, Manchester, M45BD, UK
e-mail: philipmonaghan@infrangilis.org

R. Simpson and M. Zimmermann (eds.), *The Economy of Green Cities: A World Compendium on the Green Urban Economy*, Local Sustainability 3,
DOI 10.1007/978-94-007-1969-9_22, © Springer Science+Business Media Dordrecht 2013

The race for new money and jobs in the so-called 'green economy' has come of age, as evidenced by China's latest 5-year plan (Watts 2010). China's 2011 economic blueprint outlined major initiatives to green the economy over the coming half decade including: energy efficiency and environmental services being declared 'priority industries' for the first time; a carbon intensity target being set, measuring the ratio of greenhouse gas emissions relative to Gross Domestic Product (GDP); a new environmental tax on heavy polluters to levy fees on discharges of sulfur dioxide; and a mandatory carbon trading system on a regional level.

The timing of this dramatic growth in clean technology markets was even more interesting given that the major urban centers, which would be the focus of many of these clean technologies, were simultaneously facing savage public spending cuts. Consequently, municipal authorities are looking to the private sector to fill this gap (Monaghan 2012). At the same time, there were rumblings from the UN that they were abandoning hope that a single deal on climate change could soon be reached – instead switching to proposals for a green economy on the basis that this was much more politically palatable (ENDS 2010; Goldenberg 2011).

Given this context, how are city leaders around the world attempting to capitalize on the opportunity to transform local economies from underperforming fossil fuel dependent manufacturing to more competitive low carbon sectors? One increasingly popular policy intervention is to develop a so-called 'Low Carbon Enterprise Zone' (LCEZ).

2 Clear Market Signals and Controls

The past 3 years or so has seen a remarkable rise in interest amongst local authorities across the world in the concept and practice of the so-called LCEZs (or an equivalent name, noting they may actually use a different term in their respective language while displaying similar characteristics). These zones are special economic trade areas in cities and towns within which designated clusters of low carbon industries are nurtured. In doing so, these cities are building the brand of their districts by providing a clear market signal that they are 'open for business' in the clean technology industry, whilst at the same time implementing controls that reduce the incidence of fossil fuel dependent activities. The motivations go beyond the traditional notion of greening local government to the core business of building local capabilities to attract new inward investment, regenerating areas by creating jobs and re-skilling the local workforce, and enhancing local resiliency by providing greater energy or water security. Particularly noteworthy schemes include solar energy in Baoding (China), marine energy in Liverpool (UK), energy efficient lighting in Toronto (Canada) and water efficiency in Zaragoza (Spain), as detailed in Table 1 below.

Whilst some like in Zaragoza date back 15 years and others such as in Liverpool are much more recent, it is interesting to compare and contrast their approaches in innovating and delivering success.

Table 1 Examples of LCEZs from around the world

City	Description
Baoding (China)	As part of a national to move away from heavy manufacturing to new hi-tech markets, the municipality introduced an industrial energy saving policy and regulations for businesses, which was complimented by financial incentives support for solar panel production. Baoding now has 20,000 people working in clean energy (AccountAbility and UNEP 2010)
Liverpool (UK)	A public-private alliance named The Mersey Partnership is working to develop a center of excellence in marine energy that will assist 8,000 people and especially the youth unemployed being retrained and employed in wind turbine and tidal industries over next half decade. Whilst blades tend to be produced overseas, they are assembled and maintained locally (e.g. under water cabling to connect the offshore turbines to the national electricity grid) (Regeneris and Quantum 2009)
Toronto (Canada)	The city has developed an ICT cluster to enable it to compete with Asia for a share of the US$1 billion light emitting diode (LED) technologies market (e.g. artificial street lighting can account for 13 % of local council energy bills in the UK) (AccountAbility and UNEP 2010; Monaghan 2010)
Zaragoza (Spain)	Following a demonstration project on water saving to deal with water scarcity problems that originated in 1997, a cluster on water efficiency was recently setup to allow Zaragoza to trade in the US$400 billion water market (which has an annual growth rate of 7 %). The civil society organization ECODES is helping to facilitate knowledge transfer between participants and other interested parties including demand side development through large company supply chains and corporate social responsibility programs (Zinnae 2011)

3 Cloning of Zoning or Different Approaches?

Taking the four examples listed above, one can identify a number of common and unique characteristics in these LCEZs (see Table 2 below). The most notable common feature is that they are seeking to gain a competitive advantage in low carbon markets through clusters of business capabilities and academic research. However, it is interesting to note that in some countries such as the UK, the phenomenon was originally instigated by the central government and required local government applying for ministerial designation (whereby a city or region submits a competitive bid to national government to be designated a LCEZ). The practice has since evolved to municipalities like Liverpool now also self-designating LCEZs through their own local planning frameworks.

It is also important to appreciate that seizing an economic opportunity may not have originally been their primary driver, as depicted in Table 3 below. So for instance, in the case of Zaragoza at the start it was dealing with the environmental

Table 2 Common and unique features of LCEZs

Common	Unique
• ***Competitiveness through clustering*** – particular focus on SMEs	• ***Motivation to act*** – regenerate poor neighborhoods versus energy security or water scarcity
• ***Spatial planning*** – geographically bounded areas organized by planning authority so clear market signal and less disruption	• ***Designation*** – instigated by national competition versus local self-declaration
• ***Natural resources*** – defined by utilization of local ecosystem services (e.g. the Sun, offshore wind etc.)	• ***Governance*** – multi stakeholder versus public sector only, mandatory arrangements versus voluntary alliance, with varying local constitutional powers
• ***Cross-departmental action*** – economists, accountants, planners and environmental professionals	• ***Financing*** – central grant versus local reprioritization of existing resources
• ***Barriers to change*** – officer or elected member awareness, and local opposition to certain technologies	• ***Process innovations*** – imposing a carbon compensation levy on all developers versus acting as a vocational courses broker

© Monaghan (2012)

dilemma of water shortages. The economic opportunity to trade in the US$400 billion water market was discovered later on during the course of that particular learning journey.

Yet, when it comes to these LCEZs and the green economy more generally, how are these city leaders ensuring they understand what they will gain and what they will lose as part of this great transition?

4 Winners and Losers – A Checklist to Aid City Leaders' Strategic Decisions

Whilst such zoning should broadly be welcomed, it is vital that local leaders devise a suitable model to govern such schemes. With this in mind, listed in Fig. 1 below is a set of considerations to guide cities in their decision-making process (adapted from Monaghan 2012).

For instance, if there is profit to be made from co-produced ideas (as was the case with Zaragoza) how is the local council ensuring it shares the rewards? Are the technologies involved a genuine move toward renewable energy or are they simply less polluting (for instance, incineration of non-biological domestic waste is not)? Does it alleviate poverty and local unemployment? Is this low carbon trade displacing high carbon trade and if so, what are the synergies with national growth strategies (which was a consideration that shaped Baoding's plans)? Are their innovative ways to finance the setup of these zones in the current economic climate (for instance

Table 3 Initial drivers for establishing LCEZs

	Economic			Social			Environmental		
City cluster	National growth	Regional competitiveness	Local investment	Place-making	Youth unemployment	Population migration	Energy security	Water management	Low emissions
Baoding (China)	×								
Liverpool (UK)					×				
Toronto (Canada)			×						
Zaragoza (Spain)								×	

© Monaghan (2012)

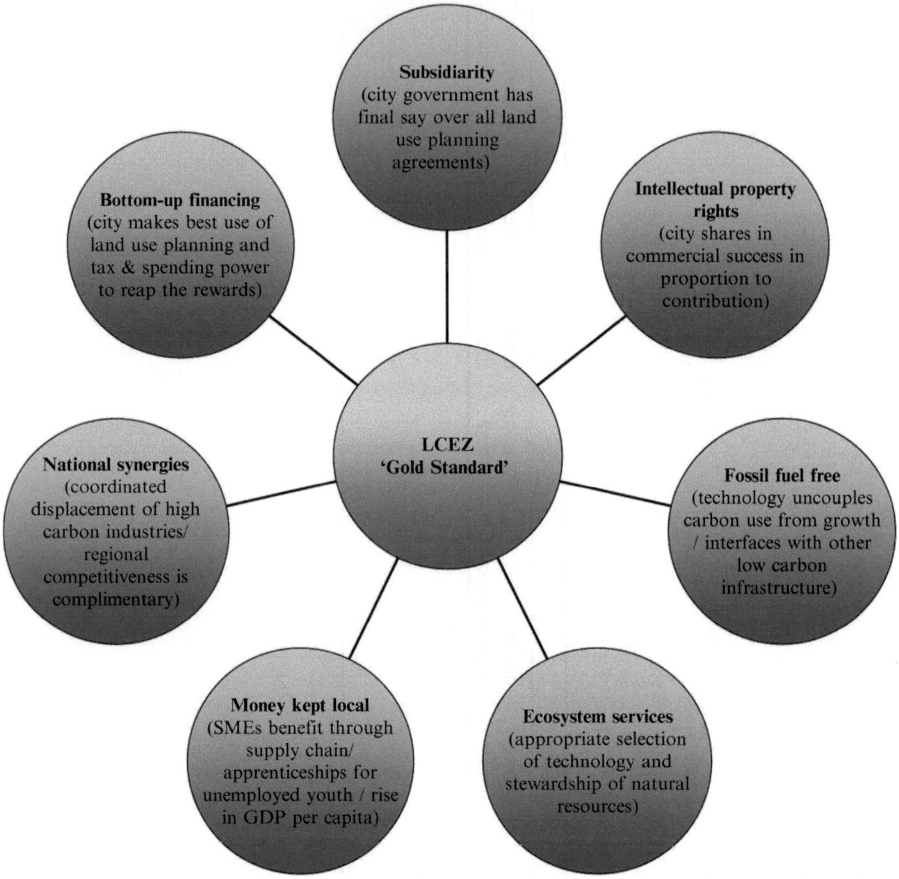

(© Monaghan, 2012)

Fig. 1 A framework for governing LCEZs

districts in Liverpool imposing a carbon compensation levy on developers to fund decentralized renewable energy)?

Clearly not all local councils will be able to transfer learning from the examples of Baoding, Liverpool, Toronto, and Zaragoza. City leaders' ability to act will be shaped by their particular context in terms of constitutional powers, the strength of the balance sheet or political appetite for change. Whilst many local authorities will have responsibilities for attracting inward investment and supporting the maintenance and creation of jobs as well as the up-skilling of a workforce, this is not universally true. Local councils in Northern Ireland, like the city of Belfast for instance, do not have responsibility for urban regeneration. At the other end of the spectrum as outlined in Box 1, California (USA) and Tokyo (Japan) are both able and willing to re-skill the local workforce and enforce low emissions sanctions on business respectively.

Box 1 Making the Best Use of Local Powers in California and Tokyo

To help stimulate the energy efficient lighting industry, the state of California has collaborated with utility companies, trade unions and electrical contractors association to develop 'journey upgrade' training for installers (McNeil and Thomas 2011). Targeting training to electricians (who must be licensed in California) and limiting participation in incentive programs to those who have achieved a standard of training has enabled contractors in the advanced light controls industry to compete on the basis of quality for new construction and retrofit developments.

Local councils in Japan have the authority to take legislative action when the national government itself has not enacted specific policies and measures towards climate change (LGA 2009). As a result some have mandated certain industries to formulate CO_2 reduction plans, amongst other measures such as local emissions trading and the buying of renewable energy bonds. For instance, Tokyo Metropolitan requires companies to formulate emission reduction plans and for retailers to place energy performance labels on electrical equipment.

5 Conclusion – Game Changer or Lemming Run for Green Growth?

LCEZs should 'in principle' be warmly applauded. They represent a fantastic opportunity to make our cities more resilient by uncoupling growth from carbon. Particularly like those springing up in cities such as Liverpool with their strong emphasis on apprenticeships and re-skilling of local unemployed youths for jobs in clean energy. This type of zoning is a process innovation that is challenging the way we think about the interface between the green economy and sustainable urban development. But other titans of business and city planners alike, who see this as some kind of snake oil remedy for their economic ills, need to think very carefully about exactly who benefits from all of this. This should be in terms of both struggling societies right now and tomorrow's generation. Good governance by city officials is vital to realizing the 'gold standard' when it comes to LCEZs.

References

AccountAbility and UNEP (2010) Climate change competitiveness index. AccountAbility, London

ENDS (2010) Cancun progress disguises post-Kyoto tensions. ENDS Report 431, p 6

Goldenberg J (2011) Too little, too late: UN chief rejects summits and seeks new way to beat global warming. The Guardian (28 January 2011: 31)

Harvey F (2011) UK left behind in race to in the green economy, Says Report. The Guardian, 29 March, p 15

LGA (Local Government Association) (2009) The climate challenge: local solutions through global learning. LGA, London

McNeil C, Thomas H (2011) Green expectations: lessons from the US green jobs market. IPPR, London

Monaghan P (2010) Sustainability in austerity: how local government can deliver during times of crisis. Greenleaf, Sheffield

Monaghan P (2012) How local resilience creates sustainable societies: hard to make, hard to break. Earthscan, London

Regeneris and Quantum (2009) The economic impact of EU and UK climate change legislation on Liverpool and Liverpool city region. Regeneris, Liverpool

Watts J (2010) China counts £130bn cost of economic growth. The Guardian, 13 May, p 7

Zinnae (2011) Urban cluster for the efficient use of water. Available via: http://www.zinnae.org/en/who-are-we/presentation. Cited 31 Dec 2011

Green Clusters and the Entrepreneurial Local Government: Portland's Economic Development Strategy

Shay Kelleher and Art von Lehe

Abstract This chapter analyzes the role of the local government in green cluster formations and city development strategies for the transition to a Green Urban Economy. Firstly, it examines the fundamentals of the urban economy including agglomeration economies but more specifically cluster formations as key aspects of an urban economy's competitiveness. It then applies these concepts to the development of green clusters as key growth sectors in the urban economy whereby cities can avail of the new business opportunities that become available as a result of increasing investments in green products and services, environmental technologies etc. Through this transition it is envisaged that a gradual shift from carbon intensive activities in favor of more environmentally sustainable economic activities will gradually come to characterize urban economies. This will ensure that cities remain competitive in the global economy by capturing market share in smart and green growth sectors. Furthermore, this chapter assesses the role of entrepreneurial local governments in creating the necessary conditions to foster green cluster development and attract green and smart growth sector investment, particularly through City Development Strategies. Finally, this chapter analyzes Portland's Economic Development Strategy as an example of how local governments can lead urban development strategies, and pursue sustainability and green cluster development as a key economic growth strategy.

Keywords Green cluster • Portland • Entrepreneurial local governance • Job creation • City development strategies

S. Kelleher (✉)
ICLEI – Local Governments for Sustainability, Bonn, Germany
e-mail: urban.research@iclei

A. von Lehe
ICLEI Oakland, California, USA
e-mail: urban.research@iclei

R. Simpson and M. Zimmermann (eds.), *The Economy of Green Cities: A World Compendium on the Green Urban Economy*, Local Sustainability 3,
DOI 10.1007/978-94-007-1969-9_23, © Springer Science+Business Media Dordrecht 2013

1 Introduction

Various economic development strategies including the concept of clusters have informed policy in urban governance and urban and regional economics for a considerable amount of time and a substantial amount of literature exists on the subject of cluster formations and agglomeration economies (see Porter 1998; Ketels 2004). This builds upon conventional theories of competiveness, economics of location (see Krugman 1991; Duffy 1995) and urban policy, particularly in the context of globalization (see Sassen 2005; Friedmann 2005). In essence, cities and regions have long competed to capture capital and investment whether in traditional manufacturing or more recently through the growth of the knowledge economy and financial services sector in which Foreign Direct Investment (FDI) has played a key role. However, urban and regional competiveness also involves the development of innovative endogenous enterprises where local innovation is an important aspect. This results in the establishment of a robust economic specialization element in cities.

Since the 2008–2009 financial crises, the term Green Economy has infiltrated both national and international economic policy discourse, as governments try to devise new ways to replenish and instigate new growth sectors in their economies in order to ensure economic viability and enhance competiveness, particularly in a developed world context (see UNEP 2011; OECD 2011). Urban areas account for the majority of their respective countries GDP[1] and for the Green Economy to be successful globally it needs to be established within our urban centers. Economic growth is fostered regionally and urban centers and metro regions are key actors in regional collaboration for economic development.[2] Cities provide considerable opportunities for a transition to the Green Economy due to factors of urban agglomeration (spillover effects, concentration of knowledge and innovation etc.) and the general structural potential of cities (see UNEP 2011; Hammer et al. 2011). However, the term Green Urban Economy has not yet been conceptualized with some notable exceptions (see ICLEI 2011, this publication).

Greening the urban economy is an important prerequisite in the transition to a global Green Economy as well as an important aspect to sustainable development. Understanding urban economic concepts of cluster formations, agglomeration economies and specialization development and policy innovation. Specific city development strategies are a growing thematic policy innovation in urban governance (see Cities Alliance 2006). By aiming to promote sustainable economic development initiatives such strategies are an essential tool in the application of green economic concepts to the urban economy, particularly in the terms of green cluster formations. Cities that specialize in green services and products, environmental technologies etc. can establish the necessary credentials for a green cluster.

[1] Half of global GDP (US$ 30,000 trillion) in 2007 came from 380 cities in developed-regions, 220 largest cities in developing-regions contributed another 10% (see McKinsey Global Institute 2011).

[2] 'Polycentricity' is also considered to be a new spatial phenomenon which are defining features of the city region of the twenty-first century (see Hall and Pain 2006).

This creates a specialized Green Urban Economy that is diverse in terms of green sector innovation and smart growth sectors, essential for maintaining economic viability. The Portland case, which will be examined in due course, is an illustrative example of how a city can pursue an economic strategy that favors sustainability and the development of green and clean tech clusters while simultaneously maintaining the city's economic edge.

2 Green Clusters

Urban green clusters can be created in two ways. Firstly, by competing for investment including FDI or through home grown endogenous business ventures where local innovation and research and development (R&D) is pivotal to the establishment of smart green clusters; it can also be a combination of both. However, any such cluster development is dependent on the establishment of the necessary conditions required for investment or innovation. Urban centers offer a favorable location for green clusters due to the presence of a highly diverse economic mix linked with their capacity to concentrate firms, R&D activities and innovation (therefore benefitting from agglomeration economies) (see UNEP 2011). Urban agglomerations and metro regions also tend to have greater endowments of human capital, a larger stock of physical capital measured by the equipment of firms and the stock of buildings and infrastructure facilities. (see Turok 2004; OECD 2006; UN-HABITAT 2008; Legates and Stout 2003). Therefore the agglomeration, structural, and synergetic potential offered by cities is an excellent opportunity for the manifestation of a Green Economy via green cluster formations.

Green clusters aim to capitalize on the new business opportunities that exist in terms of a growing market share for green products and services, due to increased investments in green infrastructure, clean technology, amongst others. Sources indicate that up to US$ 3.3 trillion has been invested in green economic activities since 2007, and this trend is set to continue ($10 trillion in investments by 2020) (see Ethical Markets Media 2012). New business models, which alter the production side of the economy to more environmentally favorable products and services, with the intention to instigate a gradual shift from carbon intensive economic activities are gradually gaining momentum. Such new industries include environmental technologies (clean tech), green products and services, renewable energy innovation etc.

Green clusters are an important shift on the production side of the green economic model, and are essential in terms of a new sustainable economic trajectory, green market development, enhancing urban competiveness and job creation. As green business is becoming more mainstream and in order to create green business hubs and clusters, urban governance needs to create the optimal conditions for capturing investment and capital, as well as policies geared towards local innovation. With this in mind, policies to enhance the capacity of cities to attract green businesses and foster smart and green growth sectors are and can be an influential force shaping local policymaking – entrepreneurial urban governance.

3 The Entrepreneurial Local Government

Regarding the urban economy, the local government needs to create the desired conditions that firms and businesses require in order to set up operations in a particular urban area. In the process, urban governance has become entrepreneurial in its actions, whereby in the age of globalization, cities compete to capture capital and inward investment in order to remain competitive. The local government is also an essential actor in fostering home grown centers of specialization. As discussed, a major aspect of the Green Urban Economy is the development of green cluster formations. Local governments can attract green economic activities that are essential for cities to remain competitive in the global economy. This is particularly true if local governments avail of the new business opportunities that exist and continue to emerge in green growth industries resulting from green investments. In creating the necessary conditions for green clusters and establishing the capabilities required for their successful formation, various initiatives can be pursued. Most importantly, these conditions are created by establishing the required statutory and policy framework, where urban governance directs policy towards specific sustainable city development strategies, which support green cluster development and other green urban economic initiatives. Sustainable economic development strategies at city level can include or be complimented by:

- Direct investment in green industries that may form a green cluster and preferential treatment for industries that result in environmental benefits.
- Incentivizing green growth industries, through financial incentives etc. and using public procurement for green market development.
- Optimizing key factors of location economics including quality of life indicators to attract investment.
- Sustainable urban planning and management i.e. urban form and energy and resource efficiency (see UNEP 2011)

When firms and companies seek investment opportunities various factors of location can influence the decision. These include a healthy natural and built environment, good public transport links, green space and a overall sustainable urban planning and management policy direction. Such urban attributes create attractive places to live and to invest while also influencing private sector decision-making. These are important prerequisites in terms of attracting firms and workers, a process which links the idea of the smart city with the sustainable city. Cities with high environmental performance coupled with smart growth sectors, including green and ICT sectors, create a signaling effect for other firms to establish operations in a particular urban area. One of the chief mitigating criterions for private sector investment depends on a strong local government commitment and vision. Creating the required regulatory and policy framework is the bedrock on which sustainable economic growth can be constructed, by which city development strategies provide a proven enabling framework. In this regard the Portland example offers an interesting insight into how a city can pursue a sustainable development strategy in order to remain economically competitive.

4 Case Study: Portland's Economic Development Strategy

In response to the 2008–2009 financial crises, Portland developed an 'Economic Development Strategy' in 2009, aiming to integrate sustainability into the fabric of the city's economy. The strategy includes a strong focus on sustainable growth in terms of green jobs, clean tech clusters and sustainable urban planning and management. Portland's strategy builds upon investments in sustainability (in economic development and public policy) to chart a new economic growth trajectory, building upon a solid history of innovative and sustainable policies. The timespan of the plan is 5 years and rests on three pillars that trace back to the concepts found at the first Rio Earth Summit in 1992:

- Sustainable job growth (economic sustainability).
- Sustainable innovation with Eco-Districts (environmental sustainability).
- Inclusive prosperity (social sustainability).

To implement the Economic Development Strategy, the Mayor established an 'Economic Cabinet', made up of leaders from all facets of the local economy including the city's target clusters which include: clean tech, software, research and commercialization, athletic and outdoor industries, and advanced manufacturing. The Economic Cabinet advises the Mayor and identifies investment opportunities. The Mayor's office is responsible for strategy updates and provides oversight on the implementation of the strategy. The Portland Development Commission (PDC), the city's economic development agency, is tasked with delivering key aspects of the Economic Development Strategy. Building on Portland's commitment to sustainability, the mission of the PDC is "to bring together resources to achieve Portland's vision of a diverse, sustainable community with healthy neighborhoods, a vibrant central city, a strong regional economy, and quality jobs and housing for all".

The creation of both the city's Economic Development Strategy and the new Bureau of Planning and Sustainability (BPS) in 2009 represents a new phase in Portland's aim to develop linkages between economic development, sustainability and green cluster formation. Portland's efforts have resulted in an estimated US$ 355–US$ 960 million in annual wages from the green building cluster in 2008. Additionally, bicycle related industries accounted for an estimated US$ 90 million in value and 850–1,150 jobs. The Economic Development Strategy helped to attract 15 new companies and retain 1,100 existing jobs and create more than 1,900 new jobs through financial assistance to 132 local businesses. In 2011 the Portland metropolitan region boasted over 27,000 jobs that produce or add value to goods and services that result in an environmental benefit. The median wage for these jobs is US$ 42,458 per year and each of these jobs produces an estimated US$ 13,952 in exports for the Portland metropolitan region's economy (see ICLEI 2012).

Furthermore, 390 businesses have obtained local, state and federal loans, grants and incentives totaling US$ 9 million. Portland also attracted the solar energy company Vestas, thin film manufacturer SoloPower and energy storage leader ReVolt to set up operations in the city. This resulted in 751 new jobs and

300 retained jobs. The city also developed capacity in the wind energy supply chain resulting in sales and services totaling US$ 2 million for local wind firms (see ICLEI 2012).

Portland's Economic Development Strategy is an illustrative example of how a local government can foster a sustainable growth industry through its entrepreneurial activities. The Economic Development Strategy was a fundamental policy and institutional framework, through which the city created the necessary conditions and capacities for the development of a green cluster in Portland, which encouraged and provided an attractive playing field for investment and growth in clean tech sectors and green products and services. Furthermore, Portland's long tradition of sustainability initiatives, particularly in terms of land-use planning, public transport strategies and a focus on public green space, renders it one of the healthiest cities in the world (see UNEP 2011). This adds to the city's green credentials and contributes to Portland's goal to become the most sustainable urban economy in the USA. Portland demonstrates that an environmental social integrity model and urban sustainable practice can be integrated within the development of robust competitive economic development strategies.

5 Conclusions

Green business and investment is increasing in market share and this newly emerging economy is gaining the attention of urban governments. Such new business opportunities have incentivized local governments in cities to pursue actions that create the optimal conditions for both fostering green business models and to attract investment for environmental technologies, green products and green services. While realizing the potential their cities offer, urban governments create the conditions that private sector actors can capitalize on. The case of Portland is an illustrative example where the city has achieved a significant drop in greenhouse gas emissions (3% below 1990 levels) while the city has developed strategies that are increasing economic development and employment. It is an encouraging example of how sustainability can be a key principle in economic growth and urban competiveness, and how cluster formations can contribute to the transition to a Green Urban Economy. Concurrently, an entrepreneurial local government is imperative to achieve this transition and should be recognized for this role. City development strategies are one example of how local governments can create a green urban advantage resulting in considerable economic advantage, a sample of the tools that local governments can use to help advance the Green Urban Economy.

References

Cities Alliance (2006) Guide to City Development Strategies: improving urban performance. Cities Alliance, UNEP and ICLEI, Washington, DC

Duffy H (1995) Competitive cities: succeeding in the global economy. E & FN Spon, London

Ethical Markets Media (2012) Green Transition Scoreboard 2012: From expanding Cleantech Sectors to emerging trends in Biomimicry. Ethical Markets Media, Saint Augustine FL
Friedmann J (2005) The world city hypothesis. In: Lin J, Mele C (eds) The urban sociology reader. Routledge, New York
Hall P, Pain K (eds) (2006) The polycentric metropolis: learning from mega-city regions in Europe. Earthscan, London
Hammer S et al (2011) Cities and green growth: a conceptual framework. OECD regional development working papers 2011/08, OECD Publishing. http://dx.doi.org/10.1787/5kg0tflmzx34-en. Cited 20 Mar 2012
ICLEI (2011) Green urban economy briefing sheet. www.iclei.org/casestudies. Cited 20 Mar 2012
ICLEI (2012) Case Study #150 Portland, Oregon, USA: building a sustainable economy on the pillars of urban innovation. www.iclei.org/casestudies. Cited 27 Mar 2012
Ketels CH (2004) European clusters. In: Structural change in European 3: innovative city and business regions. Hagbarth Publications, Bollschweil
Krugman P (1991) Increasing returns and economic geography. J Polit Econ 99(3):483–499
Legates R, Stout F (2003) The city reader. Routledge, New York
McKinsey Global Institute (2011) Urban world mapping the economic power of cities. McKinsey & Company. www.mckinsey.com/mgi. Cited 21 Mar 2012
OECD (2006) Territorial reviews: competitive cities in the global economy. OECD Publishing, Paris
OECD (2011) Towards green growth. OECD Publishing, Paris
Porter ME (1998) Clusters and the new economics of competition. Harvard Business School Press, Boston
Sassen S (2005) The urban impact of economic globalization. In: Lin J, Mele C (eds) The urban sociology reader. Routledge, New York
Turok I (2004) Cities, regions and competitiveness. Reg Stud 38(9):1069–1083
UNEP (2011) Towards a green economy: pathways to sustainable development poverty eradication. UNEP, Nairobi
UN-HABITAT (2008) State of the world cities 2008/2009: harmonious cities. Earthscan, London

Green Cities Require Green Housing: Advancing the Economic and Environmental Sustainability of Housing and Slum Upgrading in Cities in Developing Countries

Matthew A. French and Christophe Lalande

Abstract Cities throughout the developing world are in the midst of a housing crisis. Urbanization, population growth, rising economic development, the large stock of inadequate existing housing, and the need to rebuild after conflict and disasters necessitates the provision of affordable housing, the upgrading of slums, and retrofitting the existing housing on an unprecedented scale. This chapter explores these trends with a particular focus on the opportunities and challenges the green economy poses to affordable housing. Reviewing the current situation reveals that while there are many affordable housing and slum upgrading policies, programs, and projects, green economy dimensions are not sufficiently harnessed to maximize the economic and environmental opportunities that housing can provide to countries, cities, and households. The analysis highlights that there is an enormous environmental and economic potential for the greening of the housing sector in developing countries. Indeed, much of the housing stock is yet to be built and this presents a huge opportunity to build green today, reduce urban vulnerability and poverty, contribute to economic growth, and make significant environmental and economic savings in the future.

Keywords Housing • Slum upgrading • Housing policy • Economic sustainability • Developing countries

1 Introduction – The Challenges of an Urbanizing World

Half of humanity now lives in cities. By 2050, 70% of the world's population will be urban, and 95% of this urban population growth will occur in developing countries (UN-Habitat 2009a). Urbanization brings with it many positive benefits

M.A. French (✉) • C. Lalande
UN-Habitat, Nairobi, Kenya
e-mail: matthew.french@unhabitat.org; Christophe.lalande@unhabitat.org

R. Simpson and M. Zimmermann (eds.), *The Economy of Green Cities: A World Compendium on the Green Urban Economy*, Local Sustainability 3,
DOI 10.1007/978-94-007-1969-9_24,

and opportunities but it also places enormous pressure on local ecosystems, natural resources, urban land use, and the financial and human capacities of national governments and city authorities (see Davis 2005; Neuwirth 2004; Saunders 2011). Unfortunately, in most cities of the developing world these pressures combine to severely limit the sufficient provision of affordable and adequate housing leading urban dwellers to resort to living in unhealthy, overcrowded, and often relatively expensive informal housing in slums and informal settlements (Acioly and French 2012; UN-Habitat 2003). This pressure is evidenced in cities throughout Asia, Africa, Latin America and Eastern Europe. In many cities in Sub-Saharan Africa urban growth is synonymous with slum growth, meaning quite accurately that, due to a combination of poor regulatory frameworks, low incomes, inefficient private real estate markets and a lack of flexible housing finance, urbanization equals informal housing formation (Acioly 2010; UN-Habitat 2005).

The scale of this housing informality is immense with one out of every three urban dwellers in the developing world living in slum conditions (UN-Habitat 2003). Although globally the percentage of slum dwellers is decreasing,[1] the absolute number of slum dwellers is increasing and poses significant long-term challenges to the economic growth, political stability, social cohesiveness and environmental sustainability of cities. Slums are a clear manifestation of a mal-functioning housing sector that fails to provide a range of affordable housing alternatives for urban households. To solve the challenge of slums, a twin-track approach that involves supplying new affordable housing and undertaking slum upgrading programs is required. Together these can prevent the growth of new slums, address existing slums, and can significantly contribute to green economic development in cities.

At first glance the construction of new housing and the upgrading of existing slums may seem to be at odds with environmental sustainability goals. These activities suggest using more natural resources, to build more, for more people, with occupants eventually using more to maintain and operate their houses than they currently do in slums. There is certainly some basis for this assumption because in general the building sector consumes around 30% of global energy, making it one of the most resource intensive sectors of contemporary cities and one of the largest contributors to climate change (Levine et al. 2007). There exists however sufficient knowledge, expertise, and capacity to plan, design, construct, and use environmentally sustainable buildings using existing technologies and approaches. The challenge therefore is to improve the living conditions of the urban poor in a way that does not negatively impact on the environment, while at the same time improving local, regional, and national economies. Furthermore in the long term, there is enormous potential for the greening of the housing sector of developing countries precisely because much of the urban housing stock is yet to be built and this presents a huge opportunity to build green today and make significant environmental and economic savings in the future.

[1] In 1990 the percentage of the urban population in slums was 46.1. In 2010 it was 32.7%. In 1990 the absolute number of slum dwellers was 650 million but it 2010 it had risen to over 800 million. United Nations (2011a).

Over the last 50 years there have been a multiplicity of responses to addressing the housing challenge however green economy dimensions have seldom been considered (Pugh 2000, 2001). Slum upgrading programs often improve public areas through the provision of infrastructure and access to fuels and sanitation, yet they seldom consider the environmental impact of such development. Likewise, government 'social' housing projects have focused on producing large quantities of standardized affordable units often without considering the economic and environmental implications of such projects for households and cities. There is a lack of knowledge, capacity, and institutional and regulatory support for mainstreaming sustainability within slum upgrading and affordable housing programs in developing countries.

This chapter explores how housing in developing countries is an opportune setting to contribute to green economic development. It firstly outlines the conceptual underpinnings linking housing and the economy, by highlighting both the established knowledge regarding backward and forward linkages between the two, as well as the role of housing in the relatively new field of the 'green economy'. In both, housing is shown to present significant opportunities for improving the green economy of cities – that is, making them environmentally sustainable and advancing the economic development of households, settlements, cities, and national economies.

The chapter then provides a detailed situation analysis of the state of the green economy in the housing sector. It explains the dynamics of the realities on the ground: urbanization, population growth, and economic development are increasing the demand for adequate and affordable housing; the existing stock of housing needs to be retrofitted to improve its environmental efficiency; and the need to rebuild houses and residential settlements after crises, conflict, and natural disasters witnessed in many developing countries today. To support the situation analysis, the chapter then draws from international examples to communicate good practice in housing and slum upgrading policies, projects, and programs in this field. Lastly, the chapter outlines the critical blockages and constraints to scaling-up the 'greening' of the housing sector in developing countries and concludes by providing a clear outline of the key priority areas for action.

2 The Role and Contribution of Housing in Green Economic Development in Developing Countries

2.1 Housing and Sustainability

In the context of climate change mitigation, no sector is more able to produce cost effective reductions in energy consumption and greenhouse gas emissions than the building sector (Levine et al. 2007). The sector contributes up to 30% of total global annual greenhouse gas emissions and consumes up to 40% of all energy in most countries (UNEP 2009). While the specific percentage varies slightly from country to country, amongst both developed and developing countries it is generally consistent. For instance, in Indonesia about 40% of total energy consumption and associated greenhouse gas emissions are due to energy consumption in buildings which is roughly the same percentage in developed countries such as the United States at 35% total energy consumption from buildings (Utama and Gheewala 2009). Recent studies estimated the ratio of embodied energy consumption for the building sector at approximately 15–25% from the construction phase and 75–85% from the operation phase assuming a 50-year building lifespan (Nässén et al. 2007; Architecture2030 2007). These figures do not include the emissions related to the transportation of materials and other activities related to the building sector which means actual building sector emissions are certainly larger (Nässén et al. 2007).

Although the building sector is a significant contributor to global GHG emissions, it also represents the single largest source of achievable emissions reductions with current technology in the near term across the entire economy (Levine et al. 2007; Urge-Vorsatz and Novikova 2008). According to the Intergovernmental Panel on Climate Change (IPCC) Fourth Assessment Report, a 29% reduction in energy consumption in buildings is technically attainable by the year 2020 at no cost to society (Levine et al. 2007). This means money can be saved and emissions reduced when buildings are designed and constructed with efficiency and sustainability in mind, even by using currently available technologies.[2] These cost saving and attainable measures would avoid approximately five billion tons of greenhouse gas emissions over the next decade (Urge-Vorsatz and Novikova 2008). This is more than double the reductions achievable from any other sector at the same cost (IPCC 2007). This demonstrates the incredible importance that the building sector and housing have to play in a successful global strategy to mitigate climate change and towards sustainable urban development.

In developing countries the challenge is to not only address the environmental impacts of the building sector, but to balance this with the economic, social, and cultural pillars of sustainable urban development (Brown and Bhatti 2003; UN-Habitat 2012a). Sustainability is defined in the 1987 United Nations' Bruntland Commission Report as comprising economic, environmental and social dimensions, where "sustainable development is development that meets the needs of the present without compromising the ability of future generations to meet their own needs (World Commission on Sustainable Development 1987)." Cultural sustainability is also important and is therefore considered the fourth pillar of sustainability, as promoted in The Universal Declaration on Cultural Diversity: "Cultural diversity is as necessary for humankind as biodiversity is for nature" (UNESCO 2001) It is important to be "one of the roots of development understood not simply in terms of eco-

[2] Existing technology and current building design knowledge are sufficient to attain these levels of reduction. In the US residential building sector extensive studies have shown that existing technology and human resources can be successfully deployed today to create homes that are near net zero energy consumers, while remaining affordable (Parker 2009). As well, in Europe growing knowledge and new policy measures are harnessing this potential through zero carbon housing mandates to be attained by 2016 in the UK and a broader energy efficiency target of a 20% increase in efficiency by 2020 in the EU which focuses on energy efficiency labelling for buildings and other policies.

nomic growth, but also as a means to achieve a more satisfactory intellectual, emotional, moral and spiritual existence" (UNESCO 2001). Indeed in terms of housing, culture is very important as it not only links people and built environments (Rapoport 1988, 2000; Turner 1976), one criteria of 'adequate housing' is cultural adequacy (UN-Habitat and UNHCR2009).[3]

This comprehensive conception of sustainability is crucial when looking at housing in developing countries (UN-Habitat 2012b). The ever increasing need to alleviate poverty, provide adequate housing for disadvantaged peoples and provide alternatives to slums in developing countries demonstrates a critical tension between the need to increase energy access to the world's poor and simultaneously reduce emissions on the global scale. The focus must be on strategies which synergistically provide reductions in greenhouse gas emissions attributable to housing across their life cycle, climate adaptive capacity through durability and resilience to changing climatic impacts such as flooding, extreme storms and extreme heat, and simultaneously provide economic benefits and other co-benefits in the form of improved quality of life, poverty alleviation, economic development, environmental protection and improved health and safety.

In developing countries the rate of growth in the housing sector is rapidly increasing and for emerging economies such as China, India, South Africa and Brazil, the rate exceeds that of developed countries. For example, "in the next two decades, 300 million Chinese are projected to move into urban centers, and China alone will add 2 billion square meters (21.5 billion square feet) of new construction each year, doubling its building stock by 2020" (UNEP, ILO, IOE, ITUC 2008). This will mean the addition of new buildings will be greater in total floor space than the entirety of the current building sector in the United States (FGEF 2003). At this magnitude, if new housing in emerging economies fails to be the most environmentally sustainable as possible, the world faces a multi-decade lock-in of dangerous energy consumption patterning which will certainly aggravate environmental as well as economic challenges.

2.2 *Housing and the Economy*

As is the case with environmental dimensions, the economic dimensions of increasing the provision of affordable housing and upgrading existing slums present both challenges and opportunities for developing countries. The financial turmoil facing the world since 2008 has reaffirmed the close interrelation between the housing sector and the world economy. Economic growth is concentrated in cities, where housing is a "central force of sound economic development" (Kajumulo Tibaijuka 2009). Indeed, the formal and informal housing sector, including the formal and informal construction and building industries, contributes significantly to national GDP, local economic development and job creation (IMF 2008).

[3] The criteria is cultural adequacy: "housing is not adequate if it does not respect and take into account the expression of cultural identity." The other six criteria are: security of tenure; availability of services, materials, facilities and infrastructure; affordability; habitability; accessibility; and location.

Within economic housing theory, housing has traditionally been considered a consumption good (Angel 2000). Housing was seen from a social lens, as a form of welfare, where states should be motivated to address housing sector market failures for social reasons (such as health, wellbeing, etc.), not explicitly for economic reasons because housing 'tied up capital in fixed, long-term, unproductive assets'. Recent evidence, however, reaffirms that housing is an investment good and a catalyst for economic growth and development (Kajumulo Tibaijuka 2009). Housing construction, retrofitting, and maintenance has enormous potential to contribute to economic development through stimulating enterprise, and creating jobs, both directly and indirectly, in a wide range of sectors. Indeed, "investment in housing, as well as an efficient handling of housing supply, development and access, generates a high multiplier effect to the wider macro-economic and social system" (Kajumulo Tibaijuka 2009). Prioritizing investments in housing as have contributed significantly to the rapid rate of growth experienced in many countries over the last four decades. For example Singapore and Hong Kong which once had slums now have robust formal housing sectors that contribute directly and indirectly to national their economies (UN-Habitat 2011e; Kajumulo Tibaijuka 2009). Government investment in affordable housing is often the case in times of crises, for example in China and Brazil after the 2008 global financial crisis, where housing is used as a tool for economic management, macro-stabilization and national growth strategies (Yanyun Man 2011; D'Amico 2011).

Housing is economically advantageous as it has both backward and forward linkages. Backward linkages "generate wealth in the form of profits from employment, both in the construction industry itself and in the industries supplying it" (Spence et al. 1993). In many developing countries construction labor is typically unskilled and, depending on the complexity of the construction technology used, construction sector employment is a first 'port of entry' in urban areas thus providing an important safety net for rural migrants. Furthermore, residential construction is an employment multiplier; for example, in Colombia, India, Mexico and Pakistan, for every job created in residential construction, two are created elsewhere (Kajumulo Tibaijuka 2009). Forward linkages are in the form of the production of facilities required for economically productive activities. These have become increasingly apparent with the rise of 'mixed use' developments and the recognition of the role of home-based enterprises in most parts of the developing world (Tipple 1993; Kellett and Tipple 2002). When tenure is secure, low- and middle-income households are more likely to invest in housing and use them for economically productive activities, contributing to household, city and national economic activity.

Spence et al. summarize four key economic benefits of affordable housing in developing countries:

1. Housing is a stimulus to saving. It can be a vehicle to generate additional wealth and income for the poor through home-based enterprises.
2. Housing is employment generating, both on-site and through building materials production, distribution, and services.

3. Low-cost housing is more economically beneficial than high-cost housing. Investment in low-cost houses and houses which use materials manufactured in local small-scale enterprises generates significantly more employment for each unit of expenditure than luxury housing.
4. Housing has hidden employment effects. Informal construction means that a lot of construction labor is hidden, and therefore the potential of residential construction is significantly greater than is commonly realized (Spence et al. 1993).

2.3 Environmental and Economic Synergies: Housing and the Green Economy

At its core, a green economy is one that blends the environmental and economic pillars of sustainability. In 'Working towards a balanced and inclusive green economy: A United Nations system-wide perspective' (2011a) the United Nations defines the green economy at the 'visionary level' as "an economy that results in improved human well-being and social equity, while significantly reducing environmental risks and ecological scarcities." "At the operational level, the green economy is seen as one whose growth in income and employment is driven by investments that: reduce carbon emission and pollution; enhance energy and resource efficiency; [and] prevent the loss of biodiversity and ecosystem services" (United Nations 2011b).

The green economy is of relevance and importance for developing countries because it provides the opportunity to promote economic development whilst at the same time minimizing the 'lock-in effect': "The greening of infrastructure – including the buildings, energy and transport sectors, which are significant consumers of resources and emitters of greenhouse gases (GHGs) – is urgent as emerging economies and developing countries will build the bulk of their infrastructure in the next few decades (United Nations 2011b)." This awareness is shared in a recent UNEP study, Towards a green economy: pathways to sustainable development and poverty reduction, that reaffirms the fact that buildings have an 'oversized ecological footprint'; and that a green economy can make significant savings in the short and long term; can bring health and productivity benefits; increase jobs; and help developing countries achieve energy efficient building stock for decades to come (UNEP 2011).

Within the green economy there has been a strong emphasis on job creation. In the buildings sector this employment generation is in industry and service sectors that address environmental efficiencies, technologies and materials (UNEP, ILO, IOE, ITUC 2008). For example, citing a 2005 study of the ten European Union new member states – Cyprus, the Czech Republic, Estonia, Hungary, Latvia, Lithuania, Malta, Poland, Slovakia, and Slovenia – retrofitting the existing housing stock of these countries would create between 50,000 and 185,000 jobs (UNEP, ILO, IOE, ITUC 2008).

While green economy approaches in Europe and North America focus on high-technology environmental building technologies and systems, in developing countries the focus is more on broadening opportunities for unskilled construction employment and achieving economic savings in housing production through reducing the cost of materials and construction systems. It has long been argued that housing construction in developing countries should utilize locally produced materials rather than relying on imports, because this not only creates local and national jobs and stimulates local economic development, but also reduces the environmental impact of construction by reducing transport costs and therefore the embodied energy of materials (Spence et al. 1993). Studies have shown that low-cost housing development that utilizes simple, intermediate technologies generates roughly twice as much direct and indirect local employment as luxury housing development, in part because luxury housing has higher specifications and many of its components are imported (Araud 1973; Strassmann 1976); indeed, in some developing countries nearly 60% of the total value of building materials are imported (Spence et al. 1993).

Large-scale low-cost housing and slum upgrading programs offer significant opportunities for labor intensive job creation in developing countries. In slums, housing programs have the potential to be labor intensive and environmentally sustainable when using traditional building techniques, indigenous and/or recycled building materials and self-help processes (Spence et al. 1993). Indeed a move to green construction could help achieve the MDG targets, in particular Target 7d on improving the lives of slum dwellers:

> A shift away from traditional housing to green construction may provide a unique opportunity to meet these targets. Certain infrastructure costs can be bypassed by new developments in technology. For example dependence on an electricity grid may no longer be necessary with the installation of solar panels and solar water heating. By reducing energy costs, this makes the development goals more feasible for municipalities and residents. (UNEP 2011)

Green economy housing practices to produce new affordable housing also offers a unique opportunity for training and capacity building for semi- and un-skilled workers. They also provide opportunities to establish micro-enterprises for a wide range of housing inputs – direct inputs such as building materials, and indirect such as household goods. Sethuraman highlighted the multiplying effect of an initial investment in constructing informal housing units, which generate one in five more jobs than in the formal housing sector and contributes to produce six times more dwelling units (Sethuraman 1985).

Overall, this section has illustrated that to improve the housing sector in developing countries it is crucial to recognize all four pillars of sustainable urban development, rather than take a narrow sectorial view. By linking both environmental and economic pillars the green economy goes some way in adopting such an approach. The reduction of urban poverty and social vulnerability, and the improvement of slum dwellers' psycho-social wellbeing and physical living conditions should be central features in green economic development. This section has demonstrated that, at least in theory, green affordable housing and slum upgrading programs can significantly contribute to these wider aims, as well as the central underpinnings of the green economy, that of local and national economic growth, job creation, and environmental sustainability.

3 Understanding the Dynamics 'on the ground' – A Situation Analysis of the Green Economy in Housing in Developing Countries

This section explores the key trends and challenges regarding the green economy in the housing sector in developing countries. It presents a situation analysis based on the dynamic forces shaping housing demand and supply 'on the ground', and explores the current approaches to addressing these challenges with a focus on the degree to which they can be said to incorporate green economy dimensions (Fig. 1). The analysis includes brief descriptions of international examples to explain the state of the field, and communicate good practices in housing and slum upgrading policies, projects and programs in the field of green housing.

3.1 Urbanization and Population Growth

When looking at the key drivers of housing demand in developing countries, the most noticeable trend is rapid urbanization and population growth. Cities throughout Asia and Africa are growing rapidly as rural migrants move to cities in search of employment and new social, cultural, and political opportunities. Indeed, 'every day for the coming decade Asian cities will need to accommodate 120,000 new residents' (UN-Habitat 2011b). Likewise, although Africa is the least urbanized continent in the world, it is urbanizing fast and for the next two decades the urban growth rate will be above 3% per annum (UN-Habitat 2010). These dynamics cause some critical housing issues, in particular the growth of informal settlements (slums, unauthorized land subdivisions, etc.) due to the fact that the growth in the economy, employment, and in the housing sector does not match the urbanization and population growth.

Housing sector responses generally fall into two approaches. Firstly, new 'social' housing is constructed by governments, local authorities, NGOs, or housing cooperatives, with a focus on unit standardization and cost minimization. The Ethiopian Integrated Housing Development Programme (IHDP) is one such example of a government-led national housing 'social' program (Fig. 2). The program is replacing inner-city slums with multi-storey condominium housing, with a focus on local economic development, employment generation, enterprise, and poverty reduction. While the project is successful in many respects, the environmental and cultural

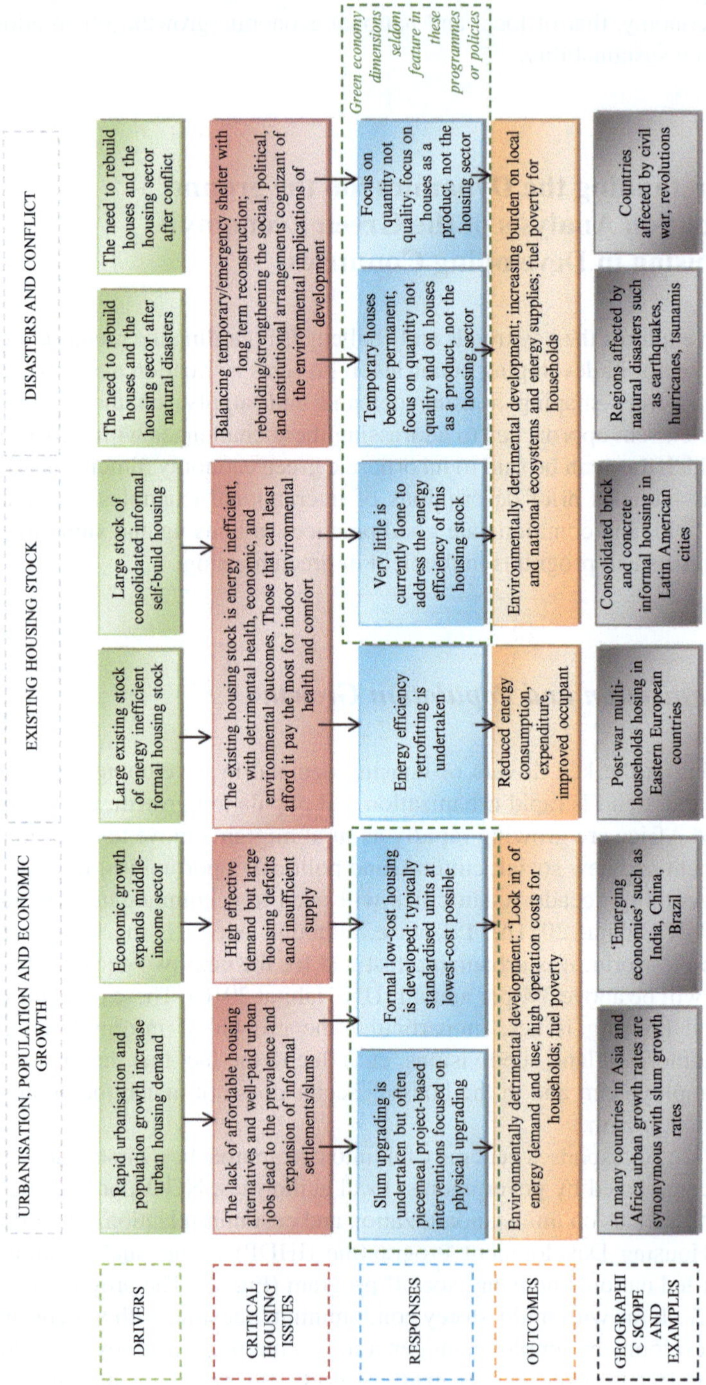

Fig. 1 Situation analysis identifying the key trends and challenges of the green economy in the housing sector in developing countries

Fig. 2 Integrated Housing Development Programme condominium units in Addis Ababa, Ethiopia (© Katherine Hegab)

sustainability of this housing program has received little attention to date (UN-Habitat 2011a). Condominium buildings are developed without consideration of such aspects as local ecosystems, micro-climatic conditions, embodied energy, operational energy use, as well as residents' cultural traditions, lifestyles and activities.

Second, upgrading of existing slums is undertaken which improves the urban environmental conditions through infrastructure and service installation. However, environmental and economic sustainability dimensions are not sufficiently mainstreamed in slum upgrading programs. While they do improve basic urban services, the resource efficiency and use of alternative environmentally sustainable infrastructure technologies is seldom considered, as is the case with housing units themselves (Fig. 3).

The case of the vertical housing project in São Paulo, Brazil (World Habitat Awards n.d.) is a good example of slum upgrading program considering economic and environmental sustainability dimensions by increasing the density and quality of the housing structures and impacting on local economic development. The project was implemented by USINA (Centre of Projects for the Built Environment) to improve the living conditions of low-income households in urban São Paulo and provide high-quality multi-story housing for low-income communities. The vertical housing project has benefited approximately 5,000 families. It is a community-led process where residents have been involved throughout all stages of the project – analysis, planning, implementation, organization, resource allocation, management and evaluation. The environmental sustainability dimension of this housing project integrated building materials and technologies that enable residents to build multi-storey housing structures, increasing density and the quality of structures. Also, the construction activities implemented contributed to the local economic development, by creating collective income generating activities related to housing construction or small scale business.

Fig. 3 Slum upgrading often involves the improvement of the physical environment, regularization of tenure, and promoting socio-economic development. This example is of a new street in Kibera, Nairobi (© Matthew French)

3.2 Economic Growth and an Expanding Middle-Income Sector

Economic growth in many developing countries is rapidly expanding the number of middle-income households who not only are demanding increased access to urban housing but also have the capacity to pay for it. In emerging economies such as Brazil, India, China, Russia and South Africa, economic growth is increasing effective housing demand as national, local, and household economies improve. Also contributing to this trend are remittances which are increasingly playing a role in raising effective demand, as foreign workers send money home and often use it for housing improvements and investment (Saunders 2011). In 2006 officially-recorded remittances to developing countries were 240 billion USD, twice as large as official aid, and nearly two-thirds of FDI in developing countries (World Bank 2008).

To meet this demand the prevailing response is large-scale new-build housing programs that seek to rapidly increase the supply of new affordable housing. In many countries, such programs are adopted rather than (or in addition to) slum upgrading because the latter is seen as ineffective in addressing the needs at scale, slums are still perceived as having no value and in need of eradication, and large programs are best at stimulating the wider economy and creating employment.

Fig. 4 Barrio Carioca in Rio de Janeiro is being constructed for households earning between 0 and 3 minimum wages as part of the Brazilian 'My House, My Life' program, which is showing promise of supplying housing to low-income households (© Matthew French)

Examples of this approach are the 'My House My Life' Program in Brazil (Gonçalves et al. 2011), the Economical and Comfortable Housing Program in China (Yanyun Man 2011), and the 'Reconstruction and Development Programme' in South Africa. Again, while the political choice to address the housing challenge is laudable and should continue, it is becoming increasingly clear that these models of development are environmentally detrimental. They do not create dense, compact, liveable cities that are carbon neutral and work with local ecosystems (Angel 2011). Instead they contribute to 'cookie-cutter' development and low-rise low-density urban sprawl – an urban model that requires large amounts of energy to produce (e.g. in laying infrastructure) and sustain (transport costs, energy costs for housing operation) (Giok Ling 2005) (Fig. 4).[4]

3.3 *Existing Housing Stock*

One aspect of the environmental sustainability of housing in developing countries that is often neglected is the existing housing stock. Eastern European and CIS countries have a large stock of multi-storey post-war housing that is in poor condition, very energy inefficient, and requires urgent upgrading. Similarly, as the

[4] In particular Chapter Six: 'Housing – relating urban basic needs to urban sustainable development', pp. 124–136.

Fig. 5 Cities throughout Latin America have a considerable stock of informally-built consolidated housing that will remain for decades to come and therefore needs retrofitting. Buenos Aires, Argentina (© Matthew French)

first developing countries to undergo urbanization, Latin American countries have largely completed their transition to urban societies (75.5% of Latin Americans live in cities) and consequently have a large stock of informal housing (UN-Habitat 2011d). Sullivan and Ward note: "throughout Latin America older (now) consolidated informal settlements … developed in the 1970s and 1980s are now fully integrated into cities such that most observers would not imagine that they had begun as shacks and squatter settlements" (Sullivan and Ward 2012). This permanent concrete and brick housing stock is the result of the gradual consolidation, upgrading and expansion over the last three decades and represents a considerable stock of housing in cities throughout the continent that will remain for decades to come. This is also true of government-built housing in Africa and Asia that has been modified and expanded (Tipple 2000) (Fig. 5).

Cato Manor Township, in Durban, South Africa is an example of environmental upgrading of existing housing. In December 2011, 30 low-income houses on one street were given a 'green upgrade'. The 30 houses were retrofitted with green building technologies by the Green Building Council of South Africa, in preparation for the UN COP-17 climate change summit in Durban with the aim to demonstrate how green housing initiatives could benefit low-income households and slum dwellers from an environmental, social, and economic perspective. Each house was provided with rainwater collection tanks, energy-saving lights, solar water heaters, and a special heat insulation cooker; the street was fitted with efficient LED streetlights. Ceiling insulation improved the thermal conditions in the low-income housing units and reduced operating costs.

Fig. 6 Retrofitted housing in Cato Manor, Durban, South Africa (© Photo courtesy of Green Building Council of South Africa. Photographer: Willem De Lange)

Since 1994, the Government of South Africa has produced 2.5 million housing units for low-income groups but environmental elements have seldom been considered. The Green Building Council of South Africa hopes that this project will have an impact in informing housing practices and contributing to formulate sustainable housing policies in South Africa. The project contributed to demonstrate a range of benefits including improving energy efficiency, occupant health, reduction in the households' carbon footprint and energy cost savings. Some households have seen their electricity bills more than halve thanks to the retrofit. The project also had a positive impact in creating jobs and training skills for the urban poor and vulnerable groups. The community of Cato Manor expressed a sense of ownership of this initiative and named its green street "Isimosezulu (meaning 'climate') COP17 Place" (Fig. 6).

Unfortunately very little is currently done in research, policy and practice to improve the sustainability of the existing housing stock in Latin America, with the exception of some cases, such as the installation of solar technology in government

Fig. 7 Upgrading post-war housing blocks in the Czech Republic (© Matthew French)

housing in Brazil (Sullivan and Ward 2012).[5] In several Eastern European countries, however, progress is being made through housing retrofitting programs. Central for these is improving the operational energy efficiency of housing through retrofitting buildings with innovative building technologies and materials. For example, two programs in the Czech Republic aim to rehabilitate multi-storey panel housing blocks with a specific focus on energy efficiency retrofitting through the installation of insulation, better-quality windows and improving construction tightness (UN-Habitat 2011c) (Fig. 7).

A similar retrofitting housing project was implemented in Bishkek, Kyrgyzstan, to provide affordable and green electric space heating for low-income households. Kyrgyzstan is one of the poorest countries in former Soviet Union and 70% of the population lives in inadequate housing conditions, often in overcrowded units without basic services and infrastructure. The capital city, Bishkek, is facing serious housing challenges due to urban growth and migration. The aim of the project was to provide affordable and environmentally sustainable housing options with seismic resistance and efficient heating solutions for low-income households in the context where there is a lack of adequate housing, unaffordable heating costs, and misuse of energy resources. The project, by decreasing the heating expenditure, is contributing to increase people's living standards. The project has reduced the environmental impacts of the existing fuel heating system and promoted the use of electric heating system and the use of traditional and locally available building materials.

[5] See also: TOR CAIXA. Sistema de Aquecimento Solar de Água – MCMV. http://downloads.caixa.gov.br/_arquivos/desenvolvimento_urbano/gestao_ambiental/tr_sas_mcmv2.pdf and Ministry of Environment 27 July 2011: http://www.mma.gov.br/sitio/index.php?ido=ascom.noticiaMMA&idEstrutura=8&codigo=6858

The project has a strong economic sustainability component, by saving up to 40% of the construction costs and up to 75% of energy costs, in comparison to the traditional brick construction technique. Reducing heating cost is crucial when over 50% of all income in a household was usually spent on space heating. The project also contributed to job creation for reed harvesting and clay gathering. Furthermore, the project was an opportunity to train and build the capacity of the workers who contributed to the construction process and to improve the health conditions of the households by changing the heating system.

3.4 Disasters and Conflict

Developing countries are disproportionately affected by war, conflict, and natural disasters, which seriously affects the housing sector and economies (International Federation of Red Cross and Red Crescent Societies 2010). Already fragile housing sectors are placed under additional and often sudden pressure to cope with physical housing and settlement reconstruction and balance this with economic, social, and political development (Guha-Sapir and Santos 2012). Prevailing responses primarily focus on the immediate need for shelter, yet temporary shelter often becomes permanent and can end up being permanent slums that inhibits long-term reconstruction and development efforts.

An example focusing on housing reconstruction in a post-conflict situation is the case of Darfur, Sudan, where over 400,000 families are currently displaced due to the conflict that broke out in February 2003 between the Sudan Liberation Movement/Army (SLM/A) and the Justice and Equality Movement (JEM). These internally displaced people are expected to integrate into new urban settlements or return to their place of origin. Their housing needs have an important impact of the environment and it is estimated that 16 million mature trees need to be cut down to produce the burnt brick they require – which, even if this quantity were available, would be an environmental catastrophe. One alternative is the production and use of stabilized soil blocks (SSB), as part of a project formulated by UN-Habitat in partnership with the Government of Sudan. A recent study highlights that the SSB technology is economically and environmentally advantageous because it is 18% cheaper than the use of burnt bricks and does not require the use of firewood (UN-Habitat forthcoming). The study also highlights the considerable potential of the SSB technology to promote jobs that are labor intensive and green. The conclusions of the assessment stressed the opportunity to establish a Building Materials Production and Construction Training Centre to enhance the housing construction, support slum upgrading initiatives, stimulate local employment and entrepreneurship, build technical capacities and favor social re-integration of IDPs and ex-militians (Fig. 8).

Experience has shown that in general there are several issues with prevailing disaster and conflict redevelopment responses, and green economy dimensions are

Fig. 8 Stabilized soil block technology in use in Sudan (© Fernando Murillo)

very weak. The construction of large quantities of environmentally-inappropriately designed mass housing units leads to environmentally detrimental development. Furthermore, often redevelopment misses the opportunity to have green economic development as a central pillar of reconstruction efforts which is also socially beneficial. Too often there is a focus on quantity, rather than quality (important for green housing) and on housing as a product. Although quantity is important, it is important to see housing as a process of engaging communities and building trust, not only a product for consumption. As a process, housing can contribute much more towards economic and social development, and in these fragile situations this is extremely beneficial.

Furthermore, such housing is often donor driven (motivating the drive for completing units quickly and at scale), and is biased towards imported 'modern' construction materials and technologies at the expense of traditional materials and technologies which are in most cases more environmentally sound. The population is too often viewed as 'beneficiaries', or as a source of free labor, and are not sufficiently engaged in the development processes of planning, designing, managing settlement reconstruction which exacerbates dependency relationships, does little to stimulate local enterprise, and destroys social capital. There is therefore a need to support long-term reconstruction from an earlier stage in post-conflict and disaster situations, and to use the housing sector to strengthen the socio-political and institutional arrangements to produce housing at a large-scale in a way that environmentally and economically sustainable.

4 Ways Forward to Scaling-Up the 'Greening' of the Housing Sector in Developing Countries

This section summarizes the current trends of housing and the green economy and highlights some of the major challenges and critical constraints to scaling-up environmentally and economically sustainable housing initiatives in developing countries. Particular emphasis is given to housing policies, legislative and institutional frameworks and capacity development dimensions, although important socio-cultural dimensions of sustainable and green housing solutions are also discussed.

4.1 Summary of Current Challenges

Seven current trends in affordable sustainable housing in the developing world can be highlighted. Firstly, there is some evidence of sustainable housing being implemented in developing countries to improve quality of life, mitigate and adapt to climate change, and contribute to economic development. However, overall, the scope and impact of these initiatives pales in comparison to the scale of the housing challenge. Furthermore, a sectorial (rather than comprehensive) approach to sustainability dominates, where environmental aspects are 'added on' to a housing project (often literally in the case of renewable energy technologies such as solar water heating or rainwater harvesting) rather than being integrated with economic, social and cultural dimensions and mainstreamed in program design, implementation, and post-build management strategies.

Second, there is enormous potential to for green housing to contribute to national, local, and household economic development and to improve environmental conditions. As much of the built environment in developing countries is yet to be built, this presents the opportunity to build green and therefore minimize the lock-in effect of an energy-intensive built environment and create healthier and more liveable cities. Developing countries need sustained economic growth and job creation and housing can play an important role in achieving this through harnessing the direct and indirect benefits of this economic activity.

Third, in developing countries there exists insufficient capacity of green housing approaches and benefits at all levels – from national governments to city authorities to urban poor households. At the national and city levels, capacities to coordinate and advance green housing efforts are limited or inconsistent due to, in many cases, insufficient knowledge; formalized policies, codes or other standards; and access to financial or material resources within developing countries. Urban households and other housing sector stakeholders are either unaware of the multiple long-term co-benefits green housing offers or are not supported to take action to directly improve the environmental and economic dimensions of housing. Many governments and municipal authorities are also reluctant to accept the climate change agenda and the urgent need to implement mitigation and adaptation measures. There is poor coordination between government departments and local and national governments to develop, fund, and monitor green housing initiatives, as well as a lack of professional and private sector capacity to plan, build and promote green housing.

Fourth, global knowledge and technologies are not reaching, or are not mainstreamed in developing countries. There exists a sharp divide between developed and developing countries in their level of awareness, value, and knowledge on green housing. In terms of technology, on-site renewable energy technologies can reduce greenhouse gas emissions, improve access to basic energy needs such as lighting, and increase adaptive capacity. Renewable energy remains a critical component for sustainable housing in terms of climate change mitigation and adaptation. But limited access in developing countries remains a challenge to increasing basic energy access that is low carbon. Advanced standards, guidelines, design tools and Life Cycle Assessment (LCA) tools, which are available in developed countries can only be utilized currently to a limited degree in developing countries.

Fifth, an opportune entry point for green housing is building design, materials and systems. The use of low cost, local and low embodied energy building materials in conjunction with bio-climatic settlement and housing design can have both economic and environmental advantages. They can improve the environmental life cycle assessment of a house, while supporting local economic development by stimulating micro-enterprise and creating direct jobs in local materials manufacture.

Sixth, green housing is more than just optimizing energy efficiency. Regionally appropriate housing design which integrates synergistic opportunities to address climate change and contribute to economic development and poverty alleviation should be carefully crafted in ways which respond to the needs and values of the occupants and the socio-cultural environment. This is particularly important because there are important social and cultural dimensions to building materials and technologies – in particular in terms of the acceptability and desirability of traditional materials and intermediate technologies – which are often perceived as 'backward' and rejected in favor of 'modern' materials and construction systems such as concrete and fired brick. Indeed 'aspirational' demands from households for socially desirable materials, technologies and lifestyles present an important challenge to mainstreaming low-technology green housing in urban areas of developing countries.

Seventh, the economic pillar is arguably the least understood aspect of sustainable housing. Economic dimensions are often viewed at the national level, in terms of the macro-economic situation and national growth, not at the level of cities, or households. Sustainable financing of large-scale housing programs remains a major challenge for scaling up affordable housing supply and upgrading existing slums, especially to avoid unintended consequences of development such as the gentrification of upgraded slums and misdirected subsidies to middle- and high-income groups.

Many stakeholders are resistant to green housing approaches due to the perceived economic disadvantages. For low-income households the initial cost of green building design can make it undesirable for uptake, even if over the life of the building costs savings can be made. Governments implementing 'social' housing and slum upgrading programs strive for quantity rather than quality, building the greatest number of units at the lowest cost. Private developers have a lack of will to include green approaches in their developments as this can be complicated, increase risk, and lower their economic return on investment.

It should also be highlighted that the sheer magnitude of urban poverty and vulnerability means long-term 'environmental sustainability' as conceived and dialogued in developed countries is of little immediate concern for many of the world's urban poor for whom more immediate concerns prevail. At the household level then, not enough is known regarding the economic values and priorities of the urban poor, and the required trade-offs between the initial capital expense for improved environmental design and the long-term economic and environmental benefits.

4.2 Key Priority Areas for Action

To address the immense challenge of sustainable housing in developing countries urgent coordinated action is required. In this section six complementary priority areas for action are outlined. It is important to note that each priority area must be underpinned by the holistic sustainability view, where green economy dimensions must be synergized with social and cultural pillars of sustainable development. This means several overarching principles for each priority area:

(a) There needs to be a specific focus on low-income groups and slum dwellers, focusing on the most vulnerable groups in the city (youth, migrants, indigenous peoples, etc.) with a specific attention to gender issues. There is a need for innovative tools to engage these groups in meaningful participation (Imparato and Ruster 2003; French 2011). This also requires supporting housing modalities and processes that suit these groups, for example self-help initiatives (for households, communities and social groups), promoting micro-enterprise, microfinance, traditional knowledge and incremental construction.
(b) In line with the enabling shelter approach, the full spectrum of housing sector stakeholders should be empowered to contribute to the greening of the housing sector, including formulating and enhancing the legal and institutional framework, developing housing policy mechanisms and instruments. Such stakeholders include national governments, local authorities, private sector, professionals, CBOs and NGOs, civil society and urban dwellers themselves (UNCHS 1988; UN-Habitat 2006).
(c) Social and cultural diversity should be respected and promoted. This implies recognizing that 'one solution does not fit all' and that different cities, countries, and regions may address green housing from different angles depending on the local socio-cultural context and aspirations.

(d) Recognize the need to make trade-offs between the different pillars of sustainability given the economic challenges facing developing countries and urban poor households, specific needs, and identified comparative advantages, for example, sometimes more urgent social dimensions are, and should be prioritized over long-term environmental concerns.

4.2.1 Improve Our Understanding of the Co-benefits and Limitations of Affordable Green Housing in Developing Countries

Research is needed because there is insufficient information available on green housing in the context of the urban poor in developing countries. While knowledge on green housing practices exists, sources of information are often fragmented. Also, field practices are not always documented, which contributes to a lack of information about sustainable housing practices. Research is needed both at the macro level – national and city – and micro level – neighborhood and household. It should investigate the environmental and economic constraints and opportunities at these levels and seek to uncover what the social and cultural barriers are to green housing, for example households' resistance to using traditional building materials, as mentioned previously. Only with an empirical understanding can pathways to scaling-up green housing solutions be forged. Importantly, research should value traditional knowledge and tap into local knowledge from communities, slum dwellers and the most vulnerable. While action research should be linked to theory development, there is a need to demonstrate the social utility of research and strengthen the linkage between field study, research, and participation in the field of sustainable housing. Research must be translated into easily accessible knowledge products, and disseminated in an easy way to stakeholders in developing countries. In particular, there is the need to provide policy-makers, target groups and end-users with practical tools and guidelines to apply existing knowledge and sustainable housing practices. Practices, policies and guides on successful sustainable housing technologies and know-how should be identified collected and shared with stakeholders and target groups according to their needs.

On-going monitoring, benchmarking and evaluation of projects are also needed to learn and scale up successful green housing solutions. If quantifiable results are to be achieved, international efforts should be increased to adapt existing tools, standards and guidelines for increased availability and usefulness in developing countries, especially for dominant modes of provision such as incremental self-built construction. There is a need to develop special tools and mechanisms to extract the best of the existing knowledge, by developing certification tools, standards and mechanisms of local knowledge for sustainable housing (for example, to deliver awards and motivate the implementation of sustainable housing practices), and by producing quick and easy adapted guides to evaluate the life cycle of traditional and low-income housing. These require developing building energy codes and databases of embodied energy in building materials so that informed decision making in housing design and construction will provide measurable results in mitigating climate change, which will necessitate defining a common agreed language on sustainable housing practices for establishing baselines and facilitating comparable evaluation of housing projects and programs.

4.2.2 Formulate and Implement Evidenced-Based Green Economy Housing Policies

Housing and urban policies that promote and stimulate an increase of green jobs and green enterprises should be implemented. This requires the associated strengthening of institutional structures and frameworks to ensure policies and codes are not only formulated but also implemented and monitored. This would involve for instance, strengthening inter-ministerial coordination between, for example, the ministry for housing, the ministry of environment, and the ministry of economic development, and, more importantly, considering the institutional framework of decentralized administration and governance. For example building codes and standards and potential tax incentives are often not part of the prerogatives of local authorities.

In most developing countries national building standards and regulations are a key area for policy reform as these pose a significant constraint to scaling up formal housing supply and unlocking the latent potential of incremental informal housing production (Payne and Majale 2004). Excessive building regulations increase development costs, further motivating urban dwellers to seek informal housing. Of course some degree of standards is necessary for health, sanitation, and security reasons, but at the lower end of the income spectrum they should not prevent the poor from realizing their right to adequate housing. Flexibility and adaptability should be used as a guide in formulating new regulations and standards.

4.2.3 Increase the Economic Attractiveness of Sustainable Housing

At the macro level, fiscal instruments for scaling up should be investigated to help increase the supply of green housing. Cities need to be able to finance and sustain green housing programs and policies, often in the face of weak governance structures and limited or non-existent taxation systems. Taxation and effective pricing policies are important but incentives to promote green housing technologies and approaches should also be encouraged. Most importantly, these incentives should reach all income levels, not only the middle- and upper-income groups. Incentives should also address the range of environmental dimensions, not only narrow energy efficiency but also water, waste, etc. and at a range of scales (not only houses but also settlements).

Incentives also need to be developed from the perspective of the urban poor. Housing finance for households to buy formal green housing (rather than settle in slums); and for households to retrofit their houses (e.g. Latin America) is urgently required. Existing micro-finance initiatives should be supported at community level and indeed scaled-up, as these have been proven to be effective mechanisms for urban poor households to improve their economic conditions.

There exists significant potential for developing countries to harness international finance mechanisms and instruments, which have to date been relatively untapped in the building sector. For example, the Clean Development Mechanism (CDM) under the Kyoto Protocol has been ineffective in supporting green housing in developing countries. Access to the CDM requires quantifiable and verifiable reductions in emissions which are additional to business as usual activities. It is often technically or economically infeasible to demonstrate this for small scale housing projects and so, linking back to priority area one, new monitoring and evaluation methodologies tailored to the special characteristics of affordable sustainable housing are needed if carbon financing is to be a viable source of financing such projects.

4.2.4 Targeted Capacity Building of All Housing Sector Stakeholders

The limited capacity of housing sector stakeholders to formulate, plan, implement and monitor green housing practices must be addressed if green housing is to be mainstreamed in developing countries. Capacity building of personnel should be accompanied by capacity development of the sector as a whole – which includes its systems, processes and institutional frameworks. Capacity building will involve technical training of built environment professionals (architects, engineers, etc.), in universities and tertiary training institutions; technical training of construction labor (through apprentices and short courses); and training for private sector personnel (housing developers, financiers, etc.). With all, there is a need for increasing Peer-to-Peer learning and establishing knowledge-sharing platforms and communities of practice in the field of green housing in developing countries. Priority should also be given to the development of tools, guidelines, and training materials and modules to build capacity of local stakeholders and policy makers in promoting and implementing sustainable and green housing practices. Capacity building activities should also explicitly raise awareness of sustainable housing and advocate for its widespread implementation.

4.2.5 Promote Networking and Exchange

There is currently a lack of exchange between developed and developing countries, as well as between developing countries themselves. An urgent need is to bridge the divide between developed and developing countries thereby linking expertise and experience. Potential actions to facilitate this are establishing a global network, and holding regional and global events such as conferences, exhibitions, trade shows, etc. The promotion of 'south-south' exchange is also required. This will enable the specificities of green housing to be explored and give countries at a similar stage of economic development, or with similar urbanization and population growth patterns the opportunity to learn from each other's experiences. Regional events, policy dialogues, field visits and exchanges are some of the many practical ways to facilitate this 'south-south' exchange. Exchange and networking can be greatly facilitated by utilizing contemporary web-based technologies to cost effectively connect with and share between stakeholders working in this sector.

5 Conclusion

The high rate of construction of new housing units in many developing countries is a unique opportunity for housing sector stakeholders to utilize a green economy approach that can tackle emissions from the housing and construction sector and create jobs and contribute to local and national economic development. The relatively new green economy concepts reflect an ongoing shift from seeing housing as a form of welfare and consumption, to an economically productive investment good and a tool that can contribute to national development and poverty reduction.

Green economy approaches however are not presently mainstreamed in housing theory and practice, especially in developing countries. There is very little emphasis on the environmental and economic sustainability of housing, and the interplay between them, especially for low-income housing. In most cases developing countries are locking themselves into environmentally detrimental housing options which also in many cases contribute less than they could to economic growth and poverty alleviation. The exception is some Eastern European countries that are taking measurable steps to improve the environmental performance of their post-war multi-storey housing stock.

The immense scale of the housing challenge and the need to urgently mitigate and adapt to a changing climate necessitates a targeted effort to increase access to affordable and environmentally sustainable housing and reduce poverty of the poorest around the world through economic development and job creation. The five key priorities outlined in this chapter are: (1) improve our understanding of the co-benefits and limitations of affordable green housing in developing countries; (2) formulate and implement evidenced-based green economy housing policies; (3) increase the economic attractiveness of sustainable housing; (4) capacity building of all housing sector stakeholders; and (5) the promotion of networking and exchange.

There is enormous potential for the green economy to thrive in the housing sector of developing countries. It is important, however, to recognize that low-income groups, slum dwellers, and the most vulnerable groups in the city should be given explicit consideration and support, and that the full spectrum of housing sector stakeholders must be engaged in the greening of the housing sector. Furthermore, it must be accepted that 'one solution does not fit all' and green housing is more than just environmental sustainability. Different cities, countries, and regions may address green housing from different angles depending on the local social, cultural and economic context, and, given this diversity, trade-offs between these pillars of sustainability will continue to have to be made given the often severe resource constraints facing developing countries and urban poor households.

Above all, transitioning to a green housing sector in developing countries will require a coordinated, committed, and sustained effort by all stakeholders – including national governments, local authorities, civil society organizations, NGOs, private sector, academia and research institutions, and urban residents themselves. From these stakeholders, champions are urgently needed to forge the way and harness the potential of these numerous actors to enact and scale-up positive change to ensure developing countries benefit environmentally and economically from housing development.

References

Acioly C (2010) The informal city and the phenomenon of slums: the challenges of slum upgrading and slum prevention. In: International New Town Institute (ed) New towns for the 21st century. The planned vs. the unplanned city. SUN Architecture, Amsterdam, pp 222–231

Acioly C, French M (2012) Housing developers: developing world. International Encyclopedia of Housing and Home. Elsevier, London

Angel S (2000) Housing policy matters: a global analysis. Oxford University Press, Oxford

Angel S (2011) Making room for a planet of cities. Lincoln Institute of Land Policy, Cambridge, MA

Araud C (1973) Direct and indirect employment effects of eight representative types of housing in Mexico. In: Araud C et al (ed) Studies on employment in the Mexican housing industry. OECD, Paris

Architecture2030 (2007) Climate change, global warming, and the built environment: Architecture 2030. Architecture 2030. Available via: http://www.architecture2030.org/home.html

Brown T, Bhatti M (2003) Whatever happened to 'housing and the environment'? Hous Stud 18:505–515

D'Amico F (2011) O Programa Minha Casa, Minha Vida e a Caixa Econômica Federal. In Costa J C et al (eds) O desenvolvimento econômico brasileiro e a Caixa: trabalhos premiados. Centro Internacional Celso Furtado de Políticas para o Desenvolvimento: Caixa Econômica Federal. Rio de Janeiro

Davis M (2005) Planet of slums. Verso, London

FGEF (2003) Environment and energy savings in the building sector in China. www.ffem.net

French M (ed) (2011) Change by design: building communities through participatory design. Urban Culture Press, Napier

Giok Ling O (2005) Sustainability and cities: concept and assessment. Institute of Policy Studies and World Scientific, Danvers

Gonçalves R, Cielici E, Castelo AM (2011) Cenários para 2012 e a confiança dos empresários da construção. Conjuntura da Construção. Fundação Getúlio Vargas/IBRE/Sinduscon-SP. São Paulo. Dez. 2011, pp 4–8

Guha-Sapir D, Santos I (eds) (2012) The economic impacts of natural disasters. Oxford University Press, Oxford

Imparato I, Ruster J (2003) Slum upgrading and participation: lessons from Latin America. The World Bank, Washington, DC

Intergovernmental Panel on Climate Change (IPCC) (2007) Climate change 2007: synthesis report. Available via: http://www.ipcc.ch/pdf/assessment-report/ar4/syr/ar4_syr.pdf

International Federation of Red Cross and Red Crescent Societies (2010) World disasters report. IFRC. Available via: http://www.ifrc.org/en/publications-and-reports/world-disasters-report/report-online/

International Monetary Fund (2008) The changing housing cycle and the implications for monetary policies, Cycle Chap. 3. In: World economic outlook, housing and the business. IMF, Washington, DC

Kajumulo Tibaijuka A (2009) Building prosperity: the centrality of housing in economic development. UN-Habitat, Nairobi

Kellett P, Tipple G (2002) Home-based enterprise and housing policy: evidence from India and Indonesia. ENHR conference, Vienna, 1–5 July 2002. Available via: http://www.dfid.gov.uk/r4d/pdf/outputs/R71388.pdf

Levine M, ÜrgeVorsatz D, Blok K, Geng L, Harvey D, Lang S, Levermore G, Mongameli Mehlwana A, Mirasgedis S, Novikova A, Rilling J, Yoshino H (2007) Residential and commercial buildings. In: Metz B, Davidson OR, Bosch PR, Dave R, Meyer LA (eds) Climate change 2007: mitigation. Contribution of working group III to the fourth assessment report of the intergovernmental panel on climate change. Cambridge University Press, Cambridge

Nässén J et al (2007) Direct and indirect energy use and carbon emissions in the production phase of buildings: an input–output analysis. Energy 32:1593–1602

Neuwirth R (2004) Shadow cities: a billion squatters, a new urban world. Routledge, London

Parker D (2009) Very low energy homes in the Unites States: perspectives on performance from measured data. Energy Build. doi:10.1016/j.enbuild.2008.11.017

Payne G, Majale M (2004) The urban housing manual: making regulatory frameworks work for the poor. Earthscan, London

Pugh C (2000) Squatter settlements: their sustainability, architectural contributions, and socio-economic roles. Cities 17(5):325–337

Pugh C (2001) The theory and practice of housing sector development for developing countries, 1950–99. Hous Stud 16(4):339–423

Rapoport A (1988) Spontaneous settlements as vernacular design. In: Patton C (ed) Spontaneous shelter. Temple University Press, Philadelphia, pp 51–77

Rapoport A (2000) Theory, culture and housing. Hous Theory Soc 17(4):145–165

Saunders D (2011) Arrival city: how the largest migration in history is reshaping our world. Windmill Books, London

Sethuraman S (1985) Basic needs and the informal sector: the case of low income housing in developing countries. International Labour Organisation (ILO), New York

Spence R, Wells J, Dudley E (1993) Jobs from housing: employment, building materials and enabling strategies for urban development. Intermediate Technology Publications, London

Straussman W (1976) Measuring the employment effects of housing in developing countries. Economic development and cultural change 24(3):623–632

Sullivan E, Ward P (2012) Sustainable housing applications and policies for low-income self-build and housing rehab. Habitat Int 36:321–323, p 313

Tipple G (1993) Shelter as workplace: a review of home-based enterprise in developing countries. Int Labour Rev 132(4)

Tipple G (2000) Extending themselves: user initiated transformations of government-built housing in developing countries. Liverpool University Press, Liverpool

Turner J (1976) Housing by people: towards autonomy in building environments. Marion Byers, London

UNCHS (1988) Global shelter strategy to the year 2000. UNCHS, Nairobi

UNEP (2009) Buildings and climate change. Summary for decision-makers. UNEP, Nairobi

UNEP (2011) Towards a green economy: pathways to sustainable development and poverty reduction. UNEP, Nairobi

UNEP, ILO, IOE, ITUC (2008) Green jobs: towards decent work in a sustainable low-carbon world. UNEP, Nairobi

UNESCO (2001) The universal declaration on cultural diversity. UNESCO, Paris

UN-Habitat (2003) The challenge of slums: global report on human settlements 2003. UN-Habitat, Nairobi
UN-Habitat (2005) Global urban observatory. Urban indicators programme, Phase III. UN-Habitat, Nairobi
UN-Habitat (2006) Enabling shelter strategies: review of experience from two decades of implementation. UN-Habitat, Nairobi
UN-Habitat (2009a) State of the world's cities 2008/2009 – harmonious cities. UN-Habitat, Nairobi
UN-Habitat (2009b) Darfur early recovery, stabilized soil blocks for sustainable urban growth. UN-Habitat, Nairobi
UN-Habitat (2010) The state of African cities 2010. UN-Habitat, Nairobi
UN-Habitat (2011a) Condominium housing in Ethiopia: the integrated housing development programme. UN-Habitat, Nairobi
UN-Habitat (2011b) Affordable land and housing in Asia. UN-Habitat, Nairobi
UN-Habitat (2011c) Affordable land and housing in Europe and North America. UN-Habitat, Nairobi
UN-Habitat (2011d) Affordable land and housing in Latin America and the Caribbean. UN-Habitat, Nairobi
UN-Habitat (2011e) Economic development and housing markets in Hong Kong and Singapore. The global urban economic dialogue series. UN-Habitat, Nairobi
UN-Habitat (2012a) Sustainable housing for sustainable cities: a policy framework for developing countries. UN-Habitat, Nairobi
UN-Habitat (2012b) Going green: a handbook of sustainable housing practices in developing countries. UN-Habitat, Nairobi
UN-Habitat (forthcoming) Economic and financial assessment of the use of Stabilized Soil Blocks (SSB) technology in Khartoum and Darfur States, Sudan. UN-Habitat, Nairobi
UN-Habitat and UNHCR (2009) The right to adequate housing. Fact Sheet No. 21. (Rev. 1) UNHCR, Geneva
United Nations (2011a) The millennium development goals report 2011. United Nations, New York
United Nations (2011b) Working towards a balanced and inclusive green economy: a United Nations system-wide perspective. Environment Management Group, Geneva
Urge-Vorsatz D, Novikova A (2008) Potentials and costs of carbon dioxide mitigation in the world's buildings. Energ Policy 36:642–661
Utama A, Gheewala SH (2009) Influence of material selection on energy demand in residential houses. Mater Des 30:2173–2180
World Bank (2008) Migration and remittances factbook. The World Bank, Washington, DC
World Commission on Sustainable Development (1987) Our common future, report on the world commission on environment and development. World Commission on Environment and Development, Published as Annex to General Assembly document A/42/427, Development and International Cooperation: Environment.
World Habitat Awards (n.d.) Self-managed vertical housing. Available via: http://www.worldhabitatawards.org/winners-and-finalists/project-details.cfm?lang=00&theProjectID=8D0EC1AA-15C5-F4C0-9947239B7C17A845
Yanyun Man J (ed) (2011) China's housing reform and outcomes. Lincoln Institute of Land Policy, New Hampshire

EcoMobility and Its Benefits in an Urban Context

Santhosh Kodukula

Abstract Cities in both the developing and developed world are at cross roads in making the right decisions for their sustainable future. The decisions taken now will frame the future of how people live and travel in cities. EcoMobility enables cities to promote travel choices and break the vicious cycle of automobile dependency. It pedals cities towards socially, environmentally and economically sound transport.

The idea of EcoMobility is to promote integration between various transport modes in a city and increase the travel options for all citizens. Studies cited in this chapter have shown that fostering EcoMobility principles has resulted in positive economic effects for many cities, in terms of job creation, increased retail volumes, rising property values, reduced accidents and health related problems. The impending issues of climate change and deteriorating air quality can therefore be affectively addressed without impeding economic effects, through the promotion of EcoMobility in cities.

Keywords EcoMobility • Economic benefits • Climate change • Air quality • Quality of life

1 Introduction

Cities are increasingly becoming dependent on personal automobiles, and being rapidly redesigned in favor of personal motorized travel. Policies majorly focus on motorization and non-motorized modes such as walking and cycling are often neglected, making them unsafe and unattractive. The automobile centered evolution of several cities in the United States lead to less dense, car dependent, resource

S. Kodukula (✉)
ICLEI – Local Governments for Sustainability, Bonn, Germany
e-mail: Santhosh.kodukula@iclei.org

R. Simpson and M. Zimmermann (eds.), *The Economy of Green Cities: A World Compendium on the Green Urban Economy*, Local Sustainability 3,
DOI 10.1007/978-94-007-1969-9_25,

inefficient and transit unfriendly agglomerations (Newman and Kenworthy 1999). The developing world cities such as Kuala Lumpur, Bangkok and Jakarta were inspired by the US and tried to build their way out of an urban transport crisis. This approach not only deteriorated the condition of the transport network but also worsened air quality, road safety, quality of life and economic activity.

Infrastructural changes such as expanding and building more roads to handle escalating motorization have been proven to generate more traffic, inducing demand rather than easing congestion. This automobile centered approach is replicated in many developing countries without a perception of the long-term ill effects of excessive automobile use. The results of excessive automobile dependency are in the form of increased congestion, increased consumption of resources and decreased road safety, wherever there is an increased average traffic speed and increased road space for vehicles (Engwicht 1993).

Urban infrastructure is expensive and needs proper planning and design for the greatest prospect of being adaptable to changed future priorities and possibilities (Stretton 1976). In this respect, two key future challenges of cities will need to be addressed: their high-energy consumption, and the heavy reliance on the fossil fuel powered motor vehicles.

One of the largest contributors to climate change in terms of CO_2 emissions is transport. In a 2004 study of 25 EU countries the World Health Organization (WHO) found that transport accounted for approximately 35% of total energy consumption. Dependence on fossil fuels has resulted in a 20% net increase of greenhouse-gas emissions since 2002, due to a rise in surface transport volumes that outweighs improvements in vehicle efficiency (WHO 2012). This proves that vehicle efficiency might not be the only or most appropriate means to tackle automobile dependency. Air quality also deteriorates with the increase in personal automobile use. More than 40 million people in 115 EU cities are exposed to air quality that is lower than the WHO air quality standards for at least one air pollutant (WHO 2012).

Present day cities need a transportation system that serves the mobility needs of all social groups in terms of promoting livability and physical activity. In addition to the social benefits, transport systems should also cause a low environmental impact compared to current transport modes and promote economic activity. This chapter attempts to define the concept of EcoMobility and the various benefits a city should receive if EcoMobility principles were promoted. The chapter elaborates the economic, social, and environmental benefits of EcoMobility and reiterates the various disadvantages of excessive dependency on personal automobiles.

2 EcoMobility as Automobile Independency

Transport and land-use policies determine a city's future transport system efficiency and diversity. Many current land-use policies unintentionally favor automobile transport over other transport modes, and some favor sprawl over more compact, multi-modal community development. Though these policies may seem justified in isolation, their cumulative impacts are harmful when they minimize transport choices.

Automobile-dependency can be defined as the unavoidable reliance on an automobile for everyday commutes. It is a principal symptom of the lack of alternatives, or when available alternatives are inefficient, unappealing and/or unsafe. An "automobile" includes cars, scooters, and motorcycles (Kodukula 2011).

Developing country cities are particularly vulnerable to the costs and risks of automobile dependency. They lack resources to build extensive highway systems and heavily subsidized parking facilities. In many developing cities, most households cannot afford an automobile, or would be financially burdened if they were forced to own and operate a car due to inadequate alternatives. On the other hand the priority given to personal automobiles is increasing in many developing countries, as they remain a status symbol for the middle and lower income groups, indicating a lack of awareness of the long-term negative effects of automobiles.

The economic costs of expanding/constructing new roads, subsidizing parking facilities, importing fuel, and dealing with accident and pollution damages are substantial, and may become even larger in the future if automobile dependency increases. Many developing countries rely on imported fossil fuel and increasing automobile dependency would further burden the country (Kodukula 2011).

A study conducted in Sydney, Australia by Diesendorf (2002) revealed that the total cost of automobile transport was 1.5 times the cost of train travel and about 2.0 times the cost of a bus travel. While all three modes receive public subsidy, the annual subsidy for automobile travel is the highest by far and is second largest (to the heavy rail) in terms of dollars spent per passenger per kilometer travelled.

In the environmental impact report for the University of California – Los Angeles's new parking structure, Shoup (2005) cites that the total external cost for each parking space is US$116.86. The external costs include the cost of the congestion (US$72.70) produced per parking space and the emission cost per parking space (US$44.16).

The 2011 Urban Mobility Report by the Texas Transportation Institute shows from research made in 439 urban areas in the United States, that congestion cost (waste of fuel and time) in 2010 was US$115 billion. A study by the same research institute in 2006 revealed that the congestion cost was US$113 billion (considering same dollar value as the 2010 figure). The average waste of fuel in 2010 was approximately 1.9 billion gallons in the 439 agglomerations (Schrank et al. 2011).

In order to reduce the costs of personal automobile dependency, a shift is required to modes of transport that provide greater benefit in terms of infrastructure and in terms of time spent commuting.

EcoMobility is a concept that attempts to shift the focus of urban transport from personal motorized modes to the integration of sustainable modes such as walking, cycling, wheeling, and public transport.

EcoMobility can be defined as a transportation system that is integrated, socially inclusive and environmentally sustainable. The main emphasis of EcoMobility is on the integration of different transport modes in a city. Integrating these modes will increase the options available for all the users in the city.

There is already a certain level of EcoMobility present in some cities. For example, many Indian cities have high levels of walking and cycling, but unfortunately these volumes are mostly limited to low income social strata. In addition, the facilities for these modes are neglected, thus reinforcing the perception that walking and cycling are related to low-income households. Many richer and economically advanced counterparts in Europe have promoted walking and cycling as a physical and social activity by creating people centered cities. This results in vibrant pedestrian areas where people spend their leisure time free from the hindrance of automobile fumes and enjoy higher road safety. Cities such as Copenhagen, Amsterdam, and Münster (to name a few), where the Gross Domestic Product (GDP) is higher than many car dependent cities, have a high bicycle usage, pedestrian activity and modal integration (Newman and Kenworthy 1999; Kodukula 2011).

As stated earlier, the major priority of EcoMobility is that the travel options available in a city are increased, chiefly by integrating the existing modes and also promoting walking, cycling, wheeling and passenging. Wheeling can be defined as the use of any non-motorized vehicle with wheels including skates, wheelchairs, tricycles, rickshaws etc. Passenging is defined as the use of any mass or public transport system running on the road or on rails.

3 EcoMobility Benefits for Cities

Promoting EcoMobility in cities will influence all the three major constituents of sustainability: the social, the economic and the environmental. The following sections will illustrate the influence of ecomobile modes on sustainability:

3.1 Economic Advantages of EcoMobility

EcoMobility has the ability to improve economic conditions by promoting green jobs through the creation of public spaces. Cities that face high automobile dependency have large economic costs in terms of built infrastructure and also in terms of costs due to traffic congestion. A study in Jakarta conducted in 2003 has shown that an estimated $68 million is wasted every year due to congestion (Asri and Hidyat 2005). This cost does not include the health impacts from increased air pollution due to congestion. A similar situation exists in major Chinese cities. In Beijing, for instance, 40% of commuters spend more than 1 h travelling per day, resulting in an economic loss of 40 million Yuan per day in 2008, while in Shanghai the GDP loss due to congestion is 10% (Hongyang 2008).

In addition to the time and fuel wasted through congestion, emission costs are also related to congestion. Vehicles stuck in congestion emit harmful gases that cause respiratory diseases and may also be fatal depending on the exposure levels and durations. Levy et al. (2010) conducted a study in 83 US Cities and concluded that congestion related emissions resulted in 4,000 premature deaths in 2000, which is approximately equivalent to a monetary value of US$31 billion (as of 2007 dollar value).

Toronto, Canada, where 71% of the population is still dependent on a personal car for commute, loses approximately CAN$3.3 billion in traffic congestion and poses a grave threat to non-motorists (CBC 2009).

The current urban transport infrastructure in many cities is completely funded from taxpayers' money. Payment is made directly by automobile users through road and vehicle tax, and indirectly by all citizens through their other taxes. Non-motorists also pay for the health and quality of life costs incurred due to the excessive use of automobiles. While motorists receive immediate benefits for their direct costs in terms of road infrastructure (wider roads, flyovers/overpasses, free or cheap parking, lower fuel prices in several countries) for their automobiles, non-motorists do not receive equal benefit from automobile infrastructure. It is therefore unfortunate that unsustainable road infrastructure such as flyovers and road expansions are also largely funded by non-motorists. This depicts an inequity in urban transport infrastructure spending. In order to reverse this trend there is a need for motorists to pay the real costs of their travel and hence economic pricing instruments are required that would enable the city to charge the motorists for the actual costs incurred on their behalf.

Studies have shown that creating commercial districts keeping pedestrians and non-motorists as main actors help to promote businesses in such areas, and will not create a negative impact on the commercial activities (Kodukula and Ross 2006; Drennen 2003). Many business owners show concern about the parking spaces available to their customers and perceive that more parking will attract more customers. Such fears have been disproved by studies applying to pedestrianization and cycling (Drennen 2003). In actual fact, businesses tend to benefit from reducing or removing parking spaces – when streets are pedestrianized and bicycle lanes provided. Equally, poor pedestrian, cycling, and public transit options can harm businesses, reducing access for foot passengers, both customers and potential employees. Improving the above conditions will increase convenience for both.

Ecomobile modes of travel also increase property values in areas of implementation and attract a wealthier class of society by promoting people friendly areas, benefiting the overall retail sales and bringing economic advantage to the commercial sector (Sermons and Seredich 2001). Businesses benefit from more attractive environments, reduced traffic speeds and increased safety for walkers. Pedestrianization encourages local residents to buy in their own neighborhoods, and attracts more customers from a wider area, improving community relations. Pedestrianization restricts car use and thereby shifts emphasis to other means of non-motorized travel (Hass-Klau 1993). Colin Buchanan and Partners (2001) assessed the impact of pedestrian improvements made to Kensington High Street, London. They found that the majority of shoppers arrived on foot (49%) and that these people made more shopping trips than drivers. They also found that total spending was dominated by public transport users (49%) and walkers (35%).

Studies have shown that households with shorter distances to non-work destinations tend to walk more often (Boarnet and Greenwald 2001). Closeness and ease of access by walking also affects the number of recreational trips (Rajamani et al. 2003). McCann (2000) found in a study that people living in a walkable and public transit oriented community spend $1,200–$6,000 less on surface transport than those people who live in auto-dependent communities.

Further, various studies in the US (TRU 2012) have show that an investment into public transportation has resulted in positive effects in terms of the economy and job creation. Weisbrod and Reno (2009) argue that investment of one billion US dollars per year in public transport operations supported the creation and maintenance of 41,000 jobs. A capital investment of one billion US dollars in public transport resulted in supporting 21,000 jobs.

This section of this chapter has showed that the efficacy attributed towards automobile oriented dependence in terms of economic advantages is debatable. There is sufficient evidence to show that diverting the investments towards an integrated transport system, catering for the pedestrians, cyclists and other non-motorists and public transport users would create a positive economic impact on society. The following sections of this chapter will detail the various social and environmental benefits derived from EcoMobility.

3.2 Social Benefits of Promoting EcoMobility

Lack of provision of infrastructure for pedestrians, cyclists, public transport users and the physically challenged translates as neglect of such people by city officials. Wide roads and ample parking spaces mean that the motorists benefit disproportionately from current levels of transport investments. Introducing EcoMobility in a city will demonstrate that the city government prioritizes needs of all people rather than a specific group, the motorists.

As previously mentioned, EcoMobility consists in integrating the various modes available in a city. For instance, people who ride a bicycle will still be able to take public transport for their daily commute, as these modes will be integrated. Similarly, pedestrians are given designated space to walk on pavements where motorized vehicles cannot intrude on their space.

In terms of road safety, promoting walking and cycling would drastically reduce the amount of road accidents. This is an important issue in many European cities as a key external cost of transport in Europe is due to accidents, estimated €158 billion a year or approximately 2.5–3% of GDP in 17 Member states (WHO 2012). Studies in Austria reveal that approximately 412 lives could be saved every year if the modal share of cycling was increased from 2 to 10%. In monetary terms these savings would equal EUR 405 million a year (WHO 2012).

Providing for pedestrians involves a redesigning of roads and streets, such as narrowing entrances, building chicanes, imposing lower speed limits, road humps, and introducing benches, trees and other street furniture, to encourage greater pedestrian use. A study of controls in 65 residential zones showed that after pedestrianizing, traffic accidents decreased by 9%, deaths by 11% and injuries by 10% and that property damage was also reduced by 5% (OECD 1990). Experiments in Nippes and Agnesviertel, both in Cologne, Germany, experienced major reduction in noise (84%) and traffic injuries (74%).

It has been found that households in communities with good public transit, pedestrian and bicycle facilities tend either to sell their cars or to reduce car usage dramatically. Residents of such communities also increase their usage of public transit, walking, and cycling. In some cases car-sharing also has increased within such communities (Boarnet and Greenwald 2001). Rabl and Nazelle (2012) conclude that if every driver switched to a bicycle for a daily commute of 5 km, doing so for 5 days a week and 46 weeks per year, they could incur a health benefit worth 1,300 €/year.

The level of transit available to households (usually measured as the number of jobs accessible by transit in 30 min), the quality of pedestrian environment (measured as the number of retail jobs accessible by 20 min of walking), and the continuity of the street network have been shown to have a significant effect on the number of cars owned in a neighborhood (Lawton 2001).

Properly planned non-motorized transport will enhance quality of life in the city by reclaiming the space from automobiles and converting them into places for pedestrians and cyclists. Curitiba, Brazil and Bogota, Colombia are both cities that have implemented world-class public transport systems and by integrating cycle use into the fabric of the city, a real reduction in automobile use has been achieved. The transformation in both of these South American cities also included increasing the public space for leisure, through reclamation of land previously used for roads.

The promotion of EcoMobility is the promotion of a whole gamut of travel options that include planning for pedestrians, cyclists, people with disabilities, public transport users as well as automobile users in an integrated manner. EcoMobility does not attempt to abolish personal automobiles from cities but rather attempts to create an optimal mix and maximum choice of all existing modes, thereby reducing excessive dependence on personal automobiles.

This section of the chapter has shown that promoting EcoMobility within cities would not only provide an economic benefit but also provide social benefits in terms of encouraging healthy living and promoting social equity through access to all income groups and people with disabilities. As shown above, EcoMobility practices have the ability to drastically reduce personal automobile use. The next section of the chapter will focus on the environmental benefits derived from fostering EcoMobility practices.

3.3 Environmental Benefits of EcoMobility

Transport can be related to various activities in an urban setting, be it leisure, the commute to work, or education. Excessive dependence on personal automobiles and hence on fossil fuels catalyzes the depletion of fossil fuel resources and accelerates climate change.

Climate change brings various changes to the ecosystem as we know it. Melting of ice caps and glaciers, rise in sea levels, droughts and floods, changes in rainfall patterns are some of these effects. Some regions of the world will also experience extreme temperatures (IPCC 2007). World CO_2 emissions from the transport sector are projected to increase by 140% from 2000 to 2050, with the biggest increase in developing countries. Particularly China, India, Latin America, and other Asian countries, a rapid rise in two-wheeled vehicles is predicted. Between 2000 and 2050, two-wheeler fuel consumption is projected to rise by more than eight times, increasing the proportion of road vehicle fuel use attributed to two-wheelers from 2 to 3% (Dalkmann and Brannigan 2007).

Lindsay (2011) shows that shifting 5% of kilometers driven from motorized vehicles to bicycles would reduce automobile travel by approximately 223 million km/year, save about 22 million liters of fuel, and reduce transport-related greenhouse emissions by 0.4%. The modal shift created by ecomobile practices in cities will directly address the issue of climate change at a city level. Promoting attractive and affordable alternatives to personal automobiles will reduce trips made on motorized modes and thereby reduce greenhouse gas emissions.

This section has shown above that adopting EcoMobility practices will have positive effect on improving the air quality as the automobile use will be greatly reduced and replaced with more environmentally and people friendly modes of transportation.

4 Conclusions

This chapter has elaborated the concept of EcoMobility as a transportation system, which is integrated, socially inclusive and environmentally sustainable. Promoting EcoMobility can be done through taking up transport projects such as pedestrianization, prioritizing bicycle travel, increasing the coverage and appeal of the public transport system and finally integrating all these modes both physically and fiscally. EcoMobility can address issues such as climate change, social equity and quality of life in cities through better urban planning. The urban environment can be transformed into lively and vibrant areas through the promotion of ecomobile transport modes.

EcoMobility does not deny the benefits provided by automobiles, nor does it suggest that cities should become completely car-free. The concept proposes an efficient transport system, offering travelers a combination of automobile, non-motorized and public transport options, with incentives to encourage users to choose the most appropriate option for each trip. This would result in an efficient and equitable transport system that maximizes benefits to users and society overall.

References

Asri DA, Hidyat B (2005) Current transportation issues in Jakarta and its impacts on environment. Proc East Asia Soc Transp Stud 5:1792–1798

Boarnet MG, Greenwald MJ (2001) The built environment as a determinant of walking behavior: analyzing non-work pedestrian travel behavior in Portland, Oregon. Center for Activity System

Analysis. Available via: http://repositories.cdlib.org/itsirvine/casa/UCI-ITS-AS-WP-01-4. Cited 1 Apr 2012

CBC (2009) Traffic jams cost Toronto $3.3B per year: OECD. Availablevia: http://www.cbc.ca/news/canada/toronto/story/2009/11/10/oecd-traffic.html. Cited 1 Apr 2012

Colin Buchanan and Partners (2001) Kensington high street study. Colin Buchanan and Partners, London

Dalkmann H, Brannigan C (2007) Module 5e: transport and climate change, sustainable transport: a sourcebook for policy-makers in developing cities. Deutsche Gesellschaft für Internationale Zusammenarbeit (GIZ), Germany

Diesendorf M (2002) The effect of land costs on the economics of urban transport systems. In: Wang KCP, Xiao G, Nie L, Yang H (eds) Traffic and transportation studies. Proceedings of third international conference on traffic and transportation studies (ICTTS2002), Guilin, China, pp 1422–1429

Drennen E (2003) Economic effects of traffic calming on urban small businesses. Department of Public Administration, San Francisco State University, San Francisco

Engwicht D (1993) Towards an ecocity: calming the traffic. Envirobook, Sydney

Hass-Klau C (1993) Impact of pedestrianization and traffic calming on retailing: a review of the evidence from Germany and UK. Transp Policy 1(1):21–31

Hongyang WU (2008) Benchmarking efficiency of sustainable urban transport in China – 7 challenges & 5 strategic priorities. CUSTReC, China Academy of Transportation Sciences-CATS, MOT

IPCC (2007) Fourth assessment report: summary for policy makers, climate change 2007: the physical science basis. Intergovernmental Panel on Climate Change (IPCC). Available via: http://ipcc-wg1.ucar.edu/wg1/docs/WG1AR4_SPM_Approved_05Feb.pdf. Cited 20 Feb 2012

Kodukula SK (2011) Rising automobile dependency: how to break the trend? Sustainable urban transport technical document # 8. Deutsche Gesellschaft für Internationale Zusammenarbeit (GIZ), Germany

Kodukula SK, Ross W (2006) Effects of pedestrianisation on the commercial and retail areas: study in Khao Road, Bangkok. World Transp Policy Pract 13(1):37–47

Lawton KT (2001) The urban structure and personal travel: an analysis of Portland, Oregon data and some national and international data. Supplementary materials: papers and analyses for the E-Vision 2000 conference. Available via: http://192.5.14.43/content/dam/rand/pubs/conf_proceedings/CF170z1-1/CF170.1.lawto.pdf. Cited 31 Mar 2012

Levy JI, Buonocore JJ, von Stackelberg K (2010) The public health costs of traffic congestion: a health risk assessment. Environ Health 9:65. Available via: http://www.ehjournal.net/content/9/1/65. Cited 1 Apr 2012

Lindsay G (2011) Moving urban trips from cars to bicycles impact on health and emissions. Aust NZ J Public Health 35(1):54–60. ISSN: 1326-0200

McCann B (2000) Driven to spend: the impact of sprawl on household transportation expenses. Available via: http://www.transact.org/PDFs/DriventoSpend.pdf. Cited 1 Apr 2012

Newman P, Kenworthy J (1999) Sustainability and cities: overcoming automobile dependence. Island Press, Washington, DC

OECD (1990) Environmental policies for cities in 1990's. OECD, Paris

Rabl A, Nazelle A (2012) Benefits of shift from car to active transport. Transp Policy 19(1):121–131. ISSN: 0967-070X. Available via: http://dx.doi.org/10.1016/j.tranpol.2011.09.008

Rajamani J, Bhat CR, Handy S, Knaap G, Song Y (2003) Assessing the impact of urban form measures in non work trip mode choice after controlling for demographic and level-of-service effects. Transportation Research Board. Available via: http://trb.metapress.com/index/R47K227J474K1902.pdf. Cited Feb 2012

Schrank D, Lomax T, Eisele B (2011) TTI's 2011 urban mobility report. Texas Transportation Institute The Texas A&M University System. Available via: http://tti.tamu.edu/documents/mobility-report-2011.pdf. Cited 1 Apr 2012

Sermons W, Seredich N (2001) Assessing traveler responsiveness to land and location based accessibility and mobility solutions. Transp Res D Transp Environ 6(6):417–428

Shoup D (2005) The high cost of free parking. American Planning Association (APA), Chicago
Stretton H (1976) Capitalism socialism and environment. Cambridge University Press, Cambridge
TRU (2012) The economic benefits of public transit: essential support for a strong economy. Data from the American Public Transit Association, Summarized by Transportation Riders United (TRU). Available via: http://www.detroittransit.org/cms.php?pageid=26. Cited 1 Apr 2012
Weisbrod G, Reno A (2009) Economic impact of public transportation investment. American Public Transportation Association, Transit Cooperative Research Program (TCRP) Project J-11. Available via: http://www.apta.com/resources/reportsandpublications/documents/economic_impact_of_public_transportation_investment.pdf. Cited 1 Apr 2012
WHO (2012) Transport and health: facts and figures. World Health Organization (WHO). Available via: http://www.euro.who.int/en/what-we-do/health-topics/environment-and-health/Transport-and-health/facts-and-fiigures/injuries2. Cited 20 Feb 2012

Assessing the Cost of Groundwater Degradation in the Urbanizing Desert Area of Wadi El Natrun

Caroline King and Boshra Salem

Abstract This chapter shows how economic models and analysis can be used to capture the cost of environmental degradation. A series of approaches to assessing the costs of environmental degradation are explored in the context of a fast-growing desert settlement on the edge of Egypt's Western Desert. First, a modeling approach focusing on the local agro-industrial economy and the loss of productivity caused by degradation in groundwater availability and quality is considered. This is based on integrating available environmental and economic models of the area. A second modeling approach isolates replacement costs associated with loss of artesianism and the failure of wells as groundwater extraction increases. Third and finally, a holistic approach to the valuation of the degradation of the desert ecosystem and its natural services is outlined. The final discussion considers the perspective for further use of these methods by local decision-makers at the Wadi El Natrun City Council, and other local authorities in rapidly developing regions.

Keywords Economic assessment • Environmental degradation • Groundwater

1 Introduction

The transition to a green economy calls for a strategic reorientation of consumption, investments and other economic activities towards enhancing resource efficiency (Pearce et al. 1989; Le Blanc 2011; UN 2012). Green economic assessments can help local decision-makers to identify the best available strategies for simultaneously

C. King (✉)
Oxford University Centre for the Environment, Oxford, UK
e-mail: caroline.king@ouce.ox.ac.uk

B. Salem
University of Alexandria, Alexandria, Egypt
e-mail: boshra.salem@dr.com

R. Simpson and M. Zimmermann (eds.), *The Economy of Green Cities: A World Compendium on the Green Urban Economy*, Local Sustainability 3,
DOI 10.1007/978-94-007-1969-9_26,

enhancing economic sustainability *as well as* protecting ecosystems and human well-being (Hassan et al. 2005; TEEB 2010). This chapter focuses on the challenge for local decision-makers to incorporate green economic modeling and assessment techniques into ongoing decision-making concerning natural resource uses. For this purpose, a case study addressing groundwater resource development and conservation is considered in the context of a rapidly developing region in North Africa.

In the water sector, pressing challenges concern the provision of water supplies for economic activities and domestic uses, while also ensuring the conservation of the ecosystems on which the future provision of such resources depends (JMP 2010; IBRD 2010). Aquifers provide natural underground water supply systems that effectively distribute water for drinking and irrigation use across vast areas without the need for construction of pipes or pumping infrastructure (WWAP 2009; WDR 2010). Where decision-makers decide to alter the function of these ecological systems, the long- and short-term consequences of their interventions require careful consideration. Extraction of the natural resource for one use may compromise its availability and quality for another (see e.g. Gisser and Sanchez 1980; Hellegers et al. 2001; Shiferaw 2008; Pulido-Velazquez et al. 2008). Over-extraction can eventually threaten the future of the system as a whole, and all those that depend upon it.

In this chapter, we consider the application of simple models for green economic assessment of land-use planning and groundwater regulation decisions in an urbanizing area in the Western Desert of Egypt.[1] Due to the rapid pace of economic development in this area, a series of such models have already been applied. These provide examples of available tools and their potential to support local decision-making in other developing regions (Slootweg 2010; Goodstadt et al. 2010). Following a brief conceptual overview, three simple approaches to economic modeling and assessment are reviewed. We demonstrate the accessibility of these tried and tested methods, and consider the outcomes generated from their application to a single selected case study area. We then provide a discussion of some of the limitations of each approach for use in this and other local development contexts. The final discussion invites local decision-makers to make use of the insights gained from this case to shape the green economic models that will work best for them.

2 The Challenge: Capturing the Cost of Environmental Degradation

Following Local Agenda 21, the principle of sustainability has been widely accepted by local decision-makers, but its operationalization has remained challenging (UN 2012). At the global level, an increasing array of conceptual tools and assessment

[1] Such models can be applied easily by local decision-makers using either back-of the envelope calculations, or with the support of widely available spreadsheet softwares, e.g. Excel.

initiatives has emerged (see an overview in UNEP 2009). These have generated one-off multi-scale global assessments incorporating economic analysis of environmental change (Hassan et al. 2005; Sukhdev 2010), as well as periodic assessments of the potential costs and benefits of adaptation to climate change (e.g. Parry et al. 2007), and the achievement of the Millennium Development Goal for environmental sustainability. These global assessments necessarily connect methodologies and statistical activities at other scales, including regional, national and local scales, and involve various sectors, including the water sector (see e.g. UNSD 2005, 2010; WWAP 2009; JMP 2010; EEA 2011; IBRD 2010). However, the use that has been made of green economic assessment tools at the local level in developing countries is still considered to be limited (Goodstadt et al. 2010).

For local decision-makers, assessing the costs of degradation can help in the selection of land and groundwater management options that best fit local aspirations towards a green economy. Green or otherwise, economic assessment is increasingly used directly by local decision-makers as an element of Environmental Impact Assessment (EIA) and Strategic Environmental Assessment (SEA) (see review in Hussain and Gundimeda 2010). It can also enable local decision-makers to inform decisions taken at other levels of government (e.g. concerning decisions relating to taxes, subsidies, trade regulations, etc.) (TEEB 2010). Emerging green economic measures of potential interest to local decision-makers include Payments for Environmental Services (PES) (Engel et al. 2008; Barbier 2011). Such measures can be applied alongside other local environmental management measures to regulate the externalities of agricultural and other land uses affecting groundwater availability and quality (Abildtrup et al. 2012).

3 The Tools: Assessing the Cost of Groundwater Degradation: A Conceptual Overview

The purpose of economic valuation is to make disparate decision outcomes affecting resources, income streams and expenditures incurred comparable to each other. Monetary units are readily understood by decision-makers and the lay public, and facilitate comparison with budgets and priorities for spending by local decision-makers (DeFries et al. 2005). A range of methodological guides have appeared to support local decision-makers in the application of emerging valuation techniques, including applications to assess the economic significance of changes in water resource conditions (TEEB 2010; Ash et al. 2010). Available approaches for valuation of environmental outcomes are commonly considered in terms of their relation to existing market values (see Table 1). Where prices do not exist, these can often be derived using available green economic methods. In some cases, naming a price may be controversial. Nevertheless, although imperfect, (green) economic valuation remains the most readily available means through which utility can be effectively quantified and compared (Bolt et al. 2005, see Kumar 2010).

Table 1 Menu of methods for valuation of changing groundwater conditions (adapted from DeFries et al. 2005; Bolt et al. 2005; Hussain and Gundimeda 2010)

Type	Methodology	Approach	Data requirements
Direct market price	Market price	Price per unit of water	Water user fees, unit price
Market alternative	Productive value	Value-added based on input to production function	Market price of produce, input prices
Market alternative	Human capital	Impact on work-days, morbidity and mortality	Health effects, wage rates, value of a life
Market alternative	Replacement cost	Cost to replace and/or treat groundwater supply	Cost of technologies
Surrogate market	Travel cost	Demand curve generated from transportation costs	Survey or participatory workshop
Surrogate market	Hedonic price	Premium for environmental quality e.g. in housing market	Market information
New market	Benefits transfer	Comparison to values generated elsewhere	Previous studies
Non-market (stated preference)	Contingent valuation	Direct questions on willingness to pay – e.g. for additional units of irrigation water	Survey or participatory workshop
Non-market (stated preference)	Choice experiments	Questions on preferred options from a range	Survey or participatory workshop
Non-market (stated preference)	Relative value	Questions on relative value of a range	Survey or participatory workshop

Note: The methods are not exclusive and may often be combined – e.g. contingent valuation can be used to assess willingness to pay replacement costs (see e.g. Tentes and Damigos 2012)

Generic challenges in valuation of environmental changes include identification of the spatial and temporal boundaries and functions of complex systems. Groundwater systems are often multi-layered, extensive and porous, frequently lacking comprehensive information on physical conditions and trends (see Puri and Aureli 2009). Groundwater users can also be diverse and unpredictable. To define and assess scenarios of future impacts from local decision-making requires information concerning their behavior, as well as other unknown market- and risk-related factors (Pascual et al. 2010). For the construction of groundwater-related infrastructure in developing regions, discount rates are usually high. Green economic assessments of future resource degradation have to face both technical and ethical considerations to rebalance immediate development priorities and resource use patterns with future sustainability. Other challenges may concern the distribution of economic effects of degradation amongst groundwater users.

Environmental models and economic models are separate tools that can be effectively combined to generate comprehensive assessments weighing up local development choices in the form of decision-making scenarios. The various components of the assessment process require different skills to conceptualize and operationalize, and may fit together in a range of different ways. Sometimes, models must be complex, in order to accurately capture and reflect environmental and social conditions. At others, simplicity may be required in order to ensure that the factors on which the decisions are based can be easily understood by non-specialist stakeholders. A stepwise process towards an economic assessment of deterioration in water quality or quantity might approach these methodological choices as follows (after Hussain and Gundimeda 2010):

- Step 1: Specify and agree on the problem
- Step 2: Identify which ecosystem services are relevant to the decision
- Step 3: Define the information needs and select appropriate methods
- Step 4: Assess the expected changes in the flow of ecosystem services
- Step 5: Identify and value the policy options
- Step 6: Assess distributional impacts of policy options

In the remainder of this chapter, we consider a series of examples of the use of green economic assessment methods in assessing the cost of groundwater degradation in a single case study. This enables the varying outcomes and recommendations from each assessment to be contrasted and compared. Limitations specific to each of the methods and to their applications in the examples provided are also identified, revealing the choices made by decision-makers at each of the six stages of the assessment.

4 The Case of Wadi Natrun: A Rapidly Growing Human Settlement in a Water Scarce Environment

Wadi Natrun is an administrative centre located in the frontier Governorate of Behayra, in the Western Desert of Egypt. The rapidly growing urban area provides a center for the administration of five surrounding villages, a number of small hamlets, and a large expanse of farm and desert land. The total population of the Wadi Natrun administrative area (Markaz) has risen from around 25,000 in 1996 to well over 70,000 people in 2006 (CAPMAS 2007).

The extent of agricultural activity is also rapidly expanding, due to increasing use of the groundwater resource for irrigation. According to the Egyptian Agricultural Census, the cultivated area around the villages and town had approximately doubled in size over the period 2000–2010, to cover an extent close to 1,000 hectares. Administrative divisions for the surrounding area had changed between census dates, but by 2010, these outlying areas added a further 3,000 ha, bringing the total cultivated area to around 4,000 ha, and still growing.

According to the 2007 population census, irrigated agriculture accounted for 55% of local economic activity. Other economic activities were in industries servicing the agricultural sector, e.g. transportation and storage (8%), retail and servicing of vehicles (10%). Intensive agricultural use of each hectare of land in this region is estimated to provide four full-time jobs for agricultural workers, plus an equal number in the service sectors (i.e. a total of eight jobs per hectare) (World Bank 2007).

However, not all of the farmland is cultivated intensively. Farming patterns present in the area also include smallscale subsistence farming, and sporadic speculative cultivation by absentee investors.

Although the climate is arid, a natural depression in the desert landscape supplies the area with drainage water from the cultivated lands in the Nile delta and any rainfall that occurs in this and the surrounding areas. Drainage water enters the study area from the East through a shallow aquifer, known as the Pleistocene, terminating in a depression containing a series of lakes. The main aquifer in the study area is the Pliocene aquifer. This is estimated to receive around 60 mm^3 of recharge per year, coming from both the Pleistocene aquifer, above, and the deeper Miocene aquifer, below (RIGW 1990; Fekry 1993). The Miocene aquifer extends West and Southwards beneath the Western Desert, and contains mostly non-renewable groundwater resources.

Since the 1960s, increasing volumes of groundwater have been extracted from this aquifer for irrigation use (Idris and Nour 1990). Groundwater resources have been extracted from the study area in excess of recharge, the water table has been falling and salinization has been observed (Ibrahim 2005; Masoud and Atwia 2011).

5 Example 1: Lost Productivity Value of Groundwater Degradation

5.1 Overview of the Assessment Method

The cost of groundwater degradation can be calculated as the value of foregone production due to degradation of the resource as an input to irrigation. This is based on the net market value of crops produced by subtracting all other input costs from the output value of the crop. The method is relatively well established in agricultural economics (see Acharya and Barbier 2000). Projections of the costs of degradation rely on the identification of a production-function relationship between the resource and its net productive value. Models can either be linear, or they may be complicated by progressive changes in the production-function. These might include changes in the value of any of the inputs to production, as well as changes in the anticipated ratio of crop production per unit of water input that may result from degradation in the condition and availability of the groundwater resource (see FAO 2002; Shiferaw 2008).

5.2 Application to Assessment of Groundwater Degradation in Wadi Natrun, 2007–2017

A production-oriented method was applied in a series of assessments of the economic impacts to be anticipated from the proposed construction of a canal bringing water from the Nile to irrigate an 80,000 ha area located to the South of Wadi Natrun

(Slootweg 2010). The assessment involved a series of impact scenarios, including environmental impacts on Wadi Natrun, which is situated directly downstream in the groundwater drainage flow path from the proposed development and economic impacts on the Egyptian economy. Following an initial feasibility study, a detailed assessment tested three alternative strategies for the conjunctive use of ground water and surface water from the proposed canal over a varying surface area. This assessment was used to refine the design of the project, and to determine any necessary measures for mitigation of its impacts. A final assessment then focused two scenarios, with- and without- the project (World Bank 2007). The scope of the assessment methods deepened progressively through the iterative assessment phases, moving from back of the envelope calculations to increasingly detailed scientific work and public consultation.

5.3 Approach to Data Collection, Environmental and Economic Analysis

A series of environmental and economic modeling techniques were integrated to support the assessment. The assessment framework, termed DRAINFRAME, drew on precedents in the drainage engineering sector. Changing groundwater levels were simulated using a groundwater flow simulation package (TRIWACO). This was combined with simulation of water scarcity and salinity effects on cropping patterns (Attia et al. 2005). Further interviews with farm managers, farm owners, investors, semi skilled laborers, skilled laborers (both residents and commuters) were conducted, and a Farmer Advisory Council (FAC) was created. Crop budget studies for ten major crops in the project area were used to develop scenarios simulating economic impacts (World Bank 2007). The valuation used economic prices for inputs and outputs, based on the international value of the commodity at border, adjusted for processing and transportation costs.[2] Job creation was estimated on the basis of available estimates of jobs created per hectare of cultivated land in agriculture and related sectors (see Sect. 4).

5.4 Outcome, Recommendations and Limitations

Effects on the natural environment in Wadi Natrun due to alteration of the drainage and recharge pattern were anticipated to be positive, and not to require further consideration under the Egyptian Natural Habitats Policy (OP 4.04), since the area had no officially protected status (World Bank 2007). The without-project scenario, showing a steady decline in production across the region was compared to the future with the project, in which local job creation would be increased. The application of

[2] Required price information is readily available for users in most developing countries through FAOSTAT and UN COMSTAT online databases.

this approach focusing on productive value builds in assumptions concerning what should be produced, and by whom. In this case, the focus on intensive agricultural export production works on an assumption that this is the optimal development strategy – leaving aside other sectors such as water bottling, hospitality and recreation. The application of the method also did not take into consideration the diversity of producers in the area, which includes small-scale and subsistence farmers, as well as export producers. A subsequent consultants' report investigated capacity-related assumptions concerning income generation by local farmers (AFD 2008).

6 Example 2: Replacement Cost Assessment of Groundwater Degradation

6.1 Overview of the Assessment Method

The replacement cost method focuses on the costs of engineering man-made substitutes for natural resources. If the objective in a green economy is to transform the relationship between society and nature, rather than to maintain it, replacement may not be a sufficient objective. Nevertheless, as a valuation method that is well established in environmental economics, this may provide a useful starting point for an assessment of the costs of groundwater degradation. The following general conditions have been suggested for use of replacement cost methods in valuing natural resources (Bockstael 2000):

- the replacement provides functions that are equivalent in quality and magnitude to the loss;
- the least costly option available for replacement is identified;
- individuals in aggregate would be willing to incur the replacement costs.

The method is readily used to capture degradation of groundwater availability and quality in terms of the marginal increase in extraction costs with falling artesianism (Croitoru and Sarraf 2010). Other possible alternative technologies for replacement would include estimation of the costs of water treatment or artificial recharge interventions.

6.2 Application to Assessment of Groundwater Degradation in Wadi Natrun, 1994–2005

This method has been used to assess groundwater degradation costs to smallholder farmers in Wadi Natrun using shallow wells for pumping water from lower depths and digging replacement wells over the past 20 years (King 2011). The assessment was developed as an academic project, using techniques in environmental assessment that have appeared in the published literature on economic assessment of groundwater degradation in developing countries. The application of the method focused on differentiation of the impacts of degradation amongst large and small farmers, using deep and shallow wells (King and Salem 2012). The objective was to illustrate the social dimension of the common pool resource management challenge (after Ostrom 1990; Sekar 2001; Shah 2009). As in the previous example, the assessment drew on surveys characterizing local farm budgets for farms of varying sizes (Yehia 2004; AFD 2008; King and Salem 2012). The focus remained on local agricultural groundwater users, nevertheless, the replacement cost approach and method could also accommodate consideration of costs to other existing groundwater users in the area, including factories, water bottling facilities, and water utilities.

6.3 Approach to Data Collection, Environmental and Economic Analysis

Each new well and increase in volumes of water pumped causes loss of artesian pressure to affect all surrounding well-owners. Furthermore, in some cases, as pressure is reduced, wells will cease to provide water altogether, and will then require replacement. An estimate of the replacement cost can therefore be based on projection of the marginal costs for increased pumping and the replacement of wells. A simple spreadsheet model was applied (after Croitoru and Sarraf 2010) to capture these replacement costs as the water table lowered over the period 1994–2005. Irrigation water requirements were multiplied by the additional fuel requirement for pumping from greater depth, and by the cost of the fuel. The number of wells replaced per year, well depths, drilling costs and pumping requirements were modeled based on local estimates generated during a field survey. The cost assessment generated was compared to smallholders` farm budgets using local market prices (King 2011).

6.4 Outcome, Recommendations and Limitations

This approach enabled trends in increasing replacement costs to smallholder farmers associated with groundwater degradation to be modeled. The trend in increasing costs of degradation to smallholders was affected by international market effects on exchange rates and fuel prices. The cost of groundwater degradation to smallholder farmers was demonstrated to exceed the incomes that they generated from agricultural production (King and Salem 2012). This was contrasted to the situation for larger farmers, whose replacement costs were equal to only a small fraction of the net value of their produce. Recommendations from this assessment to prevent the escalation of the groundwater pumping race included raising local awareness, participatory monitoring and local innovation for water-saving and recharge.

The assumption that the resource is replaceable is not always the case in areas where recharge rates are low and groundwater dates from previous more humid periods. If over-pumping continues beyond a critical threshold, the function of the system may undergo a permanent alteration, or collapse entirely. The choice of replacement technology, and definition of what is to be replaced, is critical in use of this method. If it is the functioning ecosystem as a whole that should be replaced, other natural and social components would require further consideration besides groundwater.

7 Example 3: Groundwater Degradation Effects on Ecosystem Services, 1990–2040

7.1 Overview of the Assessment Method

A growing literature on the economic assessment of ecosystem services has emerged over the past two decades (Pagiola 2008; Engel et al. 2008). Assessment of the cost of groundwater degradation using an ecosystem services framework encourages a more holistic approach, enabling consideration of a range of functions provided by the ecosystem, including various provisioning, supporting and regulating functions (see Setlhogile et al. 2011). The typical question being asked, then, is whether the total value of the mix of services provided by an ecosystem managed in one way is greater or smaller than the total value of the mix provided by that ecosystem if it were managed in another way (DeFries et al. 2005). Total Economic Value (TEV) typically disaggregates the utilitarian value of ecosystems into direct and indirect use values and non-use values, using a range of the techniques shown in Table 1. This evolving state of the art approach has been the focus of a series of recent conceptual guides emphasizing the importance of inclusive participatory approaches engaging stakeholders in the application of this method (Kumar 2010; Wittmer et al. 2010; Ash et al. 2010).

7.2 Application to Groundwater Degradation in Wadi Natrun: A Basis for Assessment

An ecosystem assessment approach has been developed to assess the significance of groundwater degradation for human security (Salem et al. forthcoming). This approach was applied by researchers at the nearby University of Alexandria, in consultation with local decision-makers, with a view to enhancing local capacities for environmental decision-making. The historic, cultural and amenity values of the area provide economic opportunities for attraction of tourists to local hotels and museums, and the production and sale of local paper, glass, cosmetic and construction products, in addition to food and water production. This assessment therefore highlights the need for qualitative consideration of existing and potential local investments in promising green economic activities and ecotourism.

7.3 Approach to Data Collection and Analysis

Ecosystem components were identified according to landcover classes visible in the area using a combination of field observation and remote sensing methodologies. This approach drew on the ecological characterization of the landcover systems in the area by students and staff at the University of Alexandria (Awad 2002). A qualitative assessment of the stocks and flows of ecosystem services in each class was outlined, as shown in Table 2, drawing on local knowledge. This was then reviewed through an expert workshop (Renaud 2008). Impacts of degradation with economic significance to be taken into consideration would include impacts on the groundwater-dependent habitats supporting desert birdlife, rare papyrus reeds (*Cyperus papyrus*) and raw materials for handmade glass-products. A participatory workshop aiming to generate a series of 'back-of-the-envelope' estimates for resource values has been proposed by researchers as a means to continue this work in a consultative manner.

7.4 Outcome, Recommendations and Limitations

The ecosystem approach engages with the complexity of human well-being and ecosystem functions and services. The application in this case is a work in progress, requiring continued local coordination and consultation. The qualitative assessment already produced, as shown in Table 2, provides a broad framework for assessment. This could be effectively combined with various techniques listed in Table 1. Inevitably, use of this approach, which requires sustained local consultation and the improvement of ecological datasets, would be a long-term undertaking, and would necessarily stretch existing local capacities. A hedonic model showing the marginal effect of resource conditions on property values in each of the landcover classes shown in Table 2. might be anticipated to stimulate a particularly lively local response, should local decision-makers wish to explore such an approach towards the creation of a green economy.

8 Discussion: What Can Be Shown Using Economic Analysis and Models to Capture the Cost of Environmental Degradation?

As illustrated in Table 1, and in the examples provided in the previous sections, local decision-makers face an array of choices concerning the selection and elaboration of green economic assessment processes. These include the selection of methods for modeling environmental changes such as groundwater degradation. These challenges are likely to require, and quite possibly stretch, available capacities. Attention to such challenges will need to be balanced with workloads that are often

Table 2 Ecosystem services in the study area and their exposure to groundwater threats (after Zahran and Willis 2009; Salem et al. forthcoming)

Ecosystem component	Ecosystem services			Vulnerabilities to groundwater degradation
	Provisioning	Supporting	Regulating	
Saline lakes	Salts for domestic and industrial uses	Habitat for birds and micro-organisms	Natural drainage area, ensuring the quality of the surrounding land for cultivation	Alteration of the quantity and quality of groundwater inflow to lakes affects water levels and chemical conditions for salt production and species habitats
Salt-marsh and half a grassland	Plants for fiber	Grazing for animals	Drainage water purification, carbon sequestration and other elements in soils, silt and plant matter	Falling groundwater tables affect the extent of salt-marsh vegetation
Sand and gravel desert land	Limited human use of wild animals for medicinal purposes and hunting	Wild plants, insects, reptiles and mammals create a food chain to break down and remove wastes	Waste decomposition and removal	Alteration of water availability in soil, rocks and gravels affects the survival of sparse vegetation and wild species
Agricultural land	Food and income	Generation of nutrient rich soil and grazing fodder for animals	Carbon sequestration and cooling effect where date palm and other trees remain under cultivation	Altered quality of water limits cultivation options. Falling groundwater table requires digging of deeper wells. Increasing salinity may also affect the diversity of cultivable fruit and vegetable species

already heavy, as well as other factors inhibiting the exploration of new decision-support opportunities from the local level – e.g. low levels of awareness, capacity and access to necessary information.

The selection of assessment models presented in the previous sections progressed from a long-established approach, based on assessment of productive value, to a more current holistic approach to ecosystem assessment. We have observed that the various modeling tools and their specific applications to our case study area generated different outcomes. These may be of more or less interest to different groups of stakeholders. When we consider these differences, some complex choices to be made by decision-makers appear. First of all, if the assessment is limited to the use of existing market values for production, as in the first example, or to current engineering costs, as in the second, will this be sufficient to enable a shift towards a green development pattern? On the other hand, if less conventional economic models are used, will these be sufficiently credible to convince skeptical stakeholders with entrenched interests of the need for a change in development patterns?

Local decision-makers will need to consider whether the cost of degradation should be assessed in relation to the agricultural economy, as in the first example, or whether other sectors should also be taken into account. In addition, should effects on the economy as a whole be assessed, or should impacts on particularly vulnerable groups such as resource dependent smallholders be singled out for attention, as in the second example? In the third example, the range of impacts is further enlarged to consider different components of the ecosystem. Selecting the right scope and timeframe for the assessment in the context of each decision to be made will require careful judgment. Our three examples included a projected future scenario, where the costs of groundwater degradation were affected by a discount rate, and a retrospective assessment where costs were exacerbated by unanticipated macroeconomic changes.

The examples provided in this chapter draw on instances where the conceptual and methodological challenges inherent in the assessment process have been overcome by local decision-makers in Wadi Natrun in various ways – e.g. through the involvement of visiting experts, students, and farmers, and through building connections amongst different local, national and international institutions, including research institutions. For local decision-makers seeking to develop further assessments in this and other locations, a six step process of scoping, local consultation and participation is underlined in Sect. 2. This process is designed to support local decision-makers in shaping the assessment approach to suit local development needs, opportunities, and aspirations for a green economy.

9 Conclusion

Our selected case study of groundwater degradation in Wadi Natrun is located in a region undergoing very rapid development under conditions of acute water scarcity. However, local decision-makers in other regions are likely to face similar challenges over the coming years, particularly where settlements are already growing and climatic conditions are dry. As demonstrated through our case study, green economic models have provided useful tools to support local decision-making in these contexts, connecting development choices to scenarios for social and environmental impact. The existing body of experience in the use of these tools in the selected case study may be considered exceptional. Nevertheless, we do not consider their application yet to have presented a perfect realization of the green economic blueprint.

All of the available models and applications considered in this chapter will continue to present conceptual and methodological challenges for local decision-makers. However, we have seen that many of these can best be resolved through a stepwise approach based on learning-by doing. Collaborative and consultative approaches involving local research institutions offer a means to build local capacities and support local decision-making. This chapter has demonstrated that green economic models are available to local decision-makers with the resolve to consider, contextualize and use them.

Acknowledgement This chapter refers to a growing body of work on environmental management and economic assessment in Wadi El Natrun. We have endeavored to refer readers to a range of available contributions amongst this literature. We wish to express additional thanks to the following institutions and individuals who have provided direct inputs to enabling this activity to take place: Officials of the Wadi El Natrun City Council, employees of the Ministry of Water Resources and Irrigation and the DRAINFRAME study team, the Egyptian Desert Research Centre Library, members of the UNU-UNESCO-IHP Case Study Team on Groundwater and Human Security (GWAHS-CS), the Universities of Alexandria and Oxford, ESRC/NERC, and finally, ICLEI, for enabling this publication. None of these institutions or individuals bears any responsibility for the content of this chapter, or any errors that it may contain.

References

Abildtrup J, Jensen F, Dubgaard A (2012) Does the Coase theorem hold in real markets? An application to the negotiations between waterworks and farmers in Denmark. J Environ Manag 93:169–176

Acharya G, Barbier EB (2000) Valuing groundwater recharge through agricultural production in the Hadejia-Nguru wetlands in northern Nigeria. Agric Econ 22:247–259

AFD (2008) West Delta Water Conservation and Rehabilitation Project – Complementary assessment of feasibility study. French Development Agency and Ministry of Water Resoruces and Irrigation, p 112

Ash N, Blanco H, Brown C, Garcia K, Henrichs T, Lucas N, Raudsepp-Hearne C, Simpson RD, Scholes R, Tomich TP, Vira B, Zurek M (2010) Ecosystems and human well-being. A manual for assessment practitioners. Covelo/Island Press, Washington/London, p 285

Attia FAR, Fahmy H, Eid MJH, Slootweg R (eds) (2005) The West Delta Water Conservation and Irrigation Rehabilitation Project (WDWCIRP) Environmental and social impacts and a framework management – plan part I: environmental and social impact assessment. Part II: environmental and social management plan – final draft. Arab Republic of Egypt Ministry of Water Resources and Irrigation and World Bank, p 86

Awad MA (2002) Land use planning of Wadi El-Natrun depression towards sustainable development. Department of Environmental Sciences. University of Alexandria, Alexandria, p 238

Barbier E (2011) The policy challenges for green economy and sustainable economic development. Nat Resour Forum 35:233–245

Bockstael N (2000) On measuring economic values for nature. Environ Science Technol 34:1384–1389

Bolt K, Ruta G, Sarraf M (2005) Estimating the cost of environmental degradation – a training manual in English, French and Arabic. World Bank, Washington, DC, p E1-79

CAPMAS (2007) Population and housing census 2006. Cairo: Central Agency for Public Mobilisation and Statistics, Arab Republic of Egypt

Croitoru L, Sarraf M (2010) The cost of environmental degradation: case studies from the Middle East and North Africa. World Bank, Washington, DC, p 168

DeFries R et al (2005) Analytical approaches for assessing ecosystem condition and human well-being. In: Hassan R, Scholes R, Ash N (eds) Ecosystems and human well-being: current state and trends: findings of the Condition and Trends Working Group. Island Press, Washington, DC, pp 37–67

EEA (2011) Europe's environment – an assessment of assessments (ed). Publications Office of the European Union. European Environment Agency, Luxembourg, p 103

Engel S, Pagiola S, Wunder S (2008) Designing payments for environmental services in theory and practice: an overview of the issues. Ecol Econ 62:663–674

FAO (2002) Agricultural drainage water management in arid and semi-arid areas. FAO irrigation and drainage paper 61. Food and Agriculture Organization of the United Nations, Rome

Fekry A (1993) Hydrogeological studies of the southern region west of the Nile Delta. Faculty of Science. Ain Shams University, Cairo

Gisser M, Sanchez DA (1980) Competition versus optimal control in groundwater pumping. Water Resour Res 16:638–642

Goodstadt V et al (2010) Spatial planning and environmental assessment. In: Wittmer H (ed) The economics of ecosystems and biodiversity for local and regional policy makers. TEEB, London, pp 105–123

Hassan R, Scholes R, Ash N (2005) Ecosystems and human well-being: current state and trends. Millennium Ecosystem Assessment. Island Press, Washington, DC

Hellegers P, Zilberman P, Ierland E (2001) Dynamics of agricultural groundwater extraction. Ecol Econ 37:303–311

Hussain S, Gundimeda H (2010) Tools for valuation and appraisal of ecosystem services in policy making. In: Wittmer H (ed) The economics of ecosystems and biodiversity for local and regional policy makers. TEEB, London, pp 41–63

Ibrahim SMM (2005) Groundwater resources management in Wadi El-Farigh and its vicinities for sustainable agricultural development. Faculty of Engineering, Irrigation and Hydraulics Department, Ain Shams University, Cairo, p 152

IBRD (2010) Sustaining water for all in a changing climate. The World Bank, Washington, DC, p 117

Idris H, Nour S (1990) Present groundwater status in Egypt and the environmental impacts. Environ Geol Water Sci 16:171–177

JMP (2010) Progress on sanitation and drinking water: 2010 update. World Health Organization and UNICEF, Geneva, p 60

King C (2011) Living with environmental change in the Endorheic Oasis Systems of the Northern Sahara. School of Geography and the Environment. University of Oxford, Oxford, p 317

King C, Salem B (2012) A socio-ecological investigation of options to manage groundwater degradation in the Western Desert, Egypt. AMBIO J Hum Environ 41:490–503. DOI 10.1007/s13280-012-0255-8

Kumar P (2010) The economics of ecosystems and biodiversity: ecological and economic foundations. TEEB/Earthscan, London/Washington, D.C

Le Blanc D (2011) Special issue on green economy and sustainable development. Nat Resour Forum 35:151–154

Masoud A, Atwia G (2011) Spatio-temporal characterization of the Pliocene aquifer conditions in Wadi El-Natrun area, Egypt. Environ Earth Sci 62:1361–1374

Ostrom E (1990) Governing the commons: the evolution of institutions for collective action. Cambridge University Press, Cambridge
Pagiola S (2008) Payments for environmental services in Costa Rica. Ecol Econ 65:712–724
Parry ML et al (2007) Contribution of working group II to the fourth assessment report of the Intergovernmental Panel on Climate Change. Cambridge University Press, Cambridge, p 976
Pascual UR et al (2010) Exploring the links between equity and efficiency in payments for environmental services: a conceptual approach. Ecol Econ 69:1237–1244
Pearce D et al (1989) Blueprint for a green economy. Earthscan, London
Pulido-Velazquez M et al (2008) Hydro-economic river basin modelling: the application of a holistic surface–groundwater model to assess opportunity costs of water use in Spain. Ecol Econ 66:51–65
Puri S, Aureli A (2009) Atlas of Transboundary Aquifers – Global maps, regional cooperation and local inventories. ISARM, UNESCO-IHP, Paris, p 328
Renaud S (2008) Groundwater and human security – case studies report of the 2nd workshop. UNU-EHS, Bonn, p 22
RIGW (1990). Hydrogeological map of Egypt scale 1:100 000 Wadi El Natrun. Research Institute for Groundwater (RIGW), Ministry of Public Works and Water Resources, p 11.
Salem B, King C, Gad M (forthcoming) Groundwater and Human Security Case Study: Egypt. In: Bigas H (ed) Groundwater and Human Security Case Studies (GWAHS-CS). United Nations University, Hamilton
Sekar C (2001) Externality effects of common property resource degradation. Indian J Agri Econ 56:346–358
Setlhogile TJ et al (2011) Economic valuation of selected direct and indirect use values of the Makgadikgadi wetland system, Botswana. Phys Chem Earth 36:1071–1077
Shah T (2009) Taming the anarchy: groundwater governance in South Asia. Resources for the Future, Washington, DC
Shiferaw (2008) Watershed externalities, shifting cropping patterns and groundwater depletion in Indian semi-arid villages: The effect of alternative water pricing policies. Ecol Econ [0921-8009] 67:327–340
Slootweg R (2010) Water transfer project influenced by ecosystem services valuation, Egypt. In TEEB Cases. The Economics of Ecosystems and Biodiversity, p 3
Sukhdev P (2010) TEEB – The economics of ecosystems and biodiversity for local and regional policy makers. TEEB, London
TEEB (2010) A quick guide to the economics of ecosystems and biodiversity for local and regional policy makers, Bonn
Tentes G, Damigos D (2012) The lost value of groundwater: the case of Asopos River Basin in Central Greece. Water Resour Manag 26:147–164
UN (2012) The future we want – zero draft. United Nations Conference on Sustainable Development (UNCSD), New York, p 19
UNEP (2009) UNEP/IPBES/2/INF/1 Gap analysis for the purpose of facilitating the discussions on how to improve and strengthen the science-policy interface on biodiversity and ecosystem services. In: Second ad hoc intergovernmental and multi-stakeholder meeting on an intergovernmental science-policy platform on biodiversity and ecosystem services, Nairobi, 5–9 October 2009, p 134. United Nations Environment Programme, Nairobi
UNSD (2005) System of environmental-economic accounting for water United Nations Statistics Division, New York
UNSD (2010) International recommendations for water statistics. UNDESA, New York
WDR (2010) Managing land and water to feed 9 billion people and to protect natural systems. In: World Bank (ed) World Development Report 2010: development and climate change. IBRD, Washington, DC, p 439
Wittmer H et al (2010) The value of nature for local development. In: Sukhdev P (ed) The economics of ecosystems and biodiversity for local and regional policy makers. TEEB, London, pp 12–27

World Bank (2007) Project appraisal document on a proposed loan in the amount of US$145 Million to the Arab Republic of Egypt for West Delta Water Conservation and Irrigation Rehabilitation Project, 22 May 2007. World Bank, Washington, DC, p 104

WWAP (2009) The 3rd United Nations World Water Development Report: Water in a changing world (WWDR-3). UNESCO Publishing, Paris, p 318

Yehia HALM (2004) Land resource assessment for sustainable agriculture development at multi spatial scale: a case study for Behira Governorate and Wadi El Natrun District, Egypt. Soil and Water Science, Faculty of Agriculture. University of Alexandria, Alexandria, p 218

Zahran MA, Willis AJ (2009) The vegetation of Egypt, 2nd edn. Springer, Dordrecht

Constructed Wetlands for the Treatment of Domestic Grey Water: An Instrument of the Green Economy to Realize the Millennium Development Goals

Gopalsamy Poyyamoli, Golda Arpudhalin Edwin, and Nandhivarman Muthu

Abstract In rapidly developing countries like India, wastewater generation has increased manifold due to an increasing population, industrialization and urbanization. This in turn has led to the deterioration of several urban water bodies and aquifers. Transition to a green economy in the water sector requires a paradigm shift from the current practices by adopting innovative technologies that provide environmental as well economic benefits. In this context, constructed wetland systems (CWs) for domestic water treatment and reuse promise to be a cost-effective alternative to conventional systems and can contribute to improved water security. This chapter discusses the potential of CWs to contribute to a green economy, the various costs and benefits associated with it, along with a case study. The case study conducted shows removal efficiency between 65 and 99% for various pollutants, which complies with established reuse standards set by the Central Pollution Control Board (CPCB). In the process, it helps to save around 47% of the overall water requirement of the household. This chapter concludes that CWs proves to be an effective instrument of a green economy.

Keywords Constructed wetlands • Water treatment • Green economy • Sustainable water management

1 Introduction

The vast majority of people in India, as in many other developing countries, still do not have access to a safe drinking water supply. In recent years, rapidly expanding population, industrial development and urbanization have exerted immense

---

G. Poyyamoli • G.A. Edwin (✉) • N. Muthu
Pondicherry University, Puducherry, India
e-mail: gpoyya9@gmail.com; edwingolda@gmail.com; m.nandhivarman@gmail.com

R. Simpson and M. Zimmermann (eds.), *The Economy of Green Cities: A World Compendium on the Green Urban Economy*, Local Sustainability 3,
DOI 10.1007/978-94-007-1969-9_27,

environmental pressures on freshwater sources. Discharging wastewater effluents rich in nitrogen and other nutrients into receiving water bodies has a number of problems including eutrophication in the receiving water bodies, and impact on human health and marine ecology. Therefore, the pollutants must be removed/remediated in order to preserve the water environment, protect the aquatic life and health of water users downstream. In turn, efficient wastewater treatment and reuse is critical for sustainable development (Al-Jayyousih 2003).

Although conventional wastewater treatment technologies have been used by municipalities and industries for over three decades, they are rather expensive for use in small communities or households of developing countries. The use of the constructed wetlands systems (CWs) in urban areas at both small and large scale is now being recognized across the world due to their good treatment performances and low construction and operating costs (Kadlec et al. 2000; Sonavane et al. 2008). It has been accepted as a low cost eco-technology alternative to conventional treatment methods (especially beneficial to small communities that cannot afford expensive treatment systems) and been increasingly accepted by the general public for reuse purposes (Green and Upton 1995; White 1995; Billore et al. 1999; Fenxia and Ying 2009; Friedler 2008). Some of the most prominent benefits of small scale CWs include water reuse and cost saving, while the large scale implementation can include additional benefits such as nutrient harvesting, habitat creation, recreational and other human use, protection of aquatic life and the health of marine ecology.

However, successful implementation of CWs in India at a small scale (either at household/community/institutional level) is still in its primitive stage because it is hindered by three principle factors: (a) lack of available investment capital, (b) insufficient knowledge about the local wetland plant species, and (c) the cost of water supplied by the local municipality is too low due to heavy subsidies which fail to motivate the users to conserve/reuse water (Edwin and Poyyamoli 2012).

Designing a standard modular system for developing countries that can be easily installed and maintained can support the creation of more green jobs, improve the environment and backup the crumbling centralized wastewater treatment infrastructure, while providing economic and social benefits (Yang et al. 2008).

2 Green Economy

A green economy is one that results in improved human well-being and social equity, while significantly reducing environmental risks and ecological scarcities. According to United Nations Environment Programme, "A green urban economy realizes opportunities to enhance human well-being and local natural resources, while reducing future costs, ecological scarcities and environmental risks." Water reclamation and greywater treatment are the two main aspects of sustainable water management and essential to realize a green urban economy. The economics of wastewater management and treatment are crucial topics in developing countries because:

- Wastewater is a significant and growing problem in many urban areas of both the developed and developing world.
- The available fresh water sources are dwindling and getting scarcer.
- Increase in fresh water pollution from human activity.
- Uncontrolled discharge of wastewater into streams and oceans, which causes a range of external costs, including cost to human health and ecosystems.
- Current wastewater treatment systems are at a rudimentary stage in developing countries and are grossly inefficient or ineffective, at a time of rapid growth.

2.1 The Potential of CWs for a Green Economy

CWs are natural systems in which wastewater is treated with wetland plants using natural processes (e.g. sedimentation, filtration, adsorption, biological degradation, volatilization, photolysis, biotic/abiotic degradation, nitrification/denitrification, microbial uptake, plant uptake, volatilization, etc.) to treat the wastewater in a controlled environment (Reed et al. 1995; Cooper et al. 1996; Constructed Wetlands Manual 1998; Gray 1999; Rai 2009). Hammer and Bastian (1989) puts it as "man-made complexes of saturated substrate, emergent and submerged vegetation, animal life and water that simulate natural wetlands for human use and benefits".

There have been several attempts by various authors to develop a vertical flow CWs that will comply with the most stringent effluent standard (i.e. 95% removal of biological oxygen demand, BOD, 90% removal of total-P and 90% nitrification) (Laber et al. 1997; Brix and Arias 2005). These and several other studies conducted in various parts of the world clearly demonstrate that such systems would be able to meet the effluent standards (e.g. Platzer 1996; Laber et al. 1997; Weedon 2003). Treating only the greywater (water from showers, laundry and kitchen) has an economic advantage over treating combined wastewater (includes blackwater from toilets) as greywater represent 70% of the total wastewater generated and has only about 10% of the nitrogen load that one would expect in the combined wastewater (Müllegger et al. 2003; Friedler et al. 2005). CWs for greywater treatment designed for single households or small communities can reclaim 70% of their wastewater and use them for various purposes including flushing toilets, for cleaning purposes, landscaping, etc. (Friedler 2008). The major disadvantages of these systems are that they require more space than conventional systems and site selection is almost always an issue due to availability of adequate land area and accessibility of the site (Sundaravadivel and Vigneswaran 2010).

CWs are especially applicable to urban areas where the collection and transport of wastewater cause 70–80% of the total costs of centralized wastewater treatment. Small scale decentralized systems can become economical if they are mass-produced using a standardized approach. In terms of the flow of investments, the centralized approach requires investment flow in building large sewerage systems whereas the decentralized approach directs investments towards production and maintenance of plants leading to more jobs (Otterpohl et al. 2003).

2.2 Costs

Using a CWs for greywater treatment proves to be more economically advantageous than treating the combined wastewater. This is mainly because treating only segregated greywater is more efficient at removing virtually all of the suspended solids and BOD, and about 80% of the Chemical Oxygen Demand (COD) after only 8 h of treatment. In general, the overall costs for wastewater treatment using natural methods can be subdivided into capital/investment costs and operational/maintenance costs. However, several of the monetary factors depend on the local situation, regulatory standards, and partly unquantifiable factors (Starkl et al. 2002).

2.2.1 Capital Costs

Major capital costs for a pilot scale CWs include land acquisition, earth moving, plastic liners to prevent groundwater contamination or infiltration, and the filter media in case of Subsurface Flow. The small scale CWs designed for single households can in most cases eliminate land costs by utilizing the free space available in the backyard or the terrace (as vertical flow CWs requires an area of 1–2 m^2 per person). The designs are usually oversized to comply with the regulatory standards and thus the costs reported in literature are usually quite large (Rousseau et al. 2008). Capital costs are highly dependent on the local situation, space availability and the decommissioning cost of the system after its life period, and cleaning/disposing cost of the filter media that remain uncertain (Rousseau et al. 2008).

2.2.2 Operational Costs

Operational costs include the costs for sampling, lab analysis, maintenance, energy, sludge disposal, cleaning, plant harvesting/weed control and labor. The maintenance costs of the CWs will be lower compared to the conventional treatment due to its low complexity, which does not require specialized skills for maintenance. It also has low consumption of energy that is often limited to pumping and disinfection of treated water.

2.2.3 Benefits

Using CWs has several advantages. It is a low cost and an eco-friendly method in treating water, which makes it a potential alternative to costly conventional techniques (Billore et al. 1999; Otterpohl et al. 2003; Trivedy 2007). Various studies have established that greywater reuse is economically, environmentally and technologically feasible for various purposes including the most common ones such as toilet flushing and landscape irrigation. There are examples of CWs that are aesthetically designed to reclaim domestic sewage for toilet flushing, landscape irrigation and aesthetic water features in Chatham County, NC (House et al. 1999). They are also designed indoors in some cases to flush the toilets in a hotel (March et al. 2004). Another case from an urban example shows how on-site greywater reuse in Israel is both acceptable to the public and is also an economically feasible solution for reducing urban water demand (Friedler 2008).

Treated water from CWs can also be used for restricted or unrestricted agricultural irrigation of crops depending on its effluent standard, and it can also serve as a tool for recharging groundwater (Emmett et al. 1996; Rousseau et al. 2008). Certain plant species have commercial value: some as ornamental plants (Belmont and Metcalfe 2003; Konnerup et al. 2009), others as raw material for various purposes including biomass fuel (Ciria et al. 2005) since the plants sequester carbon during their growth that can be used for energy production and for obtaining carbon credits. Harvested plant material can provide fibers, fodder for livestock and also can be composted for use as manure (Knight 1997).

Large scale CWs can serve as a new habitat for the local species of flora and fauna (Knight et al. 2001). Although most CWs are not designed with habitat functions in mind, they can be designed to achieve optimal wildlife potential if approached from an ecological viewpoint (Worrall et al. 1997). Large scale CWs designed from both ecological and engineering perspective can add further value for recreational and various other human use such as nature watching, walking, jogging, fishing, picnicking, relaxing and art (photography, painting) (Gearheart and Higley 1993; Knight 1997; Knight et al. 2001).

2.3 Case Study

To illustrate this, a case of a single house with five persons is considered in Puducherry, India. Here, greywater is being treated using vertical sub-surface flow CWs and the treated water is reused for toilet flushing and gardening purposes before releasing to the sewer. The complete flow diagram of the system is shown in Fig. 1.

The quantity and quality of the effluent water will in part determine how it can be reused. In general, the reuse of greywater serves three purposes:

- It reduces the fresh water requirement
- It reduces the sewage generation
- It helps close the water and nutrient cycle to a great degree at source

The system treats 350 l per day of greywater generated from showers, kitchen and laundry, while the remaining is sent to the sewer directly. The composition of the greywater generated from various source is shown in Table 1.

In primary treatment, screening and sedimentation process is used to remove the suspended solids that can be easily filtered or settled at the bottom. This is followed

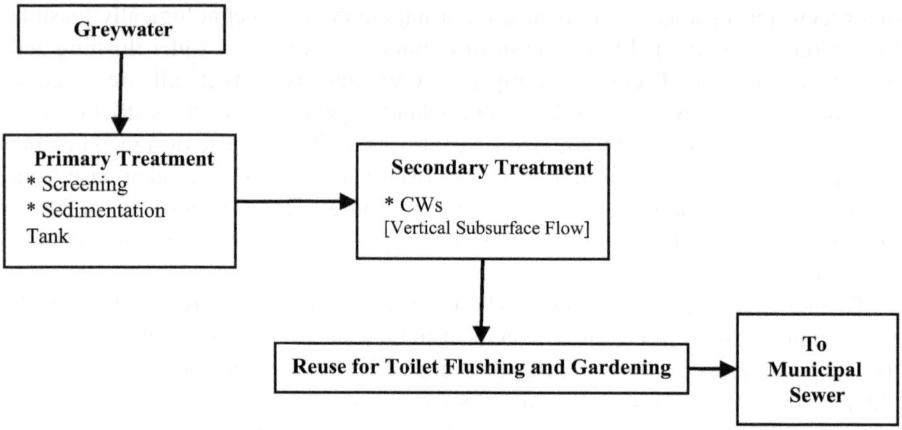

Fig. 1 Flow diagram of the greywater treatment system

Table 1 Composition of the greywater generated from various sources

Description	Percentage generated (%)	Composition
Bathroom	50–60	Soap, shampoo, hair dye, toothpaste and cleaning products
Washing	25–35	Detergents and associated chemical agents
Kitchen	10	Detergents, cleaning agents, food particles, oils, fats and other wastes

by a vertical subsurface flow CW planted with *Arundo donax* which is a freely available, fast growing indigenous plant with efficient biomass to treat the greywater through root zone treatment techniques. A collection tank of 500 l capacity is utilized to hold the treated greywater, which is then pumped to a storage tank of 500 l capacity from where it is directed to toilet flushing and gardening purpose.

The annual saving in water is calculated to be 10,500 l or about 47% as indicated in Table 2. The pollution removal efficiency of the system increased in the first 3 months before stabilizing and the laboratory analysis clearly indicate the removal efficiency of 95.2% for BOD_5, 79.2% for suspended solids, 74% total alkalinity, 81.1% for COD, 80% for turbidity, 73.2% for nitrate nitrogen, 65% for ammonium nitrogen and 99.1% for coliform bacteria.

3 Conclusion

Grey water reclamation and reuse is an alternate source of non-potable water that directly reduces the sewage flow rates and indirectly reduces the cost of treating sewage at a centralized facility. Besides the water saving advantages these systems

Table 2 Annual water saving by using treated greywater

Parameter	Before treatment and reuse	After treatment and reuse
Water source	Municipal tap water [750 lpd]	Municipal tap water [400 lpd] Greywater reuse [350 l]
Water requirement for toilet flushing and gardening	Municipal tap water [350 lpd]	Greywater reuse [350 l]
Annual cost of water	Annual expenditure on purchase of water INR 1,200/- (after subsidy from government)	Capital expenditure for greywater treatment INR 10,000/- O & M cost is ~ INR 400/- Annual expenditure on purchase of water INR 636/-
Annual water saving		10,500 l or ~ 47%

offer, they also close the water and nutrient cycles at greywater generation source itself on a local scale. The study concluded that around 47% of the water requirement can be reduced by utilizing a small scale CWs to treat and reuse greywater. For small wastewater treatment plants especially the operational costs are essential and when greywater is reused, the health/hygiene aspects also have to be considered. Frequent laboratory tests needs to be performed to ensure that the system is performing as per the guidelines set by the regulatory bodies. These systems may not be a preferable option among the urban users as long as the government provides drinking water at a huge subsidy. Though water subsidy should be used as a tool for poverty alleviation for poor families, it should not make the affordable users to underestimate the value of this precious resource and encourage wasteful use of water. As an alternative, the government should subsidize the installation of these treatment systems and make it mandatory for all group housing projects to facilitate sustainable development and help achieve the Millennium Development Goals.

References

Al-Jayyousih OR (2003) Greywater reuse: towards sustainable water management. Desalination 156(1–3):181–192

Belmont MA, Metcalfe CD (2003) Feasibility of using ornamental plants (*Zantedeschia aethiopica*) in subsurface flow treatment wetlands to remove nitrogen, chemical oxygen demand and nonylphenol ethoxylate surfactants – a laboratory-scale study. Ecol Eng 21(4–5):233–247

Billore SK, Singh N, Sharma JK, Dass P, Nelson RM (1999) Horizontal subsurface flow gravel bed constructed wetland with Phragmites karka in Central India. Wat Sci Technol 40(3):163–171

Brix H, Arias CA (2005) The use of vertical flow constructed wetlands for onsite treatment of domestic wastewater: new Danish guidelines. Ecol Eng 25:491–500

Ciria MP, Solano ML, Soriano P (2005) Role of macrophyte Typha latifolia in a constructed wetland for wastewater treatment and assessment of its potential as a biomass fuel. Biosyst Eng 92(4):535–544

Constructed Wetlands Manual (1998) Department of Land and Water Conservation, New South Wales, Vols 1 and 2, National Library of Australia

Cooper PF, Job GD, Green MB, Shutes RBE (1996) Reed beds and constructed wetland for wastewater treatment. WRC, Swindon

Edwin GA, Poyyamoli G (2012) Climate change and sustainable management of water resources. In: Leal Filho W (ed) Climate change and the sustainable use of water resources. Climate change management. Springer, Heidelberg/New York, pp 431–447

Emmett AJ, Clarke S, Howles S (1996) Conjunctive wetland treatment/aquifer storage and recovery at Regent Gardens residential development, North Field, South Australia. Desalination 106:407–410

Fenxia Y, Ying L (2009) Enhancement of nitrogen removal in towery hybrid constructed wetland to treat domestic wastewater for small rural communities. Ecol Eng 35(7):1043–1050

Friedler E (2008) The water saving potential and the socio-economic feasibility of greywater reuse within the urban sector – Israel as a case study. Int J Environ Stud 65(1):57–69

Friedler E, Kovalio R, Galil NI (2005) On-site greywater treatment and reuse in multi-storey buildings. Wat Sci Technol 51(10):187–194

Gearheart RA, Higley M (1993) Constructed open surface wetlands: the water quality benefits and wildlife benefits – City of Arcata, California. In: Moshiri G (ed) Constructed wetlands for water quality improvement. Lewis Publishers, Boca Raton, pp 561–567, Chap. 62

Gray NF (1999) Water technology: an introduction for scientists and engineers. Arnold/Wiley, London/New York, ISBN 0 340 67645 0 (pb)

Green MB, Uptown J (1995) Constructed reed beds: an appropriate technology for small communities. Water Sci Technol 32:339–348

Hammer DA, Bastian RK (1989) Wetlands ecosystems: natural water purifiers? In: Hammer DA (ed) Constructed wetlands for wastewater treatment. Lewis Publications, Chelsea, MI, pp 5–21

House CH, Bergmann BA, Stomp AM, Frederick DJ (1999) Combining constructed wet-lands and aquatic and soil filters for reclamation and reuse of water. Ecol Eng 12(1–2):27–38

Kadlec RH, Knight RL, Vymazal J, Brix H, Cooper P, Haberl R (2000) Constructed wetlands for pollution control. Process, performance, design and operation. IWA Scientific and Technical Report No.8, ISBN: 1-900222-05-1

Knight RL (1997) Wildlife habitat and public use benefits of treatment wetlands. Wat Sci Technol 35(5):35–43

Knight RL, Clarke RA, Bastian RK (2001) Surface flow (SF) treatment wetlands as a habitat for wildlife and humans. Wat Sci Technol 44(11–12):27–37

Konnerup D, Koottatep T, Brix H (2009) Treatment of domestic wastewater in tropical, subsurface flow constructed wetlands planted with Canna and Heliconia. Ecol Eng 35(2):248–257

Laber J, Perfler R, Haberl R (1997) Two strategies for advanced nitrogen elimination in vertical flow constructed wetlands. Wat Sci Technol 35(5):71–77

March JG, Gual M, Orozco F (2004) Experiences on greywater re-use for toilet flushing in a hotel (Mallorca Island, Spain). Desalination 164:241–247

Müllegger E, Langergraber G, Jung H, Starkl M, Laber J (2003) Potentials for greywater treatment and reuse in rural areas. 2nd international symposium on ecological sanitation, Lübeck

Otterpohl R, Braun U, Oldenburg M (2003) Innovative technologies for decentralised water-, wastewater and biowaste management in urban and peri-urban areas. Wat Sci Technol 48(11–12):23–32

Platzer C (1996) Enhanced nitrogen elimination in subsurface flow artificial wetlands – a multi stage concept. In: Proceedings of the fifth international conference on wetland systems for water pollution control, Universität für Bodenkultur Wien, Vienna, Austria

Rai PK (2009) Heavy metal phytoremediation from aquatic ecosystems with special reference to macrophytes. Crit Rev Environ Sci Technol 39:697–753

Reed SC, Crites RW, Middlebrooks EJ (1995) Natural systems for wastewater management and treatment. McGraw-Hill, San Francisco, ISBN 0-07-060982-9

Rousseau DPL, Lesage E, Story A, Vanrolleghem PA, De Pauw N (2008) Constructed wetlands for water reclamation. Desalination 218:181–189

Sonavane PG, Munavalli GR, Ranade SV (2008) Nutrient removal by root zone treatment systems: a review. J Environ Sci Eng 50(3):241–248

Starkl M, Ertl T, Haberl R (2002) Experiences with benchmarking of sewerage systems with a special focus on investment costs. In: University of Bradford (ed) Proceedings of the international conference on sewer operation and maintenance (CD), 26–28 Nov 2002, Bradford

Sundaravadivel M, Vigneswaran S (2010) Constructed wetlands for wastewater treatment. Crit Rev Environ Sci Technol 31(4):351–409

Trivedy RK (2007) Low cost and energy saving technologies for water and wastewater treatment. J Ind Pollut Contr 23(2):403

Weedon CM (2003) Compact vertical flow constructed wetland systems – first two years' performance. Wat Sci Technol 48(5):15–23

White KD (1995) Enhancement of nitrogen removal in subsurface flow constructed wetlands employing a 2-stage configuration, an unsaturated zone, and recirculation. Wat Sci Technol 32(3):59–67

Worrall P, Peberdy KJ, Millet MC (1997) Constructed wetlands and nature conservation. Wat Sci Technol 35(5):205–213

Yang W, Chang J, Xu B, Peng C, Ge Y (2008) Ecosystem service value assessment for constructed wetlands: a case study in Hangzhou, China. Ecol Econ 68(1–2):116–125

Decentralized Composting in Asian Cities: Lessons Learned and Future Potential in Meeting the Green Urban Economy

Dickella Gamaralalage Jagath Premakumara

Abstract Composting is a traditional agricultural practice to recycle organic waste. However, due to ongoing rapid urbanization and changing life-styles, organic waste lost its traditional link to the rural agriculture. Instead, organic waste became a serious environmental issue in developing countries. This chapter therefore discusses the experiences of two different cities in Asia (Surabaya, Indonesia, and Matale, Sri Lanka) in promoting organic waste recycling. The chapter identifies the potential of decentralized composting in realizing economic benefits such as generating incomes, creating job opportunities for the urban poor, and minimizing costs for municipal governments. The chapter also identifies environmental benefits by reducing municipal waste for landfills and greenhouse gas emission. Social benefits include the resulting improvements in the quality and coverage of waste management, quality of life, education and social capital. However, for the successful application of decentralized composting to realize a green economy, it requires strong political will and commitment and to utilize integrated policy instruments, such as economic instruments, regulations and information.

Keywords Decentralized composting • Green economy • Organic waste recycling • Municipal solid waste • Asia

1 Introduction

Solid waste management is considered to be one of the most serious environmental issues confronting urban areas in developing countries. Rapid urbanization and economic growth in Asia has resulted in a corresponding growth of solid waste, which

D.G.J. Premakumara (✉)
Institute for Global Environmental Strategies (IGES),
Kitakyushu Urban Centre, Kitakyushu, Japan
e-mail: premakumara@iges.or.jp

R. Simpson and M. Zimmermann (eds.), *The Economy of Green Cities: A World Compendium on the Green Urban Economy*, Local Sustainability 3,
DOI 10.1007/978-94-007-1969-9_28,

municipal governments find difficult to manage (Scheinberg et al. 2010). Many municipal governments in developing countries devote about 20–50% of their annual budget on solid waste management, even though 30–60% of all urban solid wastes remain uncollected and less than 50% of the population is serviced (UNEP 2009). In addition, many landfills in developing countries are poorly designed and controlled, which results in serious environmental impacts. The existing dumpsites most often reach their maximum capacity and finding new sites is becoming an increasingly difficult task, due to the scarcity of suitable land within the municipal boundaries (Zurbrugg et al. 2005). Estimates also suggest that untreated solid waste in Asia and the Pacific contribute to about 75 million equivalent tons of atmospheric carbon dioxide (CO_2) annually (UNESCAP 2007).

To address these issues, many cities in developing countries focus on establishing an integrated solid waste management system based on the 3Rs (Reduce, Reuse and Recycle) and try to reduce the amount of waste generated at the source, rather than later at the end-of-cycle (Premakumara et al. 2011). Estimates from several cities in developing countries show that as much as 20–30% of wastes generated in these cities are recycled by the informal sector through waste pickers, junk dealers and recyclers (UNESCAP 2007). A global estimate suggests that the informal sector recycling provides income and employment opportunities to about 15 million urban poor and bring savings to municipal governments in collection and transport costs of around 100 million US$ per year (Wilson 2011).

However, these recycling efforts are still limited to inorganic waste. Relatively limited or no efforts have been made to utilize the organic waste that usually comprises over 60–70% of the total waste generated in cities. This chapter therefore emphasizes the importance of decentralized composting as an approach to treat organic waste close to the source of generation. Further, organic waste can be seen as a resource rather than a cost for municipal governments, and can be viewed as an alternative to achieving a sustainable solid waste management and green urban economy in developing countries.

2 Potential of Decentralized Composting in Achieving a Green Urban Economy at the City Level

This section presents the experiences of Surabaya, Indonesia, and Matale, Sri Lanka to identify both potential and key challenges of decentralized composting in achieving a green urban economy at the city level. Each case study includes a short background, main characteristics of the composting scheme, and some lessons learned, based on the information from literature reviews, field observations and discussions with relevant stakeholders.

Table 1 Basic facts on solid waste management in Surabaya City, 2010 (Surabaya City 2011) compiled by the author

Total population	2.9 million
Total land extent	33,000 ha
Administrative districts	31
Total waste generated in the city	1,241 t per day
Waste composition	Organic 72%, paper 12%, plastic 8%, metal 1%, glass 1%, others 6%
Method of disposal	Landfill
Average cost for municipal solid waste management	23 US$ per a ton

2.1 Promotion of Decentralized Composting in Surabaya City, Indonesia

Surabaya City is the second largest city in Indonesia with a population of 2.9 million people (Surabaya City 2011). The city is confronted with the tremendous challenge of managing its solid waste in an environmentally sustainable manner. In 1994, it reached an uncontrollable state that angered citizens and lead, along with other political issues, to the dismissal of the Mayor. Subsequently, the Vice Mayor took on the responsibility of running the city (Silas 2002). Table 1 shows some basic facts on solid waste management in Surabaya City.

Since 2004, the unmanageable waste situation changed significantly by promotion of decentralized composting in Surabaya City with the technical assistance of Kitakyushu City, Kitakyushu International Techno-Cooperative Association (KITA), and the Institute for Global Environmental Strategies (IGES) in Japan under the Kitakyushu Initiative for a Clean Environment Network of the United Nations Economic and Social Commission for Asia and the Pacific (UNESCAP).

KITA from Kitakyushu City and Pusdakota (a local NGO in Surabaya) started a pilot project in a low-income neighborhood in Rungkot Lor, which covers about 1,200 households. The pilot project, which was supported by the Japan Fund for Global Environment (JFGE), educated households to separate waste at the source into organic and inorganic. The organic waste of around 0.6 t per day was then collected by Pusdakota three-times a week and transported to the nearby composting center. At the composting center, the organic waste was processed using the fermentative microorganisms, which are cultivated from locally available fermented foods (IGES 2010). This composting method was introduced by Koji Takakura, an expert from Kitakyushu City and is now widely popular as the "Takakura Method" in Surabaya City.

Using the fermentative microorganisms, Rungkot Lor's composting center produces about 3 t of compost monthly by employing two staffs from the neighborhood. Pusdakota also distributed composting baskets as a simple way to treat organic waste in households and encouraged residents to grow vegetables and herbal plants

Table 2 Basic facts of the pilot composting center in Rungkot Lor, Surabaya (KITA 2007) compiled by the author

The coverage of the pilot project	1,200 households (1,000 for compost centre and 200 for household baskets)
Amount of organic waste treated for composting	0.6 t per day
Amount of compost produced	0.1 t per day
Staff	2 person (a technician and a worker)
Initial costs for land (15,200 US$), building (800 US$), equipments (2,481) and others (919 US$)	19,400 US$
Monthly operational costs for personal (125 US$) and management (60 US$)	185 US$
Monthly income from selling compost products (0.07 US$ per kg)	210 US$
Other incomes from selling household compost baskets (1 US$ per a basket), service charges and selling of recyclable materials	15 US$

in their home gardens using their own household compost. The recyclable materials are collected separately and sold to the junk shops.

A composting center was located on land that belonged to the University of Surabaya. It was arranged that the Pusdakota would receive 190 m^2 land from the university free of charge. A part of the initial funds required for the construction of the building was received from Kitakyushu City project. Surabaya City provided the initial equipments including a shredder machine (KITA 2007). After pooling resources to cover the initial costs and materials, the compost center started to operate and produced 0.1 t of compost daily. As Table 2 shows, monthly operational costs of around 185 US$ can be easily recovered through the total monthly income of 225 US$. However, the project faced difficulties in identifying a market for the compost products and handling the seasonal fluctuation of demand for compost products (Maeda 2009). This issue was overcome by motivating farmers, schools and neighborhoods to start organic farming (Fig. 1).

The pilot project succeeded in reducing landfill waste by 300 t (85% of the total waste generated in the pilot area), which is equivalent to saving the municipal council about 6,900 US$ annually for waste management. Further, it created extra economic opportunities for community members, improved environmental education and sanitary conditions, as well as resulted in a greener and cleaner neighborhood.

After realizing the benefits of the pilot project, Surabaya City scaled-up the decentralized composting activities city-wide. Local regulation No.1/2006 on community-based solid waste management was issued and the city extended its support to the Rukun Warga (RW) – a citizens council that is responsible for waste collection within the neighborhood (*Kampung*) under the community primary collection (*Copricl*) law in 1980 (Premakumara and Maeda 2011). Surabaya City also strengthened partnerships with other stakeholders such as: the Women's Network (PKK), local NGOs, waste pickers, academic institutions, private ventures and the media. In addition, a series of awareness raising campaigns was organized covering all

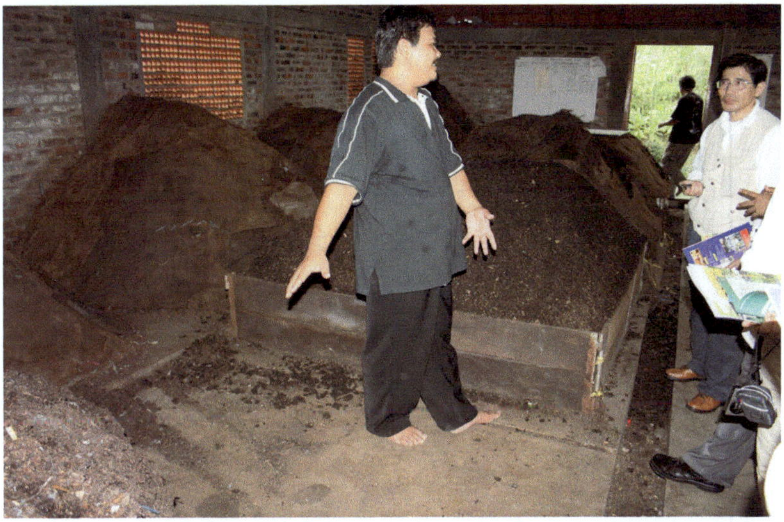

Fig. 1 Pusdakota's compost center in Rungkot Lor, Surabaya

municipal districts targeting different sectors (Surabaya City 2011). A system of community facilitators was established in each neighborhood, recruiting a team of environmental cadres (one cadre for every ten houses) based on the community leader elements. The environmental carders were trained to play an effective role of information providers in their neighborhoods on the new waste management system (UN-Habitat 2008). The municipal budget was allocated to establish additional composting centers, specifically for land, construction costs, staff, required tools and necessary trainings (Surabaya 2011). A number of neighborhood competitions, including the Green and Clean Campaign, were started in partnership with the private sector and the media to motivate and strengthen community participation, and encourage them to improve their neighborhood environment (Maeda 2009).

The enabling policy resulted in establishing at least 16 decentralized composting centers in the city to process organic waste collected from neighborhoods, markets, streets and parks. Table 3 shows how 56 workers (mostly from the surrounding low-income families) are enrolled in these compost centers producing about 480 t of compost monthly, which is used by the municipality for the maintenance of city parks and for landscaping. Green spaces in the city have also increased by 10% over 5 years (Maeda 2009) and some neighborhoods have become more popular for having grown more plants (UN-Habitat 2008).

The waste separation at the source gives higher value to recyclable materials, which are sold to junk collectors or processed by the members of PKK for handicraft. About ten small and medium sized recycling businesses promoted handicraft items that were established through private ventures. This created extra income opportunities for low-income families and waste pickers (Surabaya City 2011). A 30% waste reduction of waste deposited at the landfill was achieved by 2010,

Table 3 Basic information of the existing decentralized compost centers in Surabaya, 2010 (Surabaya City 2011) compiled by Abe and Premakumara

No.	Established	Location	Area (m^2)	Capacity of compost (t/day)	Products of compost (t/day)	Proceeding time (day)	No. of staffs	No. of shredding machine available	Source of materials used
1	1996	Uptd Bratang	391.9	3.6	1.8	20	8	2	Market, sweeping, community (cleaning day), park, papaya skin (jam factory)
2	1999	Depo Kejawen	224.4	1.8	0.9	20	3	1	Market, sweeping, community (cleaning day), papaya skin (jam factory)
3	2000	Depo Bibiskarah	117	1.05	0.45	20	1	1	Market, house, Bibis Depot
4	2005	Rumah Kompos Keputran	93	2.4	1.2	20	2	1	Market
5	2006	Rumah Kompos Lps Tenggilis Utara	60	1.2	0.6	90	2	1	House, community (cleaning day), park
6	2007	Rumah Kompos Wonorejo	273	2.1	1.05	20	3	2	Market, community (cleaning day), park
7	2007	Rumah Kompos Rungkut Asri	535.5	1.8	0.9	20	4	2	Market, community (cleaning day), sweeping, park
8	2007	Rumah Kompos Menur	207	3	1.5	20	4	2	Market, community (cleaning day), sweeping, park
9	2007	Rumah Kompos Benowo	72	1.8	0.9	20	6	2	Market, community (cleaning day), sweeping, park
10	2007	Rumah Kompos Tenggilis Rayon Taman	207	2.4	1.2	20	2	2	Market, community (cleaning day), sweeping, park

11	2008	Rumah Kompos Sonokw Ejinan	600	3.6	1.8	20	6	1	Market, community (cleaning day), sweeping, park
12	2008	Rumah Kompos Gayung Sari	64	0.9	0.45	20	1	1	Market, community (cleaning day), sweeping, park
13	2009	Rumah Kompos Putat Jaya	600	1.8	0.9	20	6	2	Market, house
14	2009	Rumah Kompos Suban Gan	600	1.8	0.9	20	5	1	Market, house
15	2010	Rumah Kompos Jamban Gan	64	0.9	0.45	20	1	1	LPS (organic waste), market, house
16	2010	Rumah Kompos Srikana	600	1.8	0.9	20	2	1	LPS (organic waste), market, house

Table 4 Basic facts on solid waste management in Matale City, 2010 (Jayaratne 2009) compiled by the author

Total population	36,695
Total land extent	860 ha
Administrative wards	13
Total waste generated in the city	32 t per day
Waste composition	Organic 84%, paper 8%, plastic 4%, metal 1%, glass 1%, others 2%
Method of disposal	Open dumping
Average cost for municipal solid waste management	17 US$ per a ton

saving the municipality solid waste management costs, which can be estimated as four million US$ annually. There has also been a reduction of greenhouse gases generated in landfills through composting in Surabaya City, which can be calculated based on the Intergovernmental Panel on Climate Change (IPCC) guidelines as about 2,800 t of CO_2 equivalent in 2010.

2.2 *Promotion of Decentralized Composting in Matale City, Sri Lanka*

Matale is a historic and important regional trade center in central Sri Lanka. It has a population of about 36,695 and generates 32 t of municipal waste per day (of which about 80% are collected and disposed by the municipal council on its open dump-site). Remaining waste are scattered along the roadsides and public spaces, creating health issues and damaging the aesthetic value of the city. Table 4 gives basic facts on the solid waste management in Matale City.

This situation provided the opportunity to Sevanatha, a local NGO, to work together with the municipal council to implement a pilot project on decentralized composting in Gongawela – a densely populated area of the city that covers 1,000 residences, commercial and market establishments (Jayaratne 2009). Matale Municipal Council, and Sevanatha established an Integrated Resource Recovery Centre (IRRC) in 2007 to treat 1 t of waste generated from within the area, using the technical approach developed by Waste Concern, an NGO in Bangladesh. The initial capital costs of the project were covered by the UNESCAP (Chularathne et al. 2007).

The IRRC is located on 600 m^2 land provided by the municipality to Sevantha on a lease basis. Households were educated to separate waste into two bins, organic and inorganic. Awareness was raised through personal visits, distribution of leaflets, and organizing meetings at neighborhoods. The collected waste was transported to the IRRC, where organic waste was treated for producing compost by applying the aerated box method. After the compost is matured in 90 days, the compost product

Table 5 Basic facts of the pilot composting center in Gongawela, Matale (Chularathne et al. 2007) compiled by the author

The coverage of the pilot project	1,000 residences, commercial and market establishments
Amount of organic waste treated for composting	1 t per day
Amount of compost produced	0.3 t per day
Staff	6 person (a technician, a driver and four workers)
Initial costs for land (3,000 US$), building (23, 000 US$), equipments (4,700 US$) and others (3,200 US$)	33,900 US$
Monthly operational costs for personal (800 US$) and management (975 US$)	1,775 US$
Monthly income from selling compost products (0.18 US$ per kg)	1,600 US$
Other monthly incomes from selling recyclables (60 US$), and collecting user charges (330)	390 US$

is sieved, packed, labeled and sold to the local markets. The recyclable materials are collected separately, stored, sorted and sold in bulk to junk shops in the city. The workforce in the IRRC consists of a technician, a driver and four workers, who are recruited from the surrounding low-income areas.

The pilot experience shows that IRRC struggled at the beginning to find a market for its compost products because of quality assurance, lack of interest, and awareness on organic agriculture among farmers and competition from subsidized chemical fertilizers. The situation gradually changed with an increasing interest of organic farming together with a growing recognition among the farmers, due to effective marketing strategies introduced by Sevanatha. The quality of the compost products was also strictly controlled under the National Standard for Compost from Municipal Solid Waste and Agricultural Waste (SLSI 1246), which was issued by the central government in 2003. It clearly states the ingredients and the nutrient (N.P.K) values of the compost product. In addition to the compost sales, the recyclable materials, which are collected separately, were sold in bulk to the junk dealers in the city. As Table 5 shows, this brought additional income to the IRRC equivalent to 60 US$ per month (Fig. 2).

However, the monthly income from sales of compost and recyclable materials (1,660 US$) alone was not enough for covering IRRC's operation costs (1,775 US$). This was overcome with an agreement between the municipality and Sevanatha. The agreement granted permission to Sevanatha to collect user charges from houses and establishments in the area. Table 6 shows the new user charge system introduced in the pilot project area. Despite difficulties at the beginning to implement user charges – as waste collection is by law a responsibility of the municipality and there is no user pay system for municipal waste management in the country- over 90% of households are now paying fees in the project area (Chularathne et al. 2007). Households have expressed their satisfaction on getting regular service and the visible improvement in the neighborhood motivated them to pay. Currently,

Fig. 2 Sevanatha's compost center in Gongawela, Matale

Table 6 New user charges system introduced in the pilot area (Chularathne et al. 2007) compiled by the author

User types	Monthly charges per unit (US$)
Residential	0.2
Commercial and pavement traders	0.9
Office and other institutions	0.4
Fruits, vegetable markets and restaurants	3.3

the collection of user charges, sales of compost and recyclables are almost enough to cover the IRRC's monthly operation costs.

During the last 3 years, the solid waste management situation in the pilot area has improved and only 10% of residual wastes now need to be placed on landfill sites. This reduces pollution from transportation and saves the municipality costs, which is estimated about 7,200 US$ annually. It also mitigates climate change by reducing methane emissions of around 350 t of CO_2 equivalent in 2010.

The success of the pilot project motivated the municipality to identify decentralized composting in its Integrated Solid Waste Management Framework in 2008 as a suitable option for treating all organic wastes generated in the city. This was further supported with the start of the Pilisaru Program, which is a national initiative of the Central Environmental Authority (CEA) to provide financial and technical assistance to local governments in implementing composting programs. As a result, a second composting center has already been established by the Matale municipality in 2010 to treat 2 t of waste per day by drawing upon the national fund. Operation of this new composting center was also handed over to Sevanatha under the legal agreement signed between the two parties for a 5-year period. The composting centers are now being scaled-up to treat 25 t of waste per day by establishing 12 IRRCs, at least one IRRC for each of the municipal ward.

3 Conclusion and Lessons Learned

In order to overcome the growing environmental problem of waste disposal, it is critical to find new approaches that promote recycling of not only inorganic wastes but also organic wastes. Organic wastes are often the larger portion of municipal wastes in developing countries. At the same time, this should be done in a way that it integrates the existing informal sector. Both case studies showed that decentralized, community-based and labor intensive composting projects could effectively help to achieve this objective.

The case studies further recognized the potential of decentralized composting in achieving environmental and social benefits with strong economic benefits. It brings environmental and health benefits through appropriate waste collection and treatment at the neighborhood level. By treating solid waste near to the source, it minimizes transportation costs, reduces the amount of wastes for landfills, prolongs the life of landfills, and saves municipal costs for landfill management. It creates new job opportunities and extra incomes for the urban poor and waste pickers. Such an approach also creates spaces for community involvement, building partnerships and social capital at the neighborhood level.

Both pilot composting centers, which are designed to service around 1,000–1,500 households and to treat between 1 and 2 t of wastes per day, prove that decentralized composting centers can be financially viable and can break even or even make profit. The operational costs of the composting center can be covered from three main regular income sources, such as collection fees from users, sale of compost products, and sales of recyclables. However, these composting centers found some difficulties at the beginning to secure a market for its compost production. This can be overcome by developing an appropriate marketing strategy during the project planning, such as establishing demonstration sites, distribution of free samples, and compost training and education. Municipal council can use compost products for city greening and landscaping, promote new life-styles on organic farming, shift subsidies towards organic fertilizers, and establish a national/local standard for compost. Improved data documentation, such as input–output tables for waste, compost, recyclables as well as monthly cost-revenue balances would also increase transparency and provide a sound basis for planning market strategies and evaluation.

Although the pilot composting centers show potential in covering operational costs, challenges were faced in finding capital to start and scale-up the approach. Both projects were established with the initial capital from local governments, national governments and international organizations. A partnership, including cost sharing, is therefore essential for the replication on a city-wide level. As seen in the case studies, composting practices have significant impacts on reducing costs for municipal governments, which can be re-invested and distributed for capital costs over time. There is also a great potential in reducing greenhouse gases in landfills by proper controlled composting centers. This allows composting centers to sell their carbon credits at the compliance market (Certified Emission Reductions, CERs, under the Clean Development Mechanism) or voluntary market (Voluntary Emission Reductions, VERs, outside the Kyoto Protocol), and get additional funding to upscale a city-wide approach. Experiences however show that it is profitable to bundle the small-scale composting centers in the city, rather than applying as a single composting center for minimizing transaction costs and time consuming application procedures.

The pilot projects have demonstrated the importance of community participation to segregate waste at the source and to pay monthly service fees for the sustainability of the composting schemes. The reasons for citizens to participate are not merely economic benefits, but also the local environmental improvements can be more important for many households. The knowledge and motivation of the households is therefore a key factor for cooperation and can be achieved through effective information, communication and education campaigns, and establishing a competition among neighborhoods.

Experience further shows that the successful implementation of decentralized composting for a green economy requires to consider a holistic approach, integrating all the elements of the composting process (waste separation, collection, transport, treatment and product utilization), and key stakeholders, who have different interests, skills and resources to participate in composting. A clear vision, strong political commitment and change in attitudes as well as technical, financial, legal and institutional capacity enabled the necessary environment for it to happen.

References

Chularathne HMU, Jayaratne KA, De Silva D (2007) Community-based Solid Waste Management Project in Matale Municipal Council, Sri Lanka. In: UNESCAP (eds) Report: Regional seminar and study visit on community-based solid waste management, 15–16 Dec 2007, Quy Nhon City, Vietnam, pp 26–56

IGES (2010) Composting for waste reduction: information kit, Hayama

Jayaratne KA (2009) UNESCAP assisted Decentralized Solid Waste Management Project: Matale Municipal Council, Sri Lanka. Expert group meeting on developing eco-efficient and sustainable urban infrastructure in Asia and Latin America, 10–12 Feb 2009, Bangkok. Available via: http://www.unescap.org/esd/environment/infra/daproject/bangkok/documents/presentations/i%20Session%209%20-%20Matale.pdf. Cited 25 June 2011

KITA (2007) Separation at source, collection and composting of waste in Surabaya, Indonesia: promoting the reduction and recycling of waste, Kitakyushu, Japan

Maeda T (2009) Reducing waste through the promotion of composting and active involvement of various stakeholders: replicating Surabaya's solid waste management model, Policy Brief 9, IGES, Japan

Premakumara DGJ, Maeda T (2011) Linking community-based composting into municipal waste management policies in Surabaya: lessons learned. HELP-O Newslett 1:7–12

Premakumara DGJ, Abe M, Maeda T (2011) Reducing municipal waste through promoting integrated sustainable waste management (ISWM) practices in Surabaya City, Indonesia. In: Villacampa Y, Brebbia CA (eds) Ecosystems and sustainable development VIII. WIT Press, Southampton, pp 457–468

Scheinberg A, Wilson DC, Rodic L (2010) Solid waste management in the world's cities. UN-Habitat by Earthscan Ltd., London/Washington, DC

Silas J (2002) Waste management policies in Surabaya: an integrated approach, Kitakyushu Initiative, Kiyakyushu. Available via: http://www.kitakyushu.iges.or.jp/successful_practices/solid_waste_management.html. Cited 27 Apr 2010

Surabaya City (2011) Community-based solid waste management as best practice in Surabaya city. In: Premakumara DGJ, Kazuyoshi H (eds) Seminar report: a networking seminar on KitaQ System composting in Asia, 29 June–01 July 2011, JICA, IGES, Kitakyushu, Japan, pp 97–104

UNEP (2009) Developing integrated solid waste management plan: training manual, Vol 4, ISWM, Osaka and Shiga

UNESCAP (2007) Report: Regional seminar and study visit on community-based solid waste management, 15–16 Dec 2007, Quy Nhon City, Vietnam

UN-Habitat (2008) Best practices: green and clean initiative in Surabaya, Indonesia. Available via: http://www.unhabitat.org/bestpractices/2008/mainview04.asp?BPID=1903. Cited 5 Nov 2011

Wilson DC (2011) Acting alone to partnerships – strategic approach for sustainable municipal waste management. Keynote speech at the CSD international conference, 16–18 Feb 2011, Tokyo, Japan

Zurbrugg C, Drescher S, Maqsood Sinha AHM, Rytz I, Enayetullah I (2005) Decentralized composting in Bangladesh, a win-win situation for all stakeholders. Resource Conservation and Recycling 43(2005):281–292. doi:10.1016/j.resconrec.2004.06.005

Biodiversity and Culture, Two Key Ingredients for a Truly Green Urban Economy: Learning from Agriculture and Forestry Policies in Kanazawa City, Japan

Raquel Moreno-Peñaranda

Abstract A truly green urban economy should promote the conservation and sustainable use of natural resources and biodiversity, inside and outside the city boundaries. Designing instruments for achieving such goals can be an arduous task for cities, as different sectors and stakeholders have divergent interests and thus a partial vision regarding what the local green economy should look like. This chapter looks at two key elements for the greening of local economies: biodiversity and culture (or biocultural diversity). By understanding the links between natural resources use and local taste and preferences, cities can more easily identify sustainable production-consumption patterns especially suitable for the locality, which, if properly scaled-up, can lead to increased sustainability and local economic development. In Kanazawa, a mid-size city in west Japan, biocultural diversity is catalyzing the local green economy. Two notable examples of how the local administration is greening the urban economy can be found in agriculture and forestry sectors. Innovative branding schemes for traditional varieties of vegetables and revitalization of traditional food culture are landmarks of local agricultural policy. In forestry, stimulus for the local crafts and traditional architecture industry, alongside native forest restoration programs are leading the way towards a greener economy.

Keywords Biodiversity • Culture • City government • Agriculture • Forestry

R. Moreno-Peñaranda (✉)
United Nations University-Institute of Advanced Studies-Operating
Unit Ishikawa/Kanazawa (UNU-IAS-OUIK)
1-1-1 Minato Mirai, Nishi-ku, Yokohama 220-8502, Japan
e-mail: Moreno@unu.edu

R. Simpson and M. Zimmermann (eds.), *The Economy of Green Cities: A World Compendium on the Green Urban Economy*, Local Sustainability 3,
DOI 10.1007/978-94-007-1969-9_29,

1 Introduction

A truly green urban economy must promote the conservation and sustainable use of ecosystems and the biodiversity they contain. Local ecosystems provide multiple benefits to urban dwellers, which often extend beyond the city boundaries. For example, urban agriculture is acknowledged for reducing footprints while increasing access to fresh produce, community building or innovative employment opportunities (Viljoen 2005; Pearson et al. 2010). Likewise, urban forests and their biodiversity provide important ecosystem services, including carbon and water management, heat island and air pollution control and multiple cultural services (McPherson et al. 1997; Nowak and Crane 2002). In a rapidly urbanizing world where cities increasingly depend on resources beyond their boundaries and local governments struggle to provide basic environmental services to citizens, ensuring that cities use their natural resources effectively and efficiently is fundamental for local wellbeing and sustainability.

Because of their competencies in sectors critical for sustainable ecosystem management – city planning, water management or greenery, among others – local governments are meant to play a key role in the urban green economy. Yet designing instruments for achieving the goals of a green economy can be an arduous task for cities, as different sectors and stakeholders have divergent interests and lack an integrative, holistic vision regarding what the local green economy should look like. In Japan, strong national government influence and marked sectorialism often results in local governments lacking ability to develop integrative policy responses to tackle local environmental challenges effectively. Yet some interesting approaches are starting to gain momentum among local governments in the country. The mainstreaming of the socio-ccological landscape concept – referring to the harmonious co-existence between societies and the ecosystems they historically inhabit- is becoming a powerful framework from which to explore sustainability and wellbeing challenges currently faced by urban areas (Japan Satoyama Satoumi Assessment 2010).

An innovative element in the socio-ecological landscape approach with the potential of helping cities to navigate the green economy transition has to do with local culture and biodiversity. By paying greater attention to the linkages between local natural resource use patterns and local culture and traditions, cities can identify sustainable local production-consumption models that are especially suitable for their locality because they lead to increased local wellbeing, sustainability and economic opportunities. Culture and biodiversity linkages are also known as biocultural diversity, a concept that refers to long term interactions between human societies and the ecosystems they inhabit, and encompasses local ecological knowledge and practices which are "a vital reservoir of experience, methods and skills that help societies to manage their resources" (UNESCO 2010). Despite increasing awareness about the implications of bio-cultural diversity on wellbeing and sustainability, the concept has yet to consolidate in the green economy approach, especially in the urban context, where it has not been explored beyond studies on rural migrants (WinklerPrins and de Souza 2005), prevalence of rural medical practices (Alves 2007), and environmental education (Shava 2010).

In Kanazawa, a mid-size city in west Japan, biocultural diversity clearly influences local policies for the green economy. Moving away from centralized control and envisioning the city as a nurturer of local culture, the city government has throughout the years incentivized cultural production. This has in turn increased the visibility of the value of local ecosystems as providers of the material goods and the creative inspiration behind local cultural production. Biocultural considerations are manifested in a variety of local policies for different economic sectors, from traditional crafts and architecture to artisanal sake or local cuisine. Agriculture and forestry are two sectors in which local policy initiatives inspired by the city's cultural traditions have a great impact on the local green economy, as both have strong implications for local wellbeing and sustainability issues (Moreno-Peñaranda 2011). In agriculture, a sector in crisis nationwide, innovative branding schemes for traditional local varieties of vegetables and revitalization of traditional food culture have resulted in the consolidation of a niche agri-food local business, with spillover effects reaching the hospitality and tourism sectors. In the forestry sector, traditionally important for the local economy but in steady decline nowadays, stimulus for the local crafts industry and traditional architecture initiatives, alongside native forest restoration programs are seeking to revitalize the production of local, high value species for the creation of local consumption-production networks.

2 Local Biodiversity and Culture: The Basis for Sustainable Agriculture and Forestry in Kanazawa

Kanazawa is a medium size city with almost half a million inhabitants located in western Japan. The city is a renowned cultural center in the country famous for its gardens, old architecture, literature, traditional crafts and cuisine. In 2009 it was designated as a UNESCO Creative City in the field of crafts. The richness of Kanazawa's culture is inextricably linked with its rich diversity of ecosystems, comprising vast extensions of alpine and broadleaf deciduous forests, fertile plateaus and plains, sand dunes, rivers, ponds, canals, a lagoon and the coast (Fig. 1). Over the years, local policies have reflected this cultural and ecological richness through different initiatives, for example uncovering the city canals and recovering much-loved firefly populations, restoring the rivers to recover traditional recreational fishing, or maintaining an incredibly lush, award winning traditional Japanese garden – Kenrokuen – hosting significant biodiversity features maintained through traditional gardening methods for over four centuries.

Agriculture and forestry are two sectors that clearly illustrate how the local government creates green economy opportunities that simultaneously tackle pressing local and global sustainability challenges of an urbanizing world. This is particularly the case for industrialized countries such as Japan, having to build sustainable and resilient agro-food systems while revitalizing rural–urban linkages. Kanazawa city includes some rural landscape and communities within its administrative boarders, where small scale farmers still make a living. Urbanization has in part been

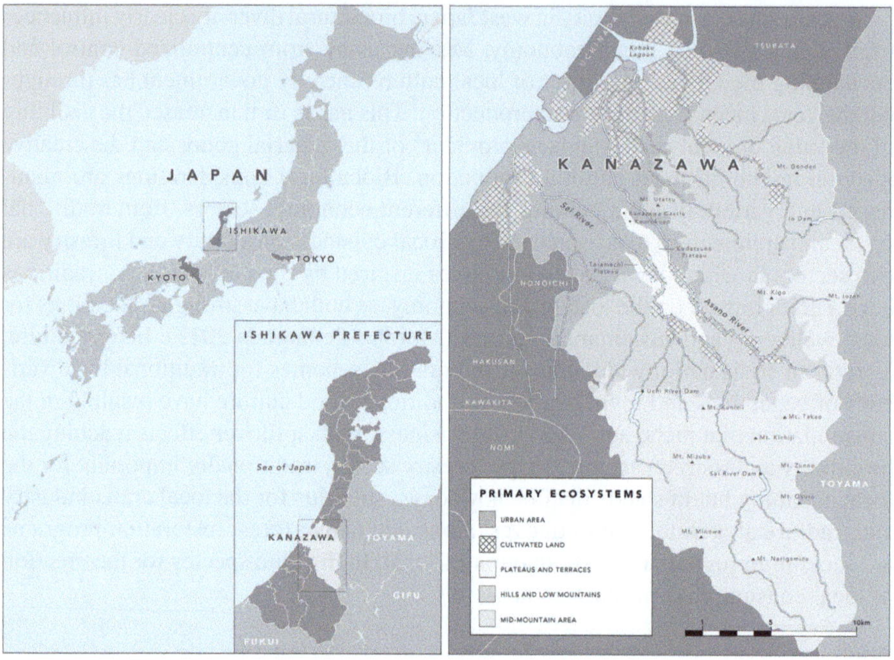

Fig. 1 Location of the city on Kanazawa in Japan (*left*); main land uses and geographical elements of the municipality (*right*). Source: United Nations University Media Center

responsible for the disappearance of such communities and with them their knowledge on sustainable natural resource management. According to the Ministry of Agriculture, Forests and Fisheries, Japan's urban and peri-urban agriculture accounts for one third of the national agricultural output, which is especially relevant for a country with very low food self-sufficiency. Yet urban agriculture is for the most part reliant on hybrids and agrochemicals, its farmers are rapidly aging and fewer young people move into the industry. Forestry also offers valuable insights as widespread abandonment of traditional practices and ageing of foresters, together with the adoption of modern construction relying on concrete, glass and steel, are behind the country's high energy footprint and rural decay.

2.1 *Agriculture: Mainstreaming Sustainable Use of Local Agro-Diversity*

As in many other Japanese cities, agriculture in Kanazawa is mainly suburban, although small cultivated patches of land can still be found inside densely urbanized districts. Around 10% of the municipal area is devoted to agriculture (4,000 ha). Paddy rice makes up 80% of the area, while grains, vegetables and orchards comprise slightly fewer than 20%. Average farmland is less than 1 ha and over half of farmers are 65 years old or older. Abandonment of fields is accelerating, especially in the inter-mountain region. Agriculture has decreased sharply in recent years. Nowadays farmers account for only 2% of local households, a decrease of over 20% over the last two decades. The value of agriculture production also decreased sharply over that period. In 2006 it was 7.1 billion yen (over 90 million USD), of which rice made up 42%, vegetables 36%, fruits 7% and ornamental plants 6%.

As in the rest of the country, modern agriculture has largely displaced traditional practices and created multiple environmental problems. Unlike in other places where the organic movement is leading the way towards more sustainable agri-food systems, in Kanazawa it has been the appeal of traditional local food varieties which has catalyzed this transformation. Before the concepts of sustainability, green economy, or sustainable agriculture were part of the development agenda, cultural values prompted local stakeholders and the government to work hand in hand for the protection and revalorization of selected local traditional varieties of vegetables – the renowned *Kaga vegetables* brand. Even two decades later, when the city is at the first stages of developing a systematic program on organic production, environmental considerations per se rank surprisingly low among the motivations behind the initiative; instead cultural values and economic considerations – namely fulfilling local consumers' taste and creating local economic opportunities – are pointed out by local authorities as the main policy drivers.

2.1.1 Kaga Vegetables: City Involvement in the Branding of Traditional Local Food Culture

Technically speaking, the term *Kaga vegetables* refers to a trademark brand comprising 15 traditional varieties of vegetables cultivated in Kanazawa prior to 1945 (Fig. 2). Cultural values have been pivotal for the creation of the varieties that today are part of the brand. The varieties have been developed throughout the centuries by local farmers, through a co-evolutionary process inextricably linked to local taste and preferences. The position of Kanazawa as a powerful political capital of rich culture during the feudal era influenced the creation of a sophisticated culinary tradition based on distinctive local products. Today, the renowned Kaga cuisine features prominently Kaga vegetables and emphasizes the broader experience of the meal, inspired greatly by the sumptuous traditional culture of the feudal times. The very term *Kaga cuisine* is the outcome of recent efforts of the City and a local stakeholder organization (the Association for the Conservation of Kaga Vegetables) for promoting Kanazawa's traditional cuisine in the hospitality business. Kaga cuisine restaurants recreate the atmosphere of old Japan; locally made Kanazawa lacquerware, together with the colorful porcelain on which the food is presented contribute to creating a carefully staged performance reflecting the intricate connection between nature and culture.

Although a total of 32 traditional local varieties of vegetables have been identified as *traditional* in Kanazawa, the 15 designated so far as Kaga vegetables are those for which production has a sufficient volume and thus is economically viable. Kaga vegetables have a prominent space in Kanazawa's retail food business and hospitality sector. Kaga vegetables are seasonal and, despite relying to a certain extent on modern agricultural inputs, are still deeply connected to the ecological features of the city's landscape, depending on specific soils, microclimates and landforms. Production encompasses different local ecosystems, and output and yield values vary greatly among varieties (Table 1). Approximately half of the city's vegetable production corresponds to Kaga vegetables. In 2008, this accounted for 1.3 billion yen (over 16 million USD).

The Importance of Local Entrepreneurship

City efforts leading to the consolidation of the Kaga vegetables brand are closely connected to local entrepreneurship, especially a local family involved in the seed business for over 150 years through their small store on one of the main communication roads crossing the city. Seeds from different parts of the country passed across their hands over the years, being the starting point of further adoption and breeding by local farmers according to the local environment and taste of the time. More recently, responding to the loss of local varieties and envisioning a business opportunity, current store owners improved seeds' stock and in 1991 established the Association for the Conservation of Kaga Vegetables, promoting the production and commercialization of seeds, vegetables and processed foods. Today the Association manages an annual budget of 5,000,000 yen (64,000 USD). The local government subsidizes 70% of the costs, while the remaining 30% is covered by the local Agricultural Cooperative and private businesses in the local agricultural, food and hospitality sectors.

Fig. 2 The 15 varieties currently under the Kaga vegetables designation, including bulb, fruit, leaf, root, stalk and tuber vegetables. Source: Kanazawa Kaga Vegetables Association

Table 1 Production of the varieties currently designated under the Kaga vegetables brand. Source: Kanazawa City Agriculture Center (2007)

Name of Kaga vegetable	Area (ha)	Production (tones)	# of producers
Summer			
Kinjiso leaves	4	72	42
Akazuiki leaves	0.3	8	9
Utsugi pumpkin	1	25	12
Eggplant	0.2	11	3
Hyacinth beans	4	24	24
Lotus root	60	672	66
Autum			
Gorojima sweet potatoes	98	2,407	65
White Gensuke daikon raddish	4	123	23
Kanazawa thick leek	0.2	4	7
Kuwai arrowhead bulbs	0.3	0.9	6
Winter			
Japanese parsley	0.4	5	5
Leaf mustard	0.1	0.3	4
Chrysanthemum leaves	0.3	3	2
Cucumber	4	620	13
Bamboo shoots	206	892	245
	382.8	4867.2	

The Kaga vegetables branding process yielded its first tangible results in 1997, when ten vegetables were designated under a local denomination of origin and sales under this newly created commercialization scheme started at the Kanazawa central wholesale market. Two years later the Kaga vegetables denomination was officially registered as a trademark brand. In 2001 the brand adopted an official logo (the mascot character "vegitan", Fig. 3), evoking the image of the varieties in a way that was designed to appeal to the local consumers. The image itself became trademark in 2009. A CD and a homepage publicizing the ten different vegetables of the brand were launched that same year. The following year two new vegetables were incorporated, and the City Major directly negotiated commercialization deals with wholesale markets in the Tokyo metropolitan area. In 2003 one additional vegetable was incorporated and promotion activities involving prestigious restaurants and retailers in the Tokyo metropolitan area were undertaken. Activities to raise the profile of Kaga vegetables and stimulate demand through the whole value chain have continued until today both locally and nationwide, disseminating information on their history and characteristics through websites, food tastings, national food expos and development and promotion of specific cooking recipes, including an innovative site for mobile phones download. Thorough the years, Kanazawa City has directly undertaken or commissioned multiple activities in support of the brand, including production subsidies, technology and research investments, educational and outreach activities.

Fig. 3 Vegitan, official image of the Kaga vegetables brand

2.2 Kanazawa Forest: Revitalizing the Urban Satoyama

With forests covering over half the municipal area, Kanazawa is also known as the "Capital of Forests". Of the city's forests, 23% are state owned 77% are privately owned. The subalpine and mountain areas to the south and east of the city are rich in relatively undisturbed deciduous forests of beech, oak and birch. At lower altitudes and on the hills adjacent to the urbanized areas, secondary forests intersect with stretches of intra-mountainous cultivated land in a mosaic characteristic of the socio-ecological production system known as *satoyama* landscape (Japan Satoyama Satoumi Assessment 2010). Satoyama is characterized by high biodiversity and ecosystem services, which have resulted through long-term interactions between societies and the ecosystems from which they depended.

Satoyama landscapes in and around cities provide many different ecosystem services essential for sustainability and local wellbeing – provisioning, supportive, regulatory and cultural. Traditionally, local forests have provided Kanazawa with abundant timber for a variety of now famous constructions such as the renowned Kanazawa castle, together with various temples, shrines, houses and urban infrastructure elements such as the bridges over the two rivers and the many canals crossing the city. Local architecture has traditionally used the variety of local forest resources to manufacture high quality products: strong and durable timbers for foundations and main pillars, lighter woods for roof beams and selected tree species for the central pillars of residential houses are an essential aesthetic element in traditional architecture. Besides construction, forests have traditionally been used as a source of energy (in the form of fuel wood and charcoal) and food (from bamboo shoots to mushrooms, in addition to hundreds of species of wild edible plants), traditional medicines, religious uses (such as the famous Kanazawa Buddhist altars), festivals and celebrations.

2.2.1 Challenges for Kanazawa Forests

Over the past few decades, traditionally managed forests have declined significantly within the city boundaries. According to Kanazawa City Forest Development Department and Kanazawa Forest Cooperative, declining economic viability of local timbers, together with depopulation and aging in mountainous areas of the municipality have resulted in lower forest production and even in significant changes in the structure and functionality of these forests (e.g. proliferation of dark overgrown vegetation and expansion of trees to the detriment of former shrub or grass species). This has in turn had a negative effect on biodiversity. There are several challenges identified by the city with regard to the viability of the local forest industry. First, most private forests are 5 ha or smaller, which makes economic profitability difficult. Second, many forest owners are no longer maintaining their forests and many do not even know where their forests are in fact located. Third, as the forestry business has become increasingly unprofitable, former forest entrepreneurs have switched to other businesses. And fourth, sharp decrease of wood and timber consumption. In Ishikawa prefecture (the broader administrative region where Kanazawa is located), consumption of timber and processed woods (both locally produced and imported) has declined by 50% over the last 30 years. Consumption of locally produced wood and timber in Kanazawa alone has fallen by 20% over the same period, mostly as a result of cheaper imports and popularization of concretization and steeling methods in construction and public works.

2.2.2 The Local Policy Response: Incentivizing Traditional Uses of Locally Grown Timber

The city government, in collaboration with forest owners, producer groups, processing and distribution industry, consumers, crafts and construction industry as well as local universities and civil society organizations related to local forest conservation is taking special measures towards reversing current trends in the local forestry sector. Like many other cities in Japan, Kanazawa City initiatives include the promotion of forests as places for recreational and educational activities, fostering the direct participation of the public in local volunteering conservation and environmental monitoring. Emphasis has been put on raising awareness about the social and cultural values of the local forests, teaching the participants how to create traditional artifacts using forest materials, or to take part in satoyama-related annual celebrations.

Yet one of the most innovative local policy initiatives for forest revitalization developed by the city government takes a significant step forward by linking incentives for the growing of high-value local tree species to stimulus for Kanazawa's traditional crafts and construction industries. The designation of Kanazawa as a Creative City for Crafts by UNESCO in 2009 rendered profound transformations in the way the local government tackled forestry issues. In early 2010, the City established a *Committee for the Promotion of Forestry in Support of Traditions*, integrated by ten members of different city government departments, experts from local universities, and local associations and stakeholders such as the Forest Owners' Association, the Wooden Fittings' Association and the Forest Industry Cooperative. The committee's mandate was to look into current timber demand and supply trends for traditional craftworks and traditional buildings in Kanazawa, identifying and discussing opportunities for the establishment of a solid local forestry industry linked to the sustainable use of local forest resources and the conservation of their biodiversity. Some of the traditional uses of local forestry products include the renowned Kanazawa lacquer ware, Buddhist altars (typically found in local homes for family praying) together with paulownia and wooden craftworks, wooden fittings, and a variety of products for traditional construction. The main goal of the Forestry in Support of Traditions initiative is to strengthen the local wood industry by means of increasing the demand of locally produced, high quality woods with a long tradition in the city, thus revitalizing the local economy (moving the city forward as a "hand craft town") while contributing to the sustainable use and biodiversity of the local forests (Fig. 4).

At the instances of the Committee, a survey of the status of local forest resources was conducted for the first time in Kanazawa. The survey showed that there are considerable untapped forest resources in the city, while highlighted problems of appropriate conservation and maintenance of high value trees. In light of the survey results, and strongly motivated by the UNESCO Creative City for Crafts designation, the establishment of a local production-consumption network for Kanazawa timbers was considered by the Committee as an effective approach to improve the status of local forests while developing local forestry-related industries.

The "Forestry in Support of Traditions" policy initiative targets demand, distribution, and supply issues simultaneously, although the main policy driver is demand. Each level encompasses particular challenges and opportunities. Regarding demand, manufacturing of craftworks and buildings has decreased in recent years, mostly because of economic factors, change of lifestyles, and influx of cheap wood products from abroad. Developing a "made in Kanazawa" brand for selected local, high quality timbers has been identified as an effective way of increasing local wood consumption by the local crafts and construction sectors, given increasing consumers' concerns regarding the sustainability of imported wood and even more so given the public appreciation for local timber because of its quality and historical tradition. So far only timbers from the renowned Kenrokuen Japanese traditional garden located in Kanazawa's historical city center are specifically branded. In terms of

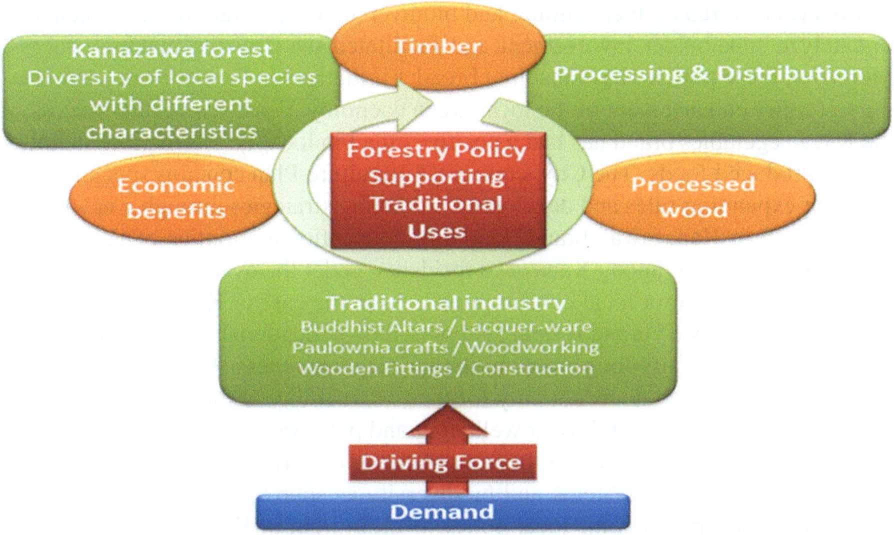

Fig. 4 Basic framework of the local forestry policy for revitalization of traditional forest uses. Source: Kanazawa City (2011)

distribution, shops and stores specializing in decorative woods are decreasing, mostly as a result of aging business owners and a lack of successors. Measures in support of these local businesses are currently being developed by the city as they are considered of strategic importance for distribution and processing of local timbers due to their historical importance in the local forestry sector. The improvement of storage facilities is also a component of the policy, given the challenge of long drying period of many local high-value woods. In terms of supply, the policy is developing special provisions for improving the growing conditions of targeted tree species, as in many cases the forests in which they are located are not properly maintained, and thus valuable trees for the "grown in Kanazawa" brand are currently being lost. Stimuli for new plantings of these targeted species in suitable areas that are currently underused are also being considered.

3 Conclusion

Culture and biodiversity are deeply linked in the development of local policies relevant for the local green economy in Kanazwa, especially in the agriculture and forestry sectors. Specific measures have been taken by the city in collaboration with relevant local stakeholders in order to mainstream sustainable local ecosystem management in connection to incentives for local economic sectors directly linked to cultural traditions.

Policies inspired by local culture and biodiversity considerations have been able to catalyze stakeholder involvement and facilitated the decision making process. Moreover, these policies have inspired broader initiatives in their sectors, contributing to further enhancement of the local green economy. For instance, the success of the Kaga vegetables brand has had a positive effect in local efforts to promote other locally produced foods. The City's current Agriculture Plan contains special provisions for expanding sales and developing marketing strategies through a newly created "grown in Kanazawa" brand, targeting conventional varieties of yet popular vegetables and fruits among Kanazawa residents, under the overall idea of "promoting local production for local consumption". As in the case of Kaga vegetables, the "grown in Kanazawa" brand is being supported by educational campaigns for local residents and especially for school children, rising awareness about the importance of consuming local products not only for the sustainability of the local industry and ecosystems but also for one's own wellbeing and pleasure.

Overall, cultural aspects have been fundamental for rising awareness about the importance of local ecosystems and biodiversity as sources of both natural goods and creative inspiration at the local policy level. In light of Kanazawa city agriculture and forestry policies, it is suggested that exploring biocultural linkages can help local governments in the design of instruments for enhancing the urban green economy, as they have the potential to catalyze sustainability and local development concerns in a synergic, integrative way.

References

Alves RRN (2007) Zootherapy goes to town: the use of animal-based remedies in urban areas of NE and N Brazil. J Ethnopharmacol 113(3)

Japan Satoyama Satoumi Assessment (2010) Satoyama-Satoumi ecosystems and human well-being: socio-ecological production landscapes of Japan – summary for decision makers. United Nations University. UNU Press, Tokyo

Kanazawa City (2011) Proposal for forest policy supporting traditional uses. Promotion committee of the forestry policy supporting traditional uses (in Japanese)

Kanazawa City Agriculture and Forestry Plan (2007) Kanazawa city government. Kanazawa, Japan (in Japanese)

McPherson EG et al (1997) Quantifying urban forest structure, function, and value: the Chicago Urban Forest Climate Project. Urban Ecosyst 1:49–61

Moreno-Peñaranda R (2011) Cities and biodiversity. In: Biodiversity in Kanazawa: through the four seasons. UNU-IAS Report. United Nations University Institute of Advanced Studies, Japan

Nowak DJ, Crane DE (2002) Carbon storage and sequestration by urban trees in the USA. Environ Pollut 116(3):381–389

Pearson C, Pilgrim S, Pretty J (eds) (2010) Urban agriculture: diverse activities and benefits for city society. Earthscan, London

Shava S (2010) Agricultural knowledge in urban and resettled communities: applications to social–ecological resilience and environmental education. Environ Educ Res 16:5–6

UNESCO (2010) A proposed joint programme of work on biological and cultural diversity lead by the secretariat of the convention on biodiversity and UNESCO. International conference on biological and cultural diversity: diversity for development, 8–10 June 2010, Montreal, Canada. Available via: http://www.unesco.org/mab/doc/iyb/icbcd_working_doc.pdf

Viljoen A (ed) (2005) Continuous Productive Urban Landscapes (CPULs): designing urban agriculture for sustainable cities. Architectural Press, Oxford

WinklerPrins AMGA, de Souza PS (2005) Surviving the city: urban home gardens and the economy of affection in the Brazilian Amazon. J Latin Am Geogr 4:107–126

An Economic Assessment of the Deforestation of Ghana's Garden City of West Africa

Jonathan Dagadu Quartey

Abstract Ghana's urban population is projected to grow from the current 52% of the total population to about 65% by 2030. Such a growth rate certainly has far reaching implications for natural environments within urban centers, particularly in the face of current economic pressures. Kumasi, Ghana's second largest city was for a long time known as the Garden city of West Africa. This status seems however to have been traded in favor of policy for urban development. In the particular case of the Kumasi forest reserve this chapter assesses the total economic use value of the forest and the factors, which influence such value through a Hedonic Price Model. The total economic use value is then compared with the user cost of the forest, to draw some useful lessons for urban forestry from economic analysis. The comparison reveals that Ghana loses a net minimum of US$ 35 million per annum in carbon credits due to the deforestation of the Kumasi Forest Reserve. The chapter therefore concludes that developing economies should adopt urban forestry as a source of funding and resources for economic development and poverty alleviation.

Keywords Kumasi Forest Reserve • Total economic value • Economic use-value • Hedonic price model • Urban forestry

1 Introduction

Forests have always contributed significantly to human welfare, and will continue to do so. The usefulness of forests to people tends to create a close association between them. Such an association can cause cities to develop in close proximity to forests

J.D. Quartey (✉)
Department of Economics, Kwame Nkrumah University of Science and Technology, Kumasi, Ghana
e-mail: jdquartey@yahoo.com

R. Simpson and M. Zimmermann (eds.), *The Economy of Green Cities: A World Compendium on the Green Urban Economy*, Local Sustainability 3,
DOI 10.1007/978-94-007-1969-9_30,

and means that urban forestry has become a reasonable and widely acceptable practice. As a result of urbanization in many developing countries, there appears to be a struggle for space between human settlements and forests. In many parts of Ghana trees have eventually been moved, but humans ended up paying the price.

A historical basis for the deforestation of the Kumasi Forest Reserve (KFR) in Ghana can be found in an ancient idea that portrayed forests as man's enemy. Overcoming the forest by pushing it as far away as possible from man's dwelling was seen as a measure of human development. The consequences of this notion for a developing economy such as Ghana, especially when compared with the realized benefits drawn by developed economies from the maintenance of urban forests, leaves much to be desired. The care and management of tree populations in urban settings should be a welcome idea for all urban communities as it improves human welfare and the urban environment.

The unsustainable nature of the traditional forestry sector in Ghana has been a source of national political debate. The excessive concentration on traditional forestry has relegated urban forestry to the background as a relatively unimportant aspect of the forestry sector. One of the costs of such neglect has been the loss of urban forests and hence the effective loss of the Kumasi, Ghana's second most urbanized city's status as "Garden City of West Africa".

Undoubtedly there has been a trade-off between the conservation of the Kumasi Forest Reserve and urban development in the region. The Kumasi Forest Reserve is currently reduced to isolated patches of trees within a site occupied by grand hotels, banks, a golf course, restaurants, road networks, workshops and other common features of urban development. To assess the cost-effectiveness and equity of such a trade-off, this chapter estimates the total economic use (developmental) value of the Kumasi Forest Reserve through a Hedonic Price Model. The value is compared with the user cost of the forest to draw some useful lessons for urban forestry from economic analysis.

2 Deforestation in Ghana

Ghana is a tropical African country located between latitudes 4.5°N and 11.5°N and longitudes 3.5°W and 1.3°E. It has a total land area of 23.85 million hectares and is divided into two main vegetation zones: a High Forest Zone, covering about a third of the land area in the South and the Savannah Zone which occupies the remaining two thirds in the North. Generally, land use activities in Ghana have been in the form of small and large scale farming, fuel wood, animal grazing, rural and urban settlements, timber extraction, cocoa production and game/park reserves (FAO 2001).

About 40% of Ghana's land area is covered by forests (9.2 million hectares), with only about 7% being of the high tropical type, all of which are found in reserves or some form of protected areas. Over the last 100 years, Ghana's forests have been under intense pressure from human and commercial activities. From 1900 to 2000 Ghana lost over 80% of its closed forest. The alarming deforestation rate of 22,000 ha per annum reported in the 1990s deteriorated to a further 115,400 ha per year between 2000 and 2005 and is now 135,395 ha per year (FAO 2011; Mongabay 2006; Tang 2010).

Deforestation and degradation of the 282 reserves and almost all non-reserve areas have been consistent. This trend has been due to a combination of factors including policy and institutional failure, urbanization, and a growing forest industry whose rate of growth the resource base cannot sustain.

The contribution of urbanization to deforestation in Ghana has not been given much attention, possibly because of the alarming nature of traditional logging and illegal causes of deforestation, which can make urban deforestation seem less significant. Another issue worth noting in this context is the almost unknown concept of urban forestry in Ghana. The next three sections provide some insight into the importance of urbanization induced deforestation in Ghana.

3 Urbanization in Ghana

Rapid urbanization has occurred in Ghana mainly due to increasing population growth, rural–urban migration and a reclassification of rural settlements as urban settlements. Working with an officially projected average urban growth rate of 3%t between 2000 and 2030, Ghana's urban population is expected to increase from 52% of the total population in 2010 to 65% of the total population by 2030 (NDPC 2010).

The implications of such an urbanization trend on Ghana are immense. Land will be needed for settlements and construction of more roads and other facilities for urban life. In 1990 the National Land use Planning Committee of Ghana estimated that each unit increase in urban population required an additional land area of 33.3 ha for additional housing, infrastructure, and related social services (FAO 2001).

This land requirement necessarily implies the clearing of substantial parts of existing off-reserve forests and forest reserves for infrastructural development, such as roads, electricity, water, schools, houses, etc. Thus more forest land stands the risk of being converted to other uses, placing more pressure on the remaining forest stock of Ghana. The next section narrows down the issue to the deforestation of Kumasi, Ghana's second largest city, and also further examines the implications for urban forestry.

3.1 Urbanization in Kumasi

Kumasi is Ghana's second largest city, located in the transitional forest zone between latitude 6.35° and 6.40°N and longitude 1.30° and 1.35°W, with a land area of about 23,415 ha. Kumasi has a population of approximately two million as indicated by national census figures of 2000. The city of Kumasi was founded in the 1680s by King Osei Tutu I to serve as the capital of the Ashanti State. Coming under British rule in 1890 and being centrally located in Ghana it became a principal transport hub and now plays a pivotal role in the distribution of goods and services in Ghana and beyond.

The Kumasi Metropolitan Assembly (KMA) is the local government authority which oversees the daily administration of the city. The city received the name

Table 1 Percentage land use change trends in Kumasi from 1986 to 2007, in hectares (derived from Tontoh 2011)

Land cover type	Forest	Wet lands	Water	Agricultural land	Urban or built up land
Percentage change	–45.30	34.20	–88.49	58.71	175.44

"Garden City of West Africa" from Queen Elizabeth II, Queen of England, when she visited the city in 1961. This was as a result of her admiration of its flora and natural beauty (KMA 2006).

The city has grown rapidly with an annual growth rate of 5.47%. Kumasi currently has about 90 suburbs, which grew from three initial communities in a concentric form to cover an area of about 10 km radius. It is estimated that 48, 46 and 6% of Kumasi are urban, peri-urban and rural respectively (KMA 2006).

3.2 *Land Use and Urbanization in Kumasi*

Types of land use in Kumasi include urban or built-up land, agricultural land, forested land, water and wetlands. Changes in land use and coverage from 1986 to 2007 are shown in Table 1 below. The conversion of various types of land use to urban or built-up land shows a 175.44% increase from 1986 to 2007. This increase has been mainly at the expense of forest land, which decreased by 45.3%, as well as water which decreased by 88.71%. The increase in agricultural land is symptomatic of the attempt to provide food for the soaring urban population.

As a result of rapid urbanization agriculture, which was a prominent type of land use in Kumasi, has seen dramatic changes in the last two decades (KMA 2006). Agricultural activities have been confined to the peri-urban areas where crop farming occurs. Increasing urbanization has led to increasing vegetable cultivation within most communities.

With the upsurge of demand for residential, industrial and commercial land versus agricultural land, about 80% of arable lands have been displaced by the construction of houses and other physical infrastructure within urban Kumasi (KMA 2006). The natural environment of Kumasi has thus been substantially altered as a result of the urban sprawl.

4 The Kumasi Forest Reserve

The Kumasi Forest Reserve (KFR), located within one kilometer from the central part of Kumasi, previously covered an area of about 1,600 ha. A pictorial view of the KFR is available from the Woods Consultancy Service (1999) Tourist map of Kumasi. The existence of a small patch of the reserve is acknowledged by the Golden Tulip Hotel, situated within the reserve area, as an attraction to its guests.

The KMA has indicated that the major land degradation issues in Kumasi result from encroachment on forest reserves, green belts, wetlands and open areas by many different kinds of developers. This trend has led to the loss of the Kumasi Forest Reserve.

Currently most literature on forests in Ghana does not recognize the existence of the KFR. This follows the forestry policy of 1948, which implicitly made provisions for the conversion of the KFR for developmental uses other than conservation. Records show that the Asantehene Osei Agyemang Prempeh II, King of Ashanti, in a letter to the Acting District Commissioner of Kumasi dated 20 October, 1947, asked for measures to stop further reservation of forest areas (Manshyia Archives 2007). Thus in the eyes of national and local authorities the KFR was only worth its land value after 1948. Since this time the forest has been left at the mercy of those who acquired land within the KFR for developmental purposes.

The KMA has also revealed that the site for Kwame Nkrumah University of Science and Technology (KNUST), formerly virgin forest, was rapidly downgraded into farm lands leading to the degradation of natural drainage and wetlands within the area.

The situation depicted above is a vivid summary of forest degradation in Kumasi and one that explains the loss of the title "Garden city of West Africa". The implementation of the Kumasi outline planning scheme from 1963 to 1988 did not lead to much improvement in this state of affairs. To obtain the desired impact the scheme should have emphasized the role of trees as a critical part of Kumasi urban infrastructure.

5 The Analytical Framework

The user cost of the Kumasi Forest Reserve signifies cost in terms of opportunities lost through deforesting the reserve. User cost is therefore a measure of benefits forgone because of the removal of forest to obtain land for urban projects. One aspect of this cost is the role the KFR could have played as a carbon store. The conversion of the reserve to other uses has given rise to a release of carbon dioxide, which has contributed to the risk of global warming. The seventh session of the Conference of the Parties to the United Nations Framework Convention on Climate Change in 2001 designated carbon sinks through afforestation and reforestation, and through forest management as eligible for credits (Sasaki and Kim 2009). This meant that types of land use that avoided carbon emissions were to be credited with the value of avoided damage, while types of land use high in carbon emissions were to be debited with the value of expected damage.

Using 1 ha of the KFR (a tropical moist forested land) as a unit of analysis and adopting related exposition of Brown and Pearce (1994) but disregarding all timescales, we can see that developmental uses of the land would be preferential to conservation:

$$(Bd - Cd) > (Bc - Cc)$$

where B equals benefits, C is costs, d is developmental uses and c is conservation uses respectively. It is worth noting that B_c is formally equivalent to the Total Economic Value (TEV) of the Conservation option comprising both use and non-use values. Introducing the element of time, we can write the equation as:

$$\sum[B_{d,t} - C_{c,t}](1+r)^{-t} > \sum[TEV - C_{c,t}](1+r)^{-t}$$

where r is the discount rate and t the length of time in years.

In the presence of properly functioning land markets, the price of land P_L should reflect the present value of expected net benefits from the development option. Conservation values however will not be reflected in land prices since they include a great deal more than the use value of the forest. Also, actual use will tend to be determined by central and local economic incentives, which have not been good shadow price reflectors (Brown and Pearce 1994). Following the above explanation the modified requirement for TEV and developmental value comparison becomes:

$$P_L > PV\ (TEV - Cc)$$

where P_V signifies present value.

The test of cost-effectiveness and equity in the deforestation of the KFR is accomplished if the right hand side (RHS) of the above inequality or any positive component of it exceeds P_L. If the RHS exceeds or is sufficiently close to P_L then a case would have been made for the conservation of the KFR. The component of the RHS we wish to compute for this test is the Carbon Storage capacity of the KFR, which will be obtained by an estimation of the carbon credit and debits of the KFR in its original form before destruction.

6 The Hedonic Model

A hedonic price model attempts to evaluate the influence of environmental factors on the prices of property. If a real-estate developer builds two identical houses – one in a high-class residential area and the other near a polluted stream – each would be priced differently due to their different environments. Thus the hedonic model seeks to attribute price difference to environmental differences. In an assessment of the determinants of land prices in or close to the Kumasi Forest Reserve, demand for land becomes linked to some characteristic features of the forest.

Few studies have examined the determinants of forest land prices using the hedonic price method (Snyder et al. 2008). When evaluating the effect of a number of local and regional characteristics on the price of forest land in Vermont, Turner et al. (1991) found that factors such as the presence of road frontage, the presence of non-forested land cover and close proximity of major roads contributed to a higher price for a plot. The study of Roos (1996), found the price of forested land in Sweden to be influenced largely by factors like size and the proportion of productive forest land on the plot. Scarpa et al. (2000) provide a further example of the hedonic model in their estimation of the non-timber value of maple-birch forests in Wisconsin. Snyder et al. (2007)

examined the influence of recreational activities and proximity features associated with forested plots in northern Minnesota through a hedonic model.

The goal of using the Hedonic model in this study was to assess the extent to which the size of plot, location, presence of trees on the plot, purpose for acquiring the plot and nearness to the KFR could influence the price of land and hence the developmental use value of the KFR. Because of the varied uses of land and the inadequacy of the valuation systems across these uses, this chapter used the value of land on or near the reserve as a proxy for the value of relevant property near the reserve. It is worth noting however that the value of land in Ghana does not reflect an absolute sale price, since the laws of Ghana do not permit an outright sale of land. In Ghana land can only be leased over a maximum period of 99 years at a time, after which ownership of land reverts back to the original owner.

The dependent variable was the "price" of the land. The independent variables for the hedonic model were assessed based on their importance in the determination of the value of a plot of land being purchased. This was done by means of a 3 point Likert scale with "very important", "somehow important" and "not important" as options. The independent variables were the size of the plot of land being sold (size), location of the plot of land within Kumasi (location), the presence of trees on the land (trees), the purpose for which the land was being purchased (use) and the proximity of the land to forest. Data on these variables were collected from 18 real-estate agents all over Kumasi for all sales of land made over the past 10 years. These agents are the main avenues through which all landed properties, both residential and commercial, are sold.

7 The Total Economic Value of the KFR

The Total Economic Value (TEV) of a forest reserve in an urban area like Kumasi is an estimate of the economic worth of the various functions of the forest. Such functions include beautification and environmental conservation on top of other economic and social benefits (Bratkovick et al. 2010). The computations of TEV were done in two different phases. One phase was based on carbon storage value of the KFR as a part component of TEV for comparison with the economic use value (P_L). The other phase was done through obtaining the expressed TEV of the current occupants of the KFR. The structure of the expressed TEV is discussed below while the carbon storage value is estimated alongside the discussion in the next section.

A face-to-face questionnaire was conducted in order to assess the value assigned to the reserve by the owners of facilities on land which it formerly occupied. This was to provide some evidence as to whether occupants attached real value to the reserve they had displaced. The questionnaire requested information on (1) the compatibility of the facility's operations with the presence of trees (COMP), (2) the intentions of the facility's owner to incorporate some trees into the setup of their premises (INTEND), (3) the acceptance of a reforestation plan within the premises (REFOR), (4) the awareness of the facility's owner that the facility is located in the former

Kumasi Forest Reserve (AWARE), (5) how much the facility's owner will be willing to pay per tree to replace displaced trees (WTP) and (6) some demographic information. Forty-four locations were originally approached and in all 39 locations cooperated. All non-cooperative business entities were excluded from the survey. Among those who cooperated two were not in a position to disclose their willingness to pay for trees and so were also excluded. A follow up questionnaire was used to ascertain the readiness of the respondent to pay the stated amount of money in reality.

8 Results

8.1 Determinants of P_L

The estate agents in Kumasi indicated that in 100% of purchases the size of the plot and its location were the most important considerations. Two location factors were of premium value to buyers: proximity to the city center and to the Kwame Nkrumah University of Science and Technology (KNUST). In the case of KNUST, this is because the university provided a premium condition due to high demand for hostel facilities by about 15,000 non-residents. Table 2 below shows the details of the results.

The use or purpose for which the land was being acquired had an influence on its price in 66.7% of purchases. The most unimportant consideration in all purchases was proximity of the plot to a forest. In 83.3% of purchases this consideration was unimportant. In cases where it was important the influence was negative, due to the belief that the forest served as a hiding place for criminals.

In 50% of all purchases the presence of trees on the plot was considered irrelevant. No purchases considered the presence of trees on the plot to be very important. These results clearly show that land value within the KFR (or for that matter any forest reserve) is equivalent to the value of the land in Kumasi, since no premium was put on the value of trees within land purchasers' desired premises. This means for the residents of Kumasi the KFR did not add value to property prices and hence was not desirable in urban Kumasi.

8.2 Expressed TEV of the KFR

To ascertain the drivers of value assigned to the trees by occupants of land located in the KFR, the information in Table 3 below was obtained through face-to-face questionnaires.

Table 3 above indicates that 87.2% of the facilities located in the KFR did not find the presence of the trees incompatible with activities within their premises. Only 33.3% of these entities however intended to plant some trees/shrubs within their premises, while over 64% objected to any reforestation plan for the area.

Table 2 Factors influencing price of land in Kumasi (author's fieldwork 2011)

	Relevance in purchases (in percentages)		
Characteristics of the plot of land	Very important	Somehow important	Not important
Size	100	0	0
Location	100	0	0
Trees	0	50	50
Use	66.7	27.8	5.6
Proximity to forest	0	16.7	83.3

Table 3 Assessment of drivers of WTP for one tree in the KFR (author's fieldwork 2011)

DRIVERS	Yes (%)	No (%)
COMP	87.2	12.8
INTEND	33.3	66.7
REFOR	35.9	64.1
AWARE	23.1	76.9

Over 76% of users of the KFR were completely unaware that their premises used to be part of a forest reserve.

Certainly the above information is not a positive sign for the success of urban forestry within the KFR region. This is confirmed by a low willingness to pay (WTP) for replacing trees, relative to the international stumpage price for such trees. This chapter used the modal WTP for one tree of Ghana cedis (GH¢) 10.50 (US$ 1.00 = GH¢ 1.60) to represent value assigned to one tree by respondents. The highest WTP given (GH¢500) was not used because the respondent failed to confirm its preparedness to pay through the follow up questions, which determine the genuineness of the WTP value. Other WTP values were also not confirmed and therefore not used.

9 Discussion

The KFR used to cover about 1,600 ha of land. Clearing it for urban developmental purposes meant the release of carbon as well as the loss of the carbon sink the forest provided. Based on computations from Houghton et al. (1987) discussed in Brown and Pearce (1994), carbon released from the KFR would be 283 t of carbon per hectare of cleared forest. This implies the production of about 452,800 t of carbon emissions through the destruction of the KFR.

Using the mean shadow price of carbon based on UK official estimates of US$ 83 per ton of carbon dioxide (Ackerman and Stanton 2010) we can see that through the destruction of the KFR, Ghana has become poorer by a minimum of US$ 37.6 million per annum. It is also worth noting that this figure excludes the carbon sequestration loss per hectare, which would be quite substantial on an annual basis. If the KFR had not been destroyed through urban policy, the forest could have earned Ghana a minimum of US$ 37.6 million per annum.

One hectare of tropical forest with the characteristic of the KFR has been estimated to contain on average about 300 individual trees belonging to a total average of about 150 different species (Oracle Educational Foundation 2011; Peters 1994). Using the modal willingness to pay (WTP) for one tree of 10.50 Ghana cedis (US$ 1.00 = 1.60 Ghana cedis (GH¢)). This brings the total WTP to GH¢ 3,150 per hectare. For the 1,600 ha of the KFR the Total WTP is GH¢ 5,040,000 = US$ 3,150,000. This means that the occupants of the KFR valued the forest much lower than the value of carbon credits the forest could have obtained for Ghana. Such a value system can be inimical to the development of urban forestry. It remains a fact however that occupants of the KFR who thought the forest close to their premises could act as a hideout for criminals may have a good point – particularly in urban areas where security networks do not function effectively due to lack of adequate resources.

The mean price of land within the reserve was GH¢18,000 per hectare. Particular types of land use however would mean that the sale of all the 1,600 ha might not be possible. Assuming an original forest coverage of about 90% (this was about 74.08% in 1986 and 40.52% in 2007, according to Tontoh 2011), from the 1,600 ha, only 1,360 ha could be offered for sale in the original state of the reserve. Land scarcity considerations would also have influenced the price of the original site of the KFR. If for the current 40.52% of forested land the price per hectare is GH¢ 18,000, signifying a scarcity of about 59.48% of the site, then the original 90% would signify the scarcity of only 10% of the land, leading to a proportional price fall, holding all other factors constant. This brings the price per hectare of originally conceived forest land within the KFR to about GH¢ 3,026.22.

The implication of this analysis is that the economic use value of the KFR (P_L) would be GH¢ 4,115,659.20, which is equivalent to US$ 2,572,287.00. Thus the main developmental gain from the economic use value for destroying the Kumasi Forest Reserve is about US$ 2.6 million. Comparing this developmental gain to the lost carbon credits of US$ 37.6 million leaves a foregone benefit of US$35 million per annum.

10 Conclusion

The deforestation of the Kumasi Forest Reserve has cost Ghana a net minimum of US$ 35 million per annum in carbon credits. Other benefits of urban forestry, namely beautification (which gave Kumasi the status of Garden City of West Africa in 1961), recreation, reduction of the urban heat island effect, reduction of storm water runoff, reduction of air pollution, reduction of energy costs through increased shade over buildings, improved wildlife habitat and the mitigation of overall urban environmental impact have also been lost through deforestation policy for urban development. Remedial measures however can be adopted to save portions of the forest, which have been legally acquired by developers (KMA 2006) but not yet destroyed.

The Achimota Forest in Accra, the capital city of Ghana, recently came to the limelight as a candidate for deforestation for urban development. Several other tropical forest areas within African cities have been targeted for destruction to create space for urban projects. A lesson must be learnt from the plight of the Kumasi Forest Reserve in the once Garden City of West Africa. Developing countries cannot afford to continue to lose money and natural resources in this way. Developing economies should embrace urban forestry as an emerging source of funding and resource for economic development and poverty alleviation.

References

Ackerman F, Stanton EA (2010) The social cost of carbon. A report for the economics of equity and the environment network. Available via: http://www.e3network.org. Cited 10 July 2011

Bratkovick S, Bowyer J, Stai S, Fernholz K, Bratkovich J, Howe J (2010) Urban forestry: an evolving discipline. Dovetail Partners Inc. Available via: www.dovetailinc.org. Cited 10 Dec 2011

Brown K, Pearce D (1994) The Economic value of non-market benefits of tropical forests: carbon storage. In: Weiss J (ed) The economics of project appraisal and the environment. Edward Elgar Publishing Limited, Hants

Food and Agricultural Organization of the United Nations (FAO) (2001) Forestry Outlook Study for Africa (FOSA) Ghana, 2nd draft. FAO, Rome, Italy

Food and Agricultural Organization of the United Nations (FAO) (2011) State of world's forests 2011. FAO, Rome, Italy

Houghton RA, Boone RD, Fruci JR, Hobbie JE, Melillo JM, Palm CA, Peterson BJ, Shaver GR, Woodwell GM, Moore B, Skole DL, Myers N (1987) The flux of carbon from terrestrial ecosystems to the atmosphere in 1980 due to changes in land use: geographical distribution of the global flux. Tellus 39B:122–39

Kumasi Metropolitan Assembly (KMA) (2006). Available via: http://kma.ghanadistricts.gov.gh/?arrow=atd&_=6&sa=5476. Cited 15 Dec 2011

Manhyia Archives (2007) Records of Kumasi Traditional Council (MAG1) – classification list. Available via: http://www.manhyiaarchives.org/mag1_list.php?code=MAG1/7. Cited 15 Dec 2011

Mongabay.com (2006) Forest protection could earn tens of millions for Ghana. Available via: http://news.mongabay.com/2006/1106-ghana.html. Cited 10 May 2010

National Development Planning Commission (NDPC) Ghana (2010) Ghana shared growth and development agenda 2010–2013. Vol 1: Policy Framework. Accra, Ghana

Oracle Educational Foundation (2011) Tropical forest. Available via: htpp://library.thinkquest.org/17456/tropicalall.html. Cited 22 Dec 2011

Peters CM (1994) Sustainable harvest of non-timber plant resources in Tropical Moist forest: an ecological primer. Available via: htpp://www.worldwidelife.org/bsp/bcn/learning/primer/engl.htm. Cited 05 Mar 2008

Roos A (1996) A hedonic price function for forest land in Sweden. Can J Forest Res 26:740–746

Sasaki N, Kim S (2009) Biomass carbon sinks in Japanese forests: 1966–2012. Forestry 82(1):2009. doi:10.1093/forestry/cpn049

Scarpa R, Buongiorno J, Hseu J, Abt KL (2000) Assessing the non-timber value of forests: a revealed-preference, hedonic model. J Forest Econ 6(2):83–107

Snyder SA, Kilgore MA, Hudson R, Donnay J (2007) Determinants of forest land prices on northern Minnesota: a hedonic pricing approach. Forest Sci 53(1):25–36

Snyder SA, Kilgore MA, Hudson R, Donnay J (2008) Influence of purchaser perceptions and intensions on price for forest and land parcels: a hedonic pricing approach. J Forest Econ 14:47–72. Available via: www.sciencedirect.com. Cited 20 Jan 2010

Tang GR (2010) Feature: forest degradation/deforestation in Ghana. Available via: News: http://news.myjoyonline.com/features/201012/57642.asp. Cited 01 May 2011

Tontoh AA (2011) A study to ascertain the growth situation of the Kumasi Metropolitan Area (KMA): a remote sensing approach. Dissertation, Kwame Nkrumah University of Science and Technology

Turner R, Newton CM, Dennis DF (1991) Economic relationships between parcel characteristics and price in the market for Vermont forestland. Forest Sci 37(4):1150–1162

Woods Consultancy Service (1999) Tourist guide in Kumasi Ashanti. Accra, Ghana

Financing a Green Urban Economy: The Potential of the Clean Development Mechanism (CDM)

Maike Sippel and Axel Michaelowa

Abstract The Clean Development Mechanism (CDM) allows entities that reduce greenhouse gas emissions through projects in developing countries to generate revenues through the sale of emission credits. Principally, city governments are well placed to set up CDM projects in the waste, building, energy and transport sectors. With the price of emission credits reaching up to 20 EUR/t CO2 CDM revenues can finance the entirety of certain waste management projects and significant shares of energy and building efficiency projects. For transport projects CDM revenues can cover a significant share of operating costs. However, to date the share of cities in CDM projects has been relatively small, which might be due to the complexity of CDM rules and some short-term orientation of city officials. Private companies within cities have been more successful. With recent reforms the CDM has become much easier. If the demand for emission credits rises sufficiently to significantly increase their price from recent lows the CDM could become a cornerstone for financing a greening of the urban economy in developing countries.

Keywords Greenhouse gas mitigation • Market mechanisms • Clean Development Mechanism • Offsets

M. Sippel (✉)
University of Applied Science, Konstanz, Germany
e-mail: maike.sippel@gmx.de

A. Michaelowa
University of Zurich, Zurich, Switzerland
e-mail: axel.michaelowa@pw.uzh.ch

R. Simpson and M. Zimmermann (eds.), *The Economy of Green Cities: A World Compendium on the Green Urban Economy*, Local Sustainability 3,
DOI 10.1007/978-94-007-1969-9_31, © Springer Science+Business Media Dordrecht 2013

1 Introduction

In order to achieve cost-effective emission reductions and promote sustainable development in developing countries, the Kyoto Protocol introduced the Clean Development Mechanism (CDM). The CDM is designed to finance greenhouse gas (GHG) emission reduction projects in developing countries. Projects and activities, which result in measurable emission reductions beyond a business-as-usual scenario, can be validated by independent auditors and registered with the CDM Executive Board of the UNFCCC to earn Certified Emission Reduction (CERs) credits.

CERs, which are denominated in tons of CO_2 (t CO_2) equivalent, can be traded and sold and used by industrialized countries to achieve their emission reduction targets under the Kyoto Protocol. Their price has fluctuated between EUR 4 and 25. The CER price has essentially been driven by the demand of companies under the EU Emission Trading Scheme, as CERs can be used to cover a part of the company compliance requirements. By the end of 2011, over 3,700 CDM projects had been registered with the UNFCCC and emission reductions from these projects are expected to result in more than two billion t CO_2 until 2012 (UNFCCC 2011).

For their transformation to Green Urban Economies, cities can benefit from the CDM. Firstly, municipalities can directly access CDM revenues for municipal GHG mitigation projects, or secondly, other local stakeholders such as private companies can set up CDM projects.

2 Priority Sectors for Greenhouse Gas Mitigation in Cities

GHG emissions from urban areas mainly originate from the following sectors: commercial and residential buildings, transport, energy supply, industry, waste and to a varying degree agriculture/forestry (e.g. Kennedy et al. 2009; Satterthwaite 2008; UN-Habitat 2011). GHG emissions from the consumption of fossil fuels dominate by far. Energy-related emissions in developing country cities are expected to increase, as their population and economy grow. The large volume of new buildings and infrastructure in cities in rapidly developing countries can either lock in climate mitigation benefits or set a high emission pathway for decades (Sovacool and Brown 2010).

Municipal action is decisive for many important GHG mitigation options in these sectors. For example, power generation and distribution for private households, public transport, as well as waste management are often organized by local authorities or municipal companies. Municipal land use and building regulations hold a strong potential to influence the urban form, density and building standards, and thus urban energy use.

3 Examples of CDM Projects in Priority Sectors

3.1 Energy Supply

The CDM can be used for all types of renewable energy and greenfield fossil fuel power plants, which are more efficient than the standard power plant conventionally built according to the host country's context. In China, for example, municipal power companies have invested heavily in hydropower and also submitted such projects to the CDM Executive Board. By late 2011 over 600 MW were subject to the validation or registered stage of the CDM process. Success rates in China have been quite high, because of their long experience with this project type in the CDM.

Another example is the district heating refurbishments in three Chinese cities, which have been registered as CDM projects. The largest of the projects is in Urumqi city with 1.2 million CERs per year expected, while projects in Qitaihe and Sanhe city forecast 0.2 million CERs per year.

3.2 Buildings

Retrofit of inefficient buildings or construction of highly energy-efficient buildings qualifies for the CDM. A "lighthouse" case for a poor municipality is the Kuyasa project in the slum of Khayelitsha, Cape Town. 2,300 simple houses of the Reconstruction and Development Programme were equipped with ceiling insulation, energy-efficient lamps and solar water heaters. The project cost EUR three million, generating about 6,000 CERs per year. The CER revenue covers about 30% of project costs. As CER revenues only flow after successful project implementation, co-financing is required for up-front investments. In the Kuyasa project there was a multi-year delay before co-financing could be found.

Several building efficiency-related CDM projects are implemented by energy service companies or owners of large commercial buildings. With a new CDM methodology, large-scale greenfield building projects have become possible since early 2011 and the first projects are expected to be submitted.

3.3 Transport

Increased motorization is a common characteristic in rapidly growing cities in developing countries (Labriet et al. 2009). Sovacool and Brown (2010) find private car use to be the dominant GHG emission source in all global cities they analyzed.

However, pressing local issues like congestion, urban air pollution, or accidents mostly dominate over global environmental concerns. Bus rapid transit (BRT) projects can address local and global issues at the same time, and can qualify as CDM projects. The BRT provides segregated bus-lanes in order to guarantee rapid movement of buses. They are relatively low in infrastructure costs (as compared to tram, light rail or metro systems) and can be built in a relatively short time. GHG emission reductions by BRT mainly results from the renewal of the bus fleet, increased capacity of buses, higher occupancy rates and modal shift.

The City of Bogotá in Columbia implemented its TransMilenio BRT as the first CDM transport project. TransMilenio is a public-private partnership. While the City of Bogotá invested to provide the required infrastructure (e.g. segregated lanes, stations, terminals), the private sector is responsible for the bus fleet, the ticket selling system, and the trunk and feeder service. CDM revenues were projected to cover 10% of the overall investment costs or one third of the investment by Bogotá (Grütter 2007). The project achieved 36% of its forecast level of CERs in its first 4 years of issuance (Sippel and Michaelowa 2009).

The TransMilenio BRT results, among others, in the following co-benefits: less local pollutants such as nitrogen dioxides (NOx) and particulate matters, less congestion, accidents, noise pollution, and improved competitiveness of the Bogota city region due to an effective, modern and attractive transit system.

With a recent revision of transport methodologies under the CDM, rail-based MRT systems have also become attractive. Two metro projects were registered in New Delhi and Mumbai, India, with a forecasted annual CER volume reaching 0.5 million and 0.2 million, respectively. Revenues can thus finance several percentage points of the operating cost of the systems.

3.4 Waste

Landfill sites are a primary source of methane, which has a high global warming potential. Landfill gas can be captured by installing gas collection pipes and tube wells. Capturing landfill gas avoids emissions of methane gas directly into the atmosphere. In many developing countries landfill gas capturing is not regulated and thus can qualify as a CDM project. The captured gas can either be flared and methane reduced to CO_2, or used for power generation. In the first case, CERs will be issued for the reduced global warming potential of CO_2 over methane (CH_4). In the second case, additional CERs will be issued according to the energy replaced from the local grid through CH_4. Due to the relatively low cost of the methane flaring equipment and subsequent revenues from CER sales, landfill gas flaring projects can become highly profitable.

An early example for CDM projects in the waste sector is the City of São Paulo in Brazil. The municipality of São Paulo wanted landfill gas capture projects to be developed on two large municipally-owned landfill sites, namely Bandeirantes and São Joao. It successfully accessed carbon finance via the CDM.

At the beginning the municipality held a bid for a 15-year concession to develop the landfill gas project. A joint venture of two Dutch and one Brazilian companies won the bid and developed a CDM project for both landfill sites. The capital costs for project development were born by this joint venture, while the CERs resulting from the project were split equally between the joint venture and the municipality of Sao Paulo (Clapp et al. 2010). Although emission reductions have been lower than expected (e.g. due to leakage of methane due to unprofessional onsite management), the city of Sao Paulo has maximized its profit from the project by auctioning its CERs through the Sao Paulo stock exchange. This procedure resulted in revenues of EUR 26 million (C40 Cities 2009). In addition to this profit the project has produced several co-benefits. They include a reduction of explosion risk on the landfill sites, reduction of odor close to the landfill sites, new job creation, and the transfer of landfill gas capturing technology to Brazil (Clapp et al. 2010).

4 General Analysis and Conclusions on Urban CDM

Despite the potential of CDM for urban priority sectors, urban mitigation activities have under-utilized the CDM. So far cities contribute to only about 1% of registered CDM projects (Sippel and Michaelowa 2009). With regards to technology and sectors, waste management projects dominate among urban CDM projects. While transport and buildings are key urban sectors regarding their share in urban GHG emissions, the CDM has not yet been able to trigger large scale emission reductions in these sectors. This is at least partly due to the nature of CDM procedures. The CDM has been traditionally technology and sector specific, and did not easily allow for holistic and non-technical mitigation approaches. This made the CDM especially challenging to implement for transport and building projects, which tend to be dispersed and mostly include different mitigation technologies, strategies and multiple stakeholders. However, recent revisions and standardizations of CDM methodologies have removed some of these obstacles and led to an increased inflow of building-efficiency as well as transport projects.

Complexity of CDM procedures is challenging for local staff, and the CDM project cycle may be longer than time-horizons of urban decision-makers. In some cases CDM projects can result in net benefits for a municipality (like in the São Paulo landfill gas project). In other cases CER revenues provide relatively small additional funding, and other solutions for large up-front investments are needed (example of TransMilenio in Bogotá and Delhi Metro).

For the future of carbon finance in cities there is first of all a need for increased awareness and capacity for carbon finance mechanisms among local decision-makers and staff. Beyond municipal action in this area, different national/global policy options exist:

First, a reformed, post-2012 CDM might include further standardization and multiple methodology approaches. Second, lessons learned from the CDM may be used to shape a new mechanism where demand for certified urban GHG emission reductions may be generated (e.g. by domestic offset schemes). In a combination of points two and three, the way may lead from the CDM Program of Activities to NAMAs (Nationally Appropriate Mitigation Actions) covering all mitigation activities within one city and eventually sectorial approaches.

Lessons learned from the CDM should also be taken into account when designing other financial mechanisms for a Green Urban Economy. A crucial issue for all forms of carbon finance is an increased willingness of governments to engage in legally binding emission commitments for a growing number of countries. The uncertainty regarding the international regime created by the failure of the Copenhagen conference in 2009 has persisted until now. Even the decisions of the Durban conference in 2011 to negotiate a treaty by 2015 with commitments for all countries from 2020 onwards has not been sufficient to stem the fall in CER prices. With CERs at EUR 4, only a small set of CDM project types remain attractive. So governments have to put policies in place to increase CER demand to show that there is a long-term future for the CDM and carbon markets in general.

References

C40 Cities (2009) CC40 Cities: Sao Paulo, Brazil, Sao Joao and Bandeirantes landfills. Accessed via http://www.c40cities.org/docs/casestudies/waste/sao-paulo-landfill.pdf. Cited 23 Nov 2009

Clapp C, Leseur A, Sartor O, Briner G, Corfee-Morlot J (2010) Cities and carbon market finance: taking stock of cities' experience with Clean Development Mechanism (CDM) and Joint Implementation (JI). OECD environment working paper. 29. OECD Publishing, Paris

Grütter JM (2007) The CDM in the transport sector, module 5d – sustainable transport: a sourcebook for policy-makers in developing cities. GTZ, Eschborn

Kennedy C, Steinberger J, Gasson B, Hansen Y, Hillman T, Havranek M, Pataki D, Phdungsilp A, Ramswami A, Villalba Mendez G (2009) Greenhouse gas emissions from global cities. Environ Sci Technol 43(19):7297–7302

Labriet M, Caldes N, Izquierdo L (2009) A review on urban air quality, global climate change and CDM issues in the transportation sector. Int J Glob Warm 1(3):144–159

Satterthwaite D (2008) Cities' contribution to global warming: notes on the allocation of greenhouse gas emissions. Environ Urban 20:539–549

Sippel M, Michaelowa A (2009) Does global climate policy promote low-carbon cities? Lessons learnt from the CDM. Center for Comparative and International Studies (CIS) working paper no 49, ETH Zurich and University of Zurich

Sovacool B, Brown M (2010) Twelve metropolitan carbon footprints: a preliminary comparative global assessment. Energy Policy 38:4856–4869

UNFCCC – United Nations Framework Convention on Climate Change (2011) CDM in numbers. Accessed via http://cdm.unfccc.int/Statistics/index.html. Cited 12 Dec 2011

UN-Habitat – United Nations Human Settlements Programme (2011) Global report on human settlements 2011: cities and climate change – policy directions. Earthscan, London

Part V
Experiences and Examples from City Governments

Introduction: Experiences and Examples from City Governments

Steven Bland

Keywords Local government • City examples • Introduction • Green urban economy

1 Introduction: Experiences and Examples from City Governments

The Green Urban Economy provides local governments and their partners a new lens with which to approach the effective governance of their jurisdictions in the context of twenty-first century environmental, social and economic challenges. It can provide new understanding, methods and tools, as well as strengthen existing activities and initiatives in support of sustainable development.

It should not come as a surprise that in cities around the world local governments are already using existing or new economic incentives and measures to develop local economic systems that directly or indirectly place a greater value on environmental amenities and ecosystem services. They have begun to demonstrate that greening urban systems is not just environmentally desirable, but economically beneficial.

In this chapter, six city case studies of action are presented. They demonstrate that local governments are already implementing measures to promote green economic development, but also the potential that still exists in urban areas for innovative, sustained and transformational action. The case studies also demonstrate that in practice, the implementation of green urban planning and economic policies will be geographically diverse, with different approaches, measures and roles being taken.

S. Bland (✉)
ICLEI – Local Governments for Sustainability, Bonn, Germany
e-mail: steven.bland@iclei.org

R. Simpson and M. Zimmermann (eds.), *The Economy of Green Cities: A World Compendium on the Green Urban Economy*, Local Sustainability 3,
DOI 10.1007/978-94-007-1969-9_32, © Springer Science+Business Media Dordrecht 2013

2 Chapters in This Part

In Germany the **City of Hannover's** governance reform, of merging environmental and economic responsibilities into one department after a separate environment department had been created years before, demonstrates the effectiveness of a specific institutional innovation which prevents silo-working and instead builds on the synergies between smart, pro-environment development and economic performance.

In **Auckland, New Zealand**, the importance of grounding any specific mechanisms, like those in Hannover, into a broad, strategic vision for a city area is clear. Only through building a locally-specific and appropriate strategic response can measures to green the economy deliver results.

The case of **Flanders, Belgium,** shows how specific urban development mechanisms can be used to bring green priorities and criteria into existing complex systems of multi-level governance. The interaction of local, regional, national and supranational (EU) policies is well explained to demonstrate the importance of situating mechanisms for green urban development into a coherent spatial framework.

The next two chapters focus on specific components for a Green Urban Economy. In **Western Canada**, the experiences and potential of local governments in altering their purchasing practices through green procurement is analyzed. In **Raleigh, USA**, the important principle of creating employment in green industries through up-skilling is highlighted through the green building training program: delivering synergetic environmental and economic improvements.

Lastly, the experience of **Kyoto, Japan** demonstrates that greening urban economies is not just a technical or scientific process, but a social and political one too. The importance of using green activities to raise awareness among citizens and to actively engage them is highlighted.

3 Cross-Cutting Themes: Commonalities in the Approaches

3.1 Local Governments Are Driving Innovation

A Green Urban Economy requires more than a business-as-usual approach. Innovation is required. Hans Mönninghoff reports that Hannover's institutional response in Hannover, Germany was the first of its kind in Europe, and this has in turn driven forward innovations in the land-use, planning, and energy sector. Chambers and Walters show how the city of Auckland, New Zealand, has prioritized key sustainability-related sectors (such as clean technologies and local energy generation) for targeted support to promote innovation. In the Flanders region of Belgium, Jan de Mulder demonstrates how integrated regional urban planning policies incentivize and encourage innovative urban projects. A set of indicators is used to measure the success of cities' efforts to create knowledge intensive and creative industry clusters. Finally, as the example of

Western Canada shows, the greening of procurement practices can also be steered to support local suppliers and industries: triggering local green innovation to products and services.

3.2 Local Governments Are Enabling Green Economic Opportunity

The case studies highlight the enabling role local governments can play in stimulating economic activity and leveraging targeted private sector investment and activity. Makoto Ueda demonstrates in the case of Kyoto, Japan, how government subsidy and incentives can drive the uptake of solar panels on domestic houses, benefitting local renewable energy companies. In Raleigh's green building training program, funding from federal government was targeted at recession-hit areas of the construction sector, enabling a truly "green shoots" recovery. If public investment is deliberately driven towards low-impact or pro-environmental industries and activities, the enabling role of local governments in stimulating economic activity can be significant.

3.3 Local Governments Are Engaging in Supra-Local Processes

Economic forces and processes are often not spatially confined. For a local authority to truly influence its economy, it needs to act and cooperate beyond city borders. In Flanders, "city-contracts" between cities and the Flemish government are used to promote an urban transition to a more sustainable model, with the transversal theme "creative and sustainable cities" playing an important role in this regard. As well as stimulating local greening processes, procurement practices are a key way local government can influence the supra-local scale of economic activity. The experiences of Canadian municipalities in greening their procurement processes demonstrate the strong potential of procurement as an agent of economic incentivizing (Kemp and Clarke report that all levels of Canadian government spend $100 billion on goods and services).

3.4 Local Governments Are Leading by Example

For years, local governments have been leading by example in the environmental field by *getting their own house in order*. It is even more important to do so now to demonstrate the economic viability of certain environmental activities and to build trust with local stakeholders. In Hannover, energy saving standards in buildings owned by the administration are 30% higher than the legal standard, and 17,000 m^2 of solar panels have been built by local private investors on rented municipal building roofs. Auckland council explicitly tries to inspire other stakeholders to act through its state of the art energy management system using real-time meters at major sites, and targeting a 30% reduction in waste by 2018. Lastly, in Kyoto, the municipality has purchased electric vehicles which are then shared with citizens and enterprises. All of these examples provide leadership and demonstrable results to inspire and encourage action by other city actors.

Conclusion

These case studies offer a snapshot of positive action taking place to integrate economic and environmental objectives at the local level. As the rest of the chapters in this book clearly demonstrate, even more radical action is required to provide for human needs without undermining the ability of ecosystems to provide for them. Business as usual urban planning and economic development strategies will not deliver global targets such as an 80% reduction in carbon emissions or the halting of biodiversity loss, nor will they secure the inclusive well-being of urban populations. Local governments and their private sector partners must be encouraged to think radically about the kind of local economy that will overcome these challenges. ICLEI looks forward to writing again in the very near future about more city cases which demonstrate this level of ambition, which are in turn supported by synergetic regional, national and global policy interactions.

Synergies Between Environment and Economy Within a City Administration – Experiences Made by the City of Hannover, Germany

Hans Mönninghoff

Abstract City administrations are typically built up in a sectorial manner, each sector led by a Director of Service, who, as a rule, is strongly focused on the respective area of responsibility. Horizontal issues have difficulties with such an administrative structure – in this respect the founding of independent environmental departments in many cities about 20 years ago was clear progress. The topic of environment thus obtained its own seat in the board of administration and with dedicated advisory councils gained a significantly higher emphasis in the city policies.

In 2005, the 520,000 people strong City of Hannover went a step further by being one of the first European Metropolitan Cities to combine the environmental and economical tasks into one department. At first this was observed critically, now however it is widely celebrated as a success story. This chapter presents examples of the effects of such a positive synergy between economic development and environmental protection.

Keywords Environment • Green economy • Hannover • Administration • Sustainable city policies

City administrations are typically built up in a sectorial manner, each sector led by a Director of Service who, as a rule, is strongly focused on the respective area of responsibility. Horizontal issues have difficulties with such an administrative structure – in this respect the founding of independent environmental departments in many cities about 20 years ago was clear progress. The topic of environment thus obtained its own seat in the board of administration and with dedicated advisory councils gained a significantly higher emphasis in the city policies.

H. Mönninghoff (✉)
Economic and Environmental Affairs, Hannover, Germany
e-mail: eea@hannover-stadt.de

R. Simpson and M. Zimmermann (eds.), *The Economy of Green Cities: A World Compendium on the Green Urban Economy*, Local Sustainability 3,
DOI 10.1007/978-94-007-1969-9_33,

In 2005, the 520,000 people strong City of Hannover went a step further being one the first European Metropolitan Cities to combine the environmental and economical tasks in one department. At first this was observed critically but meanwhile it is celebrated as a success story.

The following are examples for positive synergy effects between the topics of economy and environment, structured in five areas:

1 Municipal Land- and Construction Policies

Municipal land policy is a particularly exciting subject at the interface between economy and environment. The Economic Administration of Hannover regularly buys up areas of land which are further developed in planning and then sold for industrial settlement and housing construction. The result of this new integrated approach is that plots are not sold for the highest possible price, but that from several interested parties the contract is awarded to the one committed to realizing the highest ecological standards in terms of planned buildings and design. The standard is defined in detail in the contract of sale. Even where plots are not owned by the city authority it has a substantial influence on buildings and design through the stipulations made in urban planning contracts.

As a result of economic growth, an area of 110,000 m^2 on average was utilized annually for construction in the city over the last 10 years. In other cities, such development projects often caused substantial controversies between economy and nature conservation. The Hannover city administration however is able to guarantee to environmental organizations that commercial wasteland will be reactivated where possible and open spaces only utilized as far as is necessary. In this context, attention is paid to the conservation of environmental and climatic compensatory spaces as well as fresh air corridors in the city.

With acceptable densification of the city, planned sites for housing and business are situated close to existing public transport systems. Such denser areas therefore optimize ecological mobility, when contrasted to car dependent areas of uncontrolled urban sprawl.

Especially with new commercial areas, ecological aspects can be realized with relatively simple planning instruments. Planning authorities have directed that car parking spaces have to be laid out in a way that rain water seeps into the earth. In addition one tree for every four cars is compulsory. New flat roofs have principally to be grassed or planted with at least 5 cm of soil in Hannover – by this requirement alone, 650,000 m^2 of new grass roofs have been created in Hannover during the last 10 years. Without this measure more than 400,000 m^3 of rain water would have become surface run off. With the grass roofs a maximum of 150,000 m^3 drain away slowly – an important flood prevention measure. The remaining approximate 250,000 m^3 evaporate, contributing to the improvement of the city climate during hot summers.

Where natural resources are unfortunately lost due to construction work, investors have to pay for extensive compensatory measures in the field of nature conservation, which allow the city to realize a wide range of ecological projects. The measure of this success is that in spite of considerable building activities, Hannover was proclaimed "Federal Capital of Biodiversity" in contest with 129 other cities which was introduced by the environmental organizations and the Federal Ministry for the Environment in the year 2011.

2 The Promotion of Economic Development

This strong commitment towards environmental protection is a major advantage for the German economy. Approximately two million people are currently working in this sector and have generated a turnover of about 75 billion Euros in the year 2011. The current German share in the worldwide export of environmental protection products amounts to 15.4%.

While large companies like Siemens do not require any support on a local level, initial subsidies and loans are granted to small and medium-sized enterprises that are newly founded or expanding in the environmental sector in the framework of the municipal economic promotion. Today, it goes without saying that in the face of limited promotion funds preference is given to those enterprises, which achieve higher ecological standards in production or develop innovative ideas for new environmental products. Besides this there is for example a special contest for the foundation of enterprises in the area of climate protection.

The contacts in the framework of the municipal promotion of economic development are an outstanding "door opener" for the project "Ecoprofit" which is very successful in Hannover. In this cooperative project between business and city administration more than 120 enterprises have been intensively examined for possibilities to save money through improvement measures in the areas of water, energy, waste etc. There is an intensive benchmarking process taking place between the firms. The advantages for the environment are obvious and the enterprises gain the possibility to position themselves as part of the operational environmental protection, thus saving resources which stabilizes their economic position.

3 Management of Publicly Owned Buildings

The range of activities of the Hanoverian Economic Administration includes the management of about 600 buildings (schools, kindergartens, administration buildings etc.). In the course of the energy saving measures that are realized as part of an extensive program at the moment, energy saving standards are 30% higher than the legal standard (which is already relatively high in Germany) – a good investment in the future regarding the predicted rises in energy prices. New municipal buildings will generally be erected according to passive-house construction principles, which demand very little energy. Furthermore, the large roof areas of the municipal buildings are made available to private investors for the installation of photovoltaic panels – benefiting the communal financial budget which profits from the income generated by renting out the roofs. At the current time there are 31 such installations totaling 17,000 m^2 panel areas, which produce enough electricity for the overall requirements of 1,300 people.

4 The Energy Sector

In the area of energy, there are a number of opportunities for synergy between environment and economy in a city administration. In Hannover this has been focused in a campaign "Climate Alliance 2020" which is under the patronage of the Mayor. In this context, many large companies, housing associations, interest groups etc. took part in the discussion and development of the climate protection program. They are committed to do everything possible in their radius of action in order to reduce the emission of carbon dioxide in the City area by 40% by 2020 compared to 1990. Here lies a major challenge for the next decade. Due to economic growth, increase of housing area per inhabitant and more heavy goods vehicle and air traffic, the energy consumption in Hannover has only decreased by 10% from 1990 until today (while in many other cities it has increased), and there still remains a lot to be done to reach the 40% target.

The energy activities in Hannover are sponsored by a special finance fund named 'ProClimate'. By means of a small additional charge on the gas sold in the city and the profit made by the municipal energy utility, five million Euros have been granted annually for the last 10 years with which about 1,000 projects have been funded each year. This has included subsidies for home owners who insulate their buildings further than is required by law, as well as training for craftsmen and architects and the realization of flagship projects. Analyses have shown that a monetary support of 1 Euro makes about 10 Euros of private capital accessible so the whole matter is a considerable business development program for regional commerce, especially regarding tradesmanship.

The exact nature of the regional business development program within the energy sector can be revealed in the following figures: taken together, inhabitants, commerce and industry in Hannover spend approximately 1.5 billion Euros annually on electricity, heating and fuel. Included in this figure is an approximate 440 million Euros of procurement costs for gas, oil and coal which flow predominantly in the direction of Russia, the Near East and Australia. If it is possible in the long run to become independent from energy imports by reducing the energy consumption and by meeting the remaining demand through regenerative energy sources, the best part of this 440 million Euros will be made available to the regional economy every year. With the foreseeable increase in energy prices the effect could be even higher.

5 City Marketing

The economic department also includes city marketing. Here too, ecological topics have obtained a significantly higher emphasis during the recent years. Tourism advertising nowadays highlights the city's parks and recreational areas, which are well worth seeing. When relating to acquisition for the settlement of new companies, advertising is used to point out the high quality of life and the other "soft location factors", especially the 108 m^2 public green space per inhabitant of the city, which is an absolute top value for major German cities. Migration of high skilled labor away from Hannover is extremely low compared to other cities – a high quality of life in Hannover is one of the reasons for this.

6 Conclusion

The economic situation in Hannover does the city administration's sustainable strategies justice – Hannover has mastered the financial crisis of 2009 as well as the Euro-state debt crisis quite well so far. Sustainable strategies may not be the sole cause of this success of course – according to a German proverb: success always has many mothers and fathers. Amongst other reasons there are the many small and medium-sized enterprises which provide a stable engine of growth, and the larger technology orientated companies based in Hannover which perform extremely well in the export business. In contrast to many other German cities Hannover has obtained positive financial results in the current budget for four out of five consecutive years and in 2012, with regard to a total investment sum of 120 million Euros, only 16 million Euros will be taken from loans. Today, there are approximately 10,000 more sources of employment in the Hannover Region than in 2006, and close to 3,000 people work in the climate protection sector and are a part of this success story.

Last but not least, all the instruments and projects of integration between economy and environment described above are, of course, not primarily a result of an administrative merger as such but of the clear desire of the political majority to use such a merger to realize sustainable city policies and to set the appropriate priorities. If, for example, ecological standards are given higher priority than sales prices income may have to be relinquished. If the required standard of energy related building refurbishment is made higher than the current legal standard, expenses will be incurred at the beginning and the positive financial effects may only appear in the long run. In the long-term however, Hannover will benefit from putting such sustainability policy into practice and emphasizing a synergy between the economic and environmental sectors within the city administration. Hannover has the advantage, that for 24 years the same coalition (at present 37% Social Democrats and 21.5% Greens) has put sustainable policies into practice. This continuity of policies allows for long-term strategies, which is a basis for sustainable development.

More information: www.sustainable-hannover.de

Auckland Council: Creating a Transformational Shift Towards a Sustainable Eco-Economy

Paul Chambers and Andrew Walters

Abstract Auckland Council has set a goal for Auckland to become the world's most livable city. The Auckland Plan (the plan), the council's main plan for the future of the region, contains the blueprint for how Auckland can achieve this goal. Of paramount importance in becoming the world's most liveable city is the balance between economy and environment, consequently the councils economic development strategy (the EDS) is also critical to the plan, and to achieving the goals and themes within the plan and the strategy. The EDS includes a range of complementary strategic directions to become a vibrant, business friendly, creative world city. These key directions are supported by the following ubiquitous cross cutting themes:

- Creating a sustainable eco-economy.
- Facilitating an iwi/Māori economic powerhouse.
- Developing and enhancing an innovative rural and maritime economy.
- Supporting a diverse ethnic economy.

Auckland Council will demonstrate international best practices through its plans, policies, and leading by example. Its sustainability team will employ best practice sustainability approaches to lead by example. This will involve collaboration with the council's supply chain to achieve sustainable procurement for whole-of-life approach, best practice design and operation of the council's property portfolio. By leading by example, the Council will facilitate a wider change to a greener economy in Auckland contributing to the aspiration to be at the forefront of sustainable cities worldwide.

Keywords Promoting sustainability • Environment • Economy • Eco-economy • Eco-city

---

P. Chambers (✉) • A. Walters
Auckland Council, Auckland, New Zealand
e-mail: paul.chambers@aucklandcouncil.govt.nz; andrew.walters@aucklandcouncil.govt.nz

R. Simpson and M. Zimmermann (eds.), *The Economy of Green Cities: A World Compendium on the Green Urban Economy*, Local Sustainability 3,
DOI 10.1007/978-94-007-1969-9_34,

Fig. 1 The Auckland Plan

1 Introduction: About Auckland

Auckland's stunning natural environment, mild climate, diverse peoples and recreational opportunities ensure its place among the best cities in the world for quality of life. It is consistently ranked in the top five places to live in Mercer's Quality of Living survey, ranking third in 2011. Sophisticated urban neighborhoods are close to the countryside, beach and forests offering many a choice of lifestyles and activities (Fig. 1).

Auckland is New Zealand's largest city. It is home to 33% of the national population and accounts for 37% of the gross domestic product. Auckland is New Zealand's main commercial, industrial and educational center. This makes it the most popular destination for new migrants, and New Zealand's primary gateway to the rest of the world.

Auckland is also the most ethnically diverse region in New Zealand comprising of more than 190 different ethnic groups of which 11.1% identify as Maori. In Maori, Auckland is known as Tāmaki Makaurau, which refers to Auckland being a place that was well sought after from the earliest of times.

2 A Green Auckland

2.1 *The Future Vision*

The Auckland Plan outlines a range of visions for Auckland's Future; one such future vision is a greener Auckland where its waterways and coastlines are cleaner and healthier. Where the region is recognized for its clean air, low rates of water consumption, low greenhouse gas emissions, stringent sustainability criteria for urban development and stewardship of its natural environment. Many Aucklanders prefer to use public transport or telecommute from home, the energy supply is sustainably sourced and resilient, and households are energy efficient. The city is well endowed with tree-lined streets, and has networks of parks and protected areas with native bush and wetlands. Whilst much of the work has still to be done on how to achieve this vision, it presents an exciting glimpse into a possible future where a vision of sustainable eco-economy can become a reality (Fig. 2).

Fig. 2 Vision for a proposed Auckland waterfront development

2.2 *The Auckland Council*

Auckland Council came into being in November 2010 and was the most significant amalgamation of local governments in New Zealand's history. Comprised of a mix of city and district councils and a regional council, eight organizations were merged into one unitary authority. Not only is the Auckland Council unique in New Zealand, it is now the largest local body in Australasia, representing nearly 1.5 million people.

The council has two decision-making parts: (a) a governing body consisting of a mayor and 20 councilors, and (b) 21 local boards made up of 149 members. The governing body focuses on issues, decisions and strategies affecting the whole region, while local boards represent their communities and make decisions on local issues (Fig. 3).

Auckland Council plays a leading role in promoting sustainability in Auckland. Its operational roles include air and water quality monitoring, pest control and other bio-security services, waste collection and recycling, wastewater and stormwater treatment. The council also supports and helps fund a range of community-based environmental protection and education projects. These projects involve many people, have successful environmental outcomes in the areas of planting, waterway care, wetlands and so on, and are also important for raising environmental awareness. The council's work in areas as spatial planning and its resource consents process for certain activities and uses of natural and physical resources (as required under the New Zealand Resource Management Act) also play an important role in promoting a greener Auckland.

In terms of its spatial planning, the council promotes a vision of a quality compact urban form and an integrated public transport network. While still allowing for lifestyle

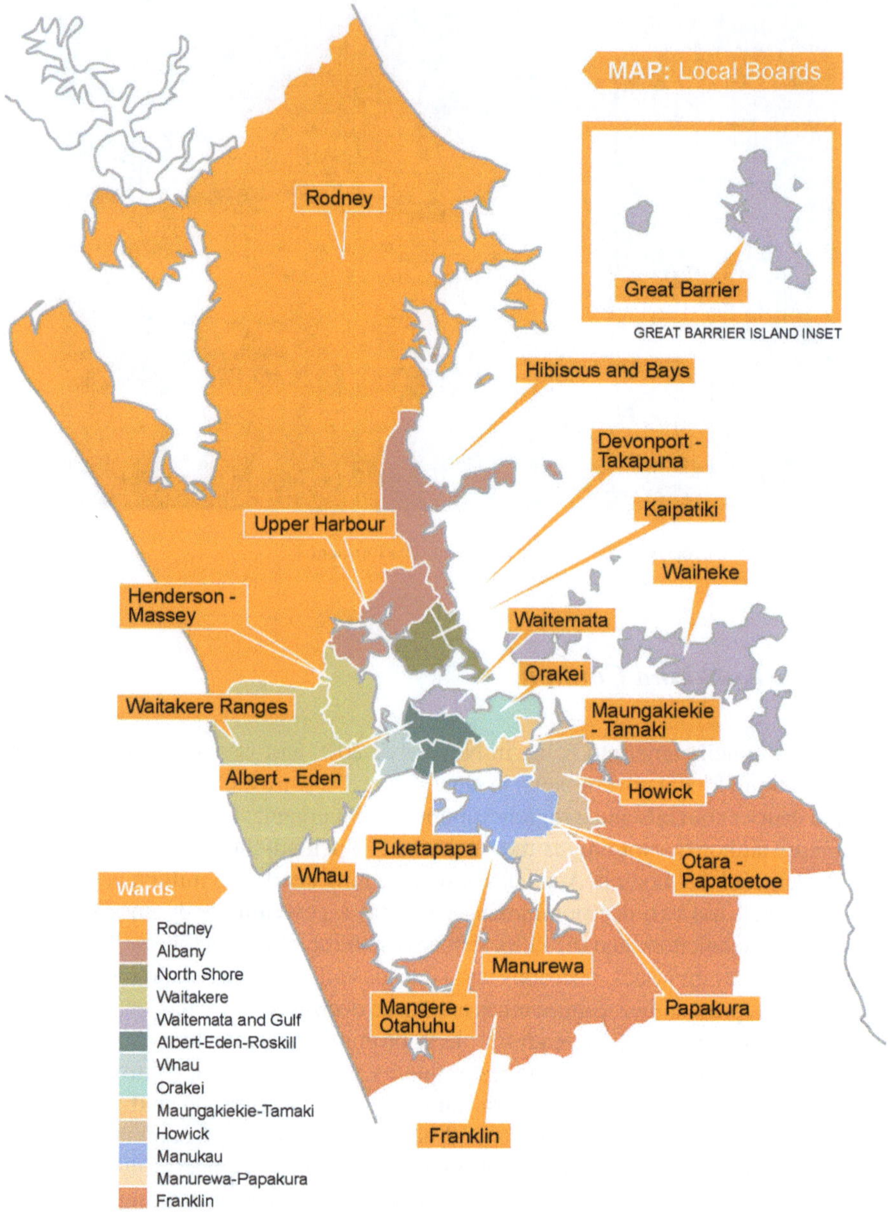

Fig. 3 Local boards of the Auckland Council

choices, future urban growth will be contained to avoid the worst effects of urban sprawl, and to maximize transport efficiencies. Well-designed and accessible intensive regional and town centers, along with the intensification of existing urban areas achieve this.

These are crucial for future environmental well-being, particularly in the light of population growth. The move toward a greener economy underlies the ability to maintain the quality of Auckland's environment over the long-term and is a priority in the council's economic development strategy.

The council also has a dedicated sustainability team that leads the council's organization-wide corporate, community and business sustainability efforts. This encompasses environmental considerations, financial efficiency, social responsibility and community involvement. Activities include quadruple bottom-line program development and implementation (that includes cultural wellbeing, alongside the economic, environmental and the social pillars of sustainability), analysis, reporting and compliance, as well as other related responsibilities to foster, develop and promote sustainability and social responsibility throughout the council's operations to ensure we lead by example.

2.3 Mayoral Vision and Targets

The Auckland Plan is the council's main future planning strategy and it aims to make Auckland the world's most livable city by 2040. It takes into account, and plans for, the additional one million people that Auckland will be home to by 2040, and the additional 400,000 dwellings needed to accommodate them. Critical to the success of the mayor's vision is the draft Auckland Plan, which embraces the economic development ideology of keeping and celebrating what's recognized as special environment assets including: three splendid harbors and associated islands, the rugged west coast and gentle east coast, a large rural environment north and south comprising the majority of Auckland's land, our volcanic landscape, and an extensive park network (Fig. 4).

Paramount to this within a sustainability context are: a commitment to green growth and environmental action, a commitment to outstanding public transport, and a radical improvement in urban living and the built environment lifting the living standards for all Aucklanders. The transformational shift to a sustainable eco-economy is summarized in the council's economic development strategy themes of creating a sustainable eco-economy, facilitating an iwi/Māori economic powerhouse, an innovative rural and maritime economy, and a diverse ethnic economy.

Auckland Council will create an environment that not only supports business sectors and platforms with clear potential now, but also enables emerging industries and expertise to develop. Auckland Council is considering areas such as clean technology, renewable energy, local energy generation, waste management, and sustainable marine farming techniques (Fig. 5).

2.4 Meeting the Targets

Auckland's challenge is to play its part in national and international efforts to reverse the trend of rising greenhouse gas (GHG) emissions. The Organization for Economic Co-operation and Development's (OECD) Economic Survey of New Zealand report (OECD 2011) identifies that New Zealand's GHG intensity of output is the second

Fig. 4 Auckland Mayor Len Brown

Fig. 5 Vision of Auckland's Queen Street

highest in the OECD (after Australia) and that New Zealand is a pioneer in planning the implementation of an emissions trading scheme that covers all sectors and gases. Auckland Council recognizes this, and the Auckland Plan has a target of reducing GHG emissions by 40% by 2040 based on 1990 levels. A preliminary estimation of Auckland's emissions indicates that Auckland accounts for approximately 14.5% of New Zealand's total (gross) GHG emissions.

Auckland's emissions profile is relatively unique when compared to the rest of New Zealand. Its largest sources of emissions are industrial purposes, transport and stationary energy for manufacturing, which combined are estimated to account for two thirds of Auckland emissions. In comparison, New Zealand's largest sources of emissions are methane (32%) and nitrous oxide (15%) from agriculture. In the rest of New Zealand, transport accounts for approximately 20% of total emissions. While Auckland's GHG emissions profile is different to the rest of New Zealand, both share an upward trend in overall emissions. Between 1990 and 2009 Auckland's

Table 1 Comparison of Auckland, New Zealand with reduction targets of selected international cities

City	Target
Auckland	40% below 1990 levels by 2040
Sydney	70% below 2006 levels by 2030
Rotterdam	50% below 1990 levels by 2025
Oslo	50% below 1990 levels by 2030
Vancouver	33% below 1990 levels by 2020 and 80% by 2050
London	34% below 1990 levels by 2020 and 80% by 2050

total (gross) emissions increased by 16.4%. As a result, the council's target to reduce GHG emissions by 40% by 2040 equates to a 57% reduction in total net emissions relative to 2009 emission levels including projected gross regional product and population growth by 2040. This represents an emissions target of 2.7 t CO_2e (Carbon Dioxide equivalent) per capita, which is comparable to targets for other leading international cities (Perry and Chambers 2010) (Table 1).

Auckland can stabilize and reverse the trend of accelerated increases in emissions by moving from a fossil fuel dependent, high energy-using, high-waste society to one that conserves energy and resources and minimizes waste. An immediate priority for Auckland Council is to investigate the viability of solar energy opportunities for water heating and electricity generation.

Auckland Council is investigating solar hot water opportunities for Auckland residents and businesses with a pilot study. In 2012, the Solar Auckland pilot offered up to 250 solar hot water systems to households and businesses at competitive commercial rates by subcontracting the work to solar hot water suppliers in accordance with the council's technical specifications.

The sustainability team has also embarked on an ambitious program to reduce organization-wide corporate emissions by establishing a carbon management tool to monitor emissions and has implemented a state of the art energy management system, utilizing real time meters for 56 of its major sites, to identify electricity, gas and water efficiency savings in its council buildings. The council also has a zero waste aspiration and has a target of 30% reduction in waste by 2018. This will be achieved through staff engagement and employment of best practice waste minimization and diversion techniques including employing in-situ composting in council buildings to treat organic waste arising within each council building. This mirrors targets set for the region in the Councils Waste Management and Minimisation Plan.

A range of environmental and resource efficiency measures will be managed by an ISO14001 compliant environmental management system, which will ensure continual improvement of the transition to a lower carbon council organization. The sustainability team has worked on a range of Environmentally Sustainable Design (ESD) benchmarks for new buildings and has drawn inspiration from the green building code Leadership in Energy and Environmental Design and the Living Building Challenge. By striving for best practice, the council aims to lead by example and inspire other Auckland businesses and organizations to transition to a sustainable eco-economy.

3 New Opportunities

There is good reason to strive towards a low carbon economy as a key component of a sustainable eco-economy. The low carbon market is the fastest growing global market, estimated to be worth £4.3 trillion by 2015 (PricewaterhouseCoopers 2009). New Zealand Trade and Enterprise notes that a low carbon economy could result in a $150 billion high-value low carbon export economy for New Zealand by 2025 (PricewaterhouseCoopers 2009). Auckland is particularly well placed to benefit from this. Adopting environmental policies can provide a longer-run competitive advantage, leading to greater investment, product differentiation and expanded export opportunities, innovation and skills development (OECD 2011). A sustainable eco-economy approach, supporting national efforts, would better position New Zealand to supply new green-based wants and technologies to the world, including emerging markets.

3.1 New Thinking

The world's most livable city goal provides a strong social, cultural, environmental and economic vision that recognizes the global reality of increasing resource constraints and increasing demand for those resources. An eco-city and eco-economy integrates ecological and economic objectives for the benefit of society as a whole. This requires new thinking and a new approach from businesses, industry, the council and central government. This new thinking is about creating a resilient, efficient economy within Auckland that is less reliant on fossil fuels and finite resources. It is also about making it easy for businesses to understand and to work with the council's commitment to sustainability. Developing a sustainable eco-economy has the potential to be a major component of creating an innovative, internationally focused economy that supports New Zealand's 'clean and green' brand.

3.2 New Approaches

Encouraging the adoption of 'greener' technologies and better environmental management practices will require a broad range of approaches by all of government and industry. Market drivers, including new procurement and certification processes in the public and private sectors, can be strong incentives for business to move in this direction. Similarly, regulatory requirements (e.g. the New Zealand Emissions Trading Scheme, and national environmental standards) and other industry-specific requirements can be drivers of change. Education and information can also be powerful change agents as they spur more informed decision-making (Green Growth Advisory Group 2011).

3.3 New Networks

The council can have a major positive influence on businesses and internationalization through its network of international partner cities. The council is engaged in formal relationships with 19 partner cities and countries worldwide. These relationships are with some of New Zealand's most important trading partners including Australia, China, Europe, Japan, Korea, the Pacific Islands and the United States. These networks offer significant opportunities for closer economic relations, and provide a conduit for Auckland and New Zealand businesses and sectors to form closer connections with our current and future markets for goods and services.

3.4 New Initiatives

To ensure Auckland is globally connected and engaged, the council will build alliances with international cities and networks. Where possible, the council's international relationships will be used to achieve economic and cultural outcomes by 'opening doors' and leveraging business and investment attraction opportunities. Examples include facilitating projects, enquiries, official visits and delegations between Auckland and its partners.

The economic development strategy and Auckland Plan proposes a number of actions to transition towards an eco-economy. In conjunction with key stakeholders, we will seek opportunities to deliver catalytic projects that fast-track the adoption of new and emerging low carbon technologies. We will promote and support the development of low energy and energy self-sufficiency zones, and identify opportunities to advance business sustainability solutions and support demonstration projects that illustrate sustainable innovation and value (e.g. showcasing sustainability initiatives that could be implemented by the Auckland Council and its council controlled organizations).

Auckland will benefit from the strategic alignment of fusing innovation and sustainability. As part of the Auckland Waterfront redevelopment, a 3.6 ha site has been identified as a potential location for an Innovation Precinct in Wynyard Quarter that, within 20 years, will be a major Asia-Pacific hub and showcase of innovation in the high-tech industries. It will create, attract and grow world-class research, talent and technology-based ventures. The innovation precinct is also an enabling technology/platform for serving all the key strategic sectors.

The emergence of green jobs and a low carbon economy will also require new skills. Ensuring that education and training providers include sustainability elements in their programs is critical to match supply with likely demand. We will work with key stakeholders to ensure that education and training providers include sustainability elements in their programs, which is critical to match supply with likely demand.

4 Conclusion

Everyone has a role to play in transitioning Auckland to a sustainable eco-economy. Auckland Council recognizes the urgency of the issue and is committed to working with government, businesses and local communities to identify an integrated set of policies and actions to achieve Auckland's targets. Particular areas of focus will be investing in public and rapid transit initiatives, new and emerging technologies (e.g. advanced metering and smart grid networks) and identifying areas where the biggest gains can be made. The council also has a significant opportunity to lead by example as a catalyst for change and to articulate a long-term vision and plan for sustainability in business and the economy.

The council and central government wish to accelerate the capacity for businesses to innovate sustainably. Eco-innovation can be fostered across a range of sectors by hosting and collaborating with businesses, organizations and research institutions to build sustainable innovation capabilities and develop a strong eco-economy based on them. Targeting sustainability-orientated sectors to help grow their market share and to improve the city's international competitiveness are key elements of this strategy. Examples include renewable and distributed energy (to retain our carbon friendly brand), alternative transport, green building (to grow our experience and skill base) and aspects of the food and beverage sector (to increase our exports). We will also seek to foster emerging sectors with high growth potential such as eco-tourism and eco-fashion.

References

Auckland Council (2011a) The Auckland plan. Auckland Council, Auckland, http://www.aucklandcouncil.govt.nz/EN/planspoliciesprojects/plansstrategies/theaucklandplan/Pages/theaucklandplan.aspx, ISBN 978-0-473-21410-4

Auckland Council (2011b) The economic development strategy. Auckland Council, Auckland

Green Growth Advisory Group (2011) Green growth issues for New Zealand, Greening New Zealand's Growth December 2011. Ministry of Economic Development, http://www.med.govt.nz/sectors-industries/environment/pdf-docs-library/Greening%20New%20Zealands%20Growth.pdf

Perry R, Chambers P (2010) Carbon now and carbon futures – a systems and performance based approach to reducing GHG emissions in the Auckland region

PricewaterhouseCoopers (2009) A clean economy vision for New Zealand in 2025. NZTE, Auckland

The Organisation for Economic Co-operation and Development (OECD) (2011) Economic Survey of New Zealand report. OECD, 2011: Economic Survey of New Zealand 2011.

URS (2011) Carbon now – regional GHG inventory projections (revised) client report for Auckland Council, Auckland

Sustainable Urban Development Policy in the Region of Flanders

Jan De Mulder

Abstract Since 2000 the Flemish authorities have developed a coherent policy framework for urban development. In 2003 a White Paper was published after which the previous (2004–2009) and present (2009–2014) Flemish governments took up these challenges. The main objectives are formulated in the two Policy Notes on Urban Development of 2004 and 2009. This policy development at the Flemish regional level has to be framed within the broader European developments (in particular the 2007 Leipzig Charter). The major issues of this urban policy are also incorporated in the two overarching regional strategies (Strategy for Sustainable Development 2006, 2011) and the Flanders in Action Plan (adopted in 2006). Both include elements for a green urban economy. The focus of the current urban policy initiative (2009–2014) is on cities as engines for societal, economic and cultural innovation, however without neglecting the urban vulnerabilities. The core issue remains the fragile balance between the living and centre-functions of cities, which needs to be re-enforced for the city dwellers. The urban policy initiative contains four major instruments: (1) the "cities fund", (2) the "city-contract", (3) the "urban renovation projects" and (4) the "city monitor". This chapter aims to provide an overview with a focus on outcomes and current challenges.

Keywords Regional policy • Urban policy • Planning • Sustainable development

1 Introduction

Belgium is Europe's most urbanized nation and has a high per capita Gross Domestic Product (Hammer et al. 2011). Belgium's cities are situated in the three constitutionally established regions: Flanders the Dutch-speaking region in the north, the

J. De Mulder (✉)
Permanent Representation of Belgium to the EU/Flemish Representation,
Wetstraat 61-63, 1040 Brussels, Belgium
e-mail: Jan.demulder@bz.vlaanderen.befl

R. Simpson and M. Zimmermann (eds.), *The Economy of Green Cities: A World Compendium on the Green Urban Economy*, Local Sustainability 3,
DOI 10.1007/978-94-007-1969-9_35,

bilingual city region of Brussels (the capital), and Wallonia the French-speaking region in the south, which includes a German speaking community. Belgium's federal structure influences the way urban policy is structured and implemented. Building on previous state reforms, the last political agreement from 11th October 2011 has further devolved urban policies to these regions.

At the same time, urban policies at the federal and (Flemish) regional level have to be framed within the broader European development of urban policies of the past decade, such as the Leipzig Charter on Sustainable European Cities adopted in 2007 and more recently the Toledo Declaration adopted in June 2010 by the EU Ministers responsible for urban matters. The Toledo Declaration reaffirms the previous EU commitments on enhancing integrated urban development policies, which are important elements to implement the Europe 2020 Strategy and the European Sustainable Development Strategy. This chapter focuses exclusively on the urban policy development in Flanders.

2 The Flemish Urban Policy

The current Flemish urban policy as reflected in the latest Policy Brief 2011–2012 of the regional Minister for Cities illustrates the tendency towards a more integrated planning approach, both at the regional and the city levels (Van Den Bossche 2011). A closer look at the contents of the past and present urban policy reveals no radical changes. The predominant social concerns that pushed the development of a Flemish urban policy were gradually broadened to incorporate newer tendencies such as ecological objectives. The latter however are often legally required by the acquis communautaire of the European Union. In general, the Flemish urban policy covers the broad objectives of sustainable development including its institutional dimension like enhancing participatory decision making. In this sense Flemish urban policy already includes a "green economy" dimension. But given the different definitions of "green economy" one may wonder how new this latest international concept is and even how far its ambitions reach (Hammer et al. 2011). Evidence based policy making requires tangible results. Urban policy developments are found in a range of planning approaches and initiatives. In order to illustrate achievements, the Flemish urban policy has developed and applies a specific monitoring tool, which uses a whole range of indicators.

2.1 Genesis of a Flemish Urban Policy

Contrary to other EU member states, the political attention for urban policies remained very low in Belgium and Flanders until the electoral victories of the extreme-right wing party in the 1990s (Loopmans et al. 2003). As a result, the more established parties finally paid more attention to deprived urban areas and their dwellers, and started the development of an urban policy in Flanders. This policy focused, more than in other countries, on the living conditions of the poor in disadvantaged neighborhoods.

A first response was the regional legislation on 'regeneration areas' followed in 1996 by the 'Sociaal Impulsfonds' (SIF). In 1999 the federal 'Grootstedenbeleid' came into being, a fund which subsidizes programs aimed at the improvement of deprived neighborhoods in the larger cities of all Belgian regions.

At the beginning of this century the need for an overarching long-term policy vision for the development of Flemish cities was acknowledged by the Flemish Government. This initiated a 2-year consultation process involving more than a thousand experts and representatives from stakeholder groups. They discussed a number of fundamental issues on the basis of 14 reports covering all knowledge on the urban dynamics in Flanders. The main outcome of this process was a White Paper guided by the key question: "What is the best way of governing cities, using and transforming them to achieve all sorts of social, economic, cultural and political possibilities, so that Flanders can be in the forefront of sustainable development?" The White Paper (Boudry et al. 2005) referred to the concept of sustainable development and reflects an integrated and multilevel approach:

> Local aspects must be seen together with the global aspects and the global aspects should be localized within a perspective of sustainable development. The city will only be able to meet this challenge if it can decipher the basic components of its own DNA, and make use of them. This is a matter of approaching the urban character on the basis of an open attitude, cohesion, density, diversity, social justice and democracy. For us, these are the elements of sustainability. The city and the urban character form the start of the social and political restructuring (ibid., p. 20).

The White Paper stated that Flanders needed a detailed social debate on a new urban policy. Such policy is necessary as the city and its urban character are not outdated phenomena. On the contrary, they can serve as the basis of a dynamic, future-oriented and sustainable development. The White Paper elaborates the concept of "urban policy" as it might be developed and implemented at the local level.

> In this sense, the urban policy becomes the test and conclusion of every good social project. This point of view is based on an interpretation of a number of important contemporary developments and requires a new perspective of urban space and urban culture. A change of attitude is needed (ibid., p. 86).

Such urban policy defines and supports the role of the changing city in an enlarged urban area, as part of the Flemish city grid. It requires its own policy level i.e. the town council in the broadest sense of the term includes: municipal government, city government, forms of consultation in urban regions, etc. (in consultation with the inhabitants of the city, the users of the city, the local social and management organizations and the private sector). The urban policy focuses on the development and implementation of an urban program. At this local level it is also important to achieve greater coordination to strengthen both the specific and the sustainable character.

The urban policy concept contains also programmatic elements:

> Therefore the urban programme is, in the first place, an inspirational vision for the future, in which cohesion on the basis of (local) diversity is projected and the city is situated in the world and in the country. Like the urban character, the urban programme is based on an interaction between a spatial and a socio-cultural dimension.
>
> It is built-up on the basis of different elements:
>
> (a) its own 'collection' (history, heritage, sources, national identity, etc.).
> (b) the existing and desired social practices (social, economic, cultural and political).
> (c) the creative processes.
> (d) an expression, an image, a position.
> (e) the main strategic lines for policy programs.
>
> The programme is based on clear principles: social justice, sustainable development, participatory democracy, a comprehensive approach and transparent government. Planning and design lead to the concrete implementation of the programme (ibid., p. 97).

The White Paper was published in 2003 and was a driving force for a new urban policy in the region of Flanders. The model of urban development Flanders deployed since 2000 and synthesized in the White Paper is derived from two strands within planning literature. The first strand starts from a model agreed upon in much of the French literature and urban development as an interaction between '*projet de ville*' and '*projet urbain*'. The '*projet de ville*' encompasses a global vision on the future of a whole city or town. Different '*projet de villes*' materialize such a vision, but are integrated within the local socio-economic and built fabric. The second strand is derived from a rather Anglo-Saxon contingent of literature that combines experiences of 15 years of strategic spatial planning with insights on the level of design by research (Vermeulen 2009).

The previous (2004–2009) and present (2009–2014) Flemish governments took up the challenges formulated in the White Paper. Its approaches were initiated and implemented. In the course of the years evaluation exercises gave input for some amendments towards an improved effectiveness.

2.2 *Objectives of the Urban Policy Initiative*

The main objectives are formulated in the two Flemish Policy Notes on Urban Development of 2004 and 2009. The former was a more concise document and gave a rather general overview. It described the regional urban policy as part of the more overarching urban policy that included the input and participation from societal stakeholders, local governments, private actors, citizens, inhabitants and consumers. The regional administration needed to re-orientate towards coaching and supporting cities. The regional urban policy had to contribute to the Flemish urban policy by innovation, investments, communication, knowledge-building and data collection. It also had to strengthen the involvement of citizens and the capacities of local administrations.

The focus of the current urban policy (Policy Note on Urban Development 2009–2014) is on the strength of cities as engines for societal, economic and cultural innovation, while not neglecting urban vulnerabilities. The core issue remains the fragile balance between the living and center-functions of cities and this balance needs to be re-enforced for the city dwellers.

The vision "to work on sustainable and creative cities in a green and dynamic urban region" is translated into nine "strategic" goals or principles:

- Making living in a city more attractive.
- Developing a spatial urban territorial policy.
- Cities work towards open and warm societies.
- Cities become important centers of creativity and entrepreneurship.
- Cities become green.
- A smart mobility for cities.
- Cities offer more opportunities for participation and involvement of their inhabitants.
- A stronger and more effective regional urban policy in partnership with good governed cities.
- All (13) policy domains of the Flemish administration support the cities.

2.3 Current Policy Frameworks

The major outlines and issues of this urban policy development have also been included in other strategies and plans or programs that were developed and adopted by the Flemish Government. Under the current Flemish Government there are two guiding integrated strategies: the 2nd regional Strategy for Sustainable Development (adopted in 2011, its predecessor in 2006) (VSDO-2 (2011)) and the Flanders in Action Plan (adopted in 2006, ViA (2006)) which is also the basis for the present Government Agreement (2009–2014). Both strategies cover all policy domains for which the Flemish Government is responsible. In 2006, the Flemish Government initiated 'Vlaanderen in Actie' (ViA) – Flanders in Action. All societal partners and stakeholders were engaged in round-table discussions with the aim of developing a new future plan for Flanders. ViA is constructed around seven broad developments or changes (called "breakthroughs") which were identified as necessary components for the achievement of this plan. One concerns the development of Flanders into a green and dynamic urban region. The new plan, called Pact 2020, was endorsed by the Flemish Government in 2008–2009 and is the mainframe for the current overall Flemish policy. The objective of this strategy is very ambitious. By 2020 Flanders aims to perform in the top league of the top regions in Europe, economically, socially and environmentally.[1] Pact 2020 consists of 20 objectives with clear targets figures.

[1] The Research Centre of the Flemish Government published in February 2011 a benchmark of Flanders compared to Denmark, Baden-Württemberg, Bayern, The Basque Country, Midi-Pyrénées, Oost-Nederland, West-Nederland, Zuid-Nederland, Finland, Sweden, North West (UK), Eastern (UK), South East (UK), South West (UK), Scotland. It does not contain indicators on urban policy issues.

Available at: http://vlaandereninactie.be/wp-content/uploads/2011/02/Flanders_Outlook_2011.pdf

The implementation of this Pact should enable progress within five principal domains: (1) Higher prosperity and welfare; (2) A competitive and sustainable economy; (3) Higher employment figures (decent jobs, longer average career terms); (4) A high quality standard of living; (5) An efficient and effective public administration. The progress towards the 20 targets is annually monitored. The implementation of the seven ViA breakthroughs is also being monitored twice a year.

The ViA "breakthrough" for "A green and dynamic urban region" contributes to five Pact 2020 objectives (i.e. innovation, energy, eco-efficiency, environment, and nature) and is implemented by no less than 68 key projects. The concept of 'smart cities' is a central issue and involves a number of initiatives concerning renewable energy, 'Smart grids', electrical cars, sustainable building and living, and resource efficiency. Most of these projects are necessary in the light of EU requirements and targets (e.g. moving towards a low carbon society). Also new planning approaches, improved social housing, and enhanced development of brownfields are incorporated in the range of projects.

Most key projects of this ViA breakthrough (and others too) are also considered as implementation tracks of the 2nd Strategy for Sustainable Development (VSDO 2). Contrary to ViA which is only a political initiative, the VSDO 2 is a legally required strategy adopted by the Flemish Government on 29 April 2011. This strategy is based on the fundamental transition management concept. The VSDO 2 includes 40 actions in order to realize the transition objectives. A particular action concerns sustainable cities and is directly linked and aimed at the existing Flemish urban policy approaches. The action aims to strengthen the sustainability requirements within the urban policy approaches. Furthermore in July 2011 the Flemish Government agreed to enhance ViA by incorporating the transition management concept. Like VSDO 2 ViA aims also at a system innovation which requires transition management approaches. For that reason the breakthrough on decisive governance was broadened with an additional intersecting theme on creative and sustainable cities.[2]

A closer look at both integrated strategies reveals strong interlinkages and interactions. In addition many building blocks have been put in place for implementing sustainable urban policies through a number of sector plans, programs and projects, such as on environmental and nature conservation planning, climate change policies, spatial planning, housing, mobility, energy etc. Obviously all these elements have an impact when elaborating and applying the specific instrumental (or "vertical") urban policy approaches as provided for by the Flemish Government (also by the federal and EU levels). Furthermore this variety of policy approaches poses huge co-ordination challenges (Beuselinck and Verhoest 2008). Some of these elements are definitely important when focusing on greening the urban economy through specific projects or other policy initiatives at the local level. A particular, mainly "environmental" sector instrument that was introduced by the Flemish administration to enhance the greening of the local level concerns the "Co-operation covenant for local authorities"

[2] Additional financing (220.000,-euro) for innovative urban projects is announced in the Policy Brief 2011–2012 of the Flemish Minister for Urban Policy.

(Paredis 2011; LNE 2008). At present its future application after 2013 remains doubtful. As the planning burden for local authorities became too heavy, the necessity to reduce top-down planning and reporting requirements or even supplementary policy tools was acknowledged at the regional level. For this reason the regional government decided in 2011 that municipalities and cities should work in the future with a single strategic plan (Vermoesen 2011; Price and Heyligen 2012).

3 Urban Policy Instruments

The current Flemish urban policy is implemented by four major, legally established instruments: (1) the "cities fund", (2) the "city-contract", (3) the "urban renovation projects" and (4) the "city monitor".[3] The set of urban policy instruments and approaches reflects very clearly the description of core policy levers for green growth in cities (Hammer et al. 2011).[4] The total regional budget for the implementation of these instruments [5] was 137 million euro in 2009 and 142 million euro in 2010.

3.1 The Cities Fund

This fund[6] became operational on 1 January 2003 and was created to support the largest 13 Flemish cities (center cities) in a structural way in order to:

1. Increase the liveableness of the city (also at neighborhood level).
2. Cope with the growing tendency towards a dual society.
3. Improve the quality of democratic governance.

For 2011, the total amount was almost 119 million euro.[7] The two largest cities (Antwerp and Ghent) received almost 75% of this amount. All 13 cities had an agreement for the period 2003–2007 and have a second one for the period 2008–2013.

---

[3] See: http://www.thuisindestad.be/startpagina-2010.html

[4] "(...) be synthesised into ***four core policy levers***, or types of policy instruments, that can be employed by sub-national authorities: (i) regulatory authority; (ii) government spending; (iii) financial incentives; and (iv) information and advocacy. These levers can be applied internally (i.e. toward internal government operations, facilities, and staff) or externally (i.e. toward the public, local businesses, or others in a position to take action to help implement the city's goals) (Hammer 2009)" (Hammer et al. o.c. at p. 73).

[5] This budget is part of the budget for the Agency for the Interior, which belongs to the Public Governance policy domain, one of the 13 policy domains of the Administration of the region of Flanders.

[6] Legal basis is the Flemish decree of 13 December 2002 and Executive Orders of 27 January 2003 and 11 December 2009.

[7] The Flemish Community Commission in Brussels receives 13.2 million euro (in addition to the amount for the 13 Flemish cities).

Each multi-annual agreement indicates the city objectives and how the resources of the fund will be used to achieve these objectives. A visitation commission reviewed the functioning of this fund in 2005. The results of this evaluation had an impact on the negotiations for the agreements for the period 2008–2013. More attention was paid to formulate societal effects and the choice of indicators in order to monitor these developments. A new review is being scheduled for the end of 2013 with a visitation organized in the course of 2011. For each city a report has been drafted and the visitation commission published a synthesis report. The Flemish Government started discussions on the review in 2012.

3.2 *The City-Contract*

By the end of 2007 the Flemish Government had signed a city contract with each of the 13 center cities. These contracts cover 6 years (period 2007–2012) and went through a mid-term review in 2009. Such contracts reflect a city's vision on sustainable development and its financial and administrative support from the Flemish regional Government. These contracts contain engagements from the Flemish Government as well as from the city. Each contract has two chapters: A general one with a central theme on "living" and a more specific chapter related to the particular city. The contracts must be implemented by the end of 2012 in accordance with the agreed timeframes.

The Flemish Government introduced this "contract-instrument" on request from the cities as they favored a more inclusive approach of "urban projects" as proposed by the regional government. These contracts are means for a more efficient and "customer-oriented" multi-level governance within the Flemish region to include new themes (suggested top-down or bottom-up) and embed structural operations of the Flemish administration.

Within the ViA context, the city contracts are considered as a tool for partnership between the regional and local authorities and as a way to realize a transition process. The ViA breakthroughs and the additional transversal theme on creative and sustainable cities are also triggers for input in the preparation of the next generation of city contracts. In order to organize a smooth transfer between the two generations, the Flemish Minister for Urban Policy will announce a new working method in 2012.

3.3 *The Urban Renewal Projects*

The Flemish Government subsidizes urban projects with an innovative character.[8] These projects should address different functions (e.g. synergies between living and

---

[8] Based on the decree of 22 March 2002 and Executive Orders of 16 March 2007 and 17 Augustt 2010.

working) and also leverage and enhance local dynamics. Projects in the 13 center cities as well as in 21 other provincial cities are eligible for financing. The Flemish government is subsidizing urban renovation projects on the basis of seven quality criteria related to spatial impact, planning and design, participation and public-private partnership. The selection criteria are ambitious and demanding. The projects must serve as a lever for a town or a district of a town and significantly improve the general quality of life of a district. They must be established in consultation with the inhabitants of the district and the local social partners, and must be able to demonstrate a minimum financial contribution of 30% from the private sector. One of the crucial aspects of urban projects is the management role of the town itself. The town must ensure that in addition to economic and financial profitability, the project also achieves social added value. The subsidies for urban renewal have not only provided financial and qualitative incentives for urban projects, but also generate a positive dynamic and the necessary recognition.

In recent years there has been a striking leap forward in terms of the quality of urban renovation projects, particularly at the level of the quality of the design and planning (Loeckx et al. 2009). This leap forward in terms of quality is related to several factors (e.g. the expertise, available competences) but the most important factor is undoubtedly the method of 'concept funding'. Concept funding is designed to help smaller towns that often have less planning capacity and project experience than the larger towns. In recent years several smaller towns presented far-reaching and potentially innovative projects, but lacked the quality to attract project subsidies. 'Concept funding' can be used in these cases. Projects which are seen as 'nearly very good', 'potentially excellent' or 'very important' can make an appeal for knowledge, experience, talent and financial means made available in order to improve such project proposals. "Experts" with experience in urban projects are appointed in locally composed teams, who formulate a more focused description of the issues and select a design agency to be in charge of a "design study". This explores the urban planning, infrastructural and architectural possibilities for the development of a high quality urban space. Some of the funding is also used for advice on communication, participation and the development of efficient forms of public-private partnership. The latest call for projects happened in the last quarter of 2011. The budget for 2012 is 12.7 million euro (Flemish Parliament 2011). The evaluation and selection is done by an external and multidisciplinary jury. These projects have a positive impact on the neighborhoods. The current system will be further evaluated and adapted to improve the quality and flexibility (Block et al. 2011a).

3.4 The City Monitor

The City Monitor is a learning tool at strategic level for all stakeholders involved or interested in urban policies. The first edition was published in 2004 and since then it has been continuously developed and adapted. The figures, which the City Monitor provides, indicate how livable the Flemish cities are and how sustainable their development is. This is not only used as a way to measure the situation and progress but also a communication instrument.

Indicators covered by the City Monitor were revised in 2010. The City Monitor now contains some 216 indicators concerning 11 policy fields (of which one deals with entrepreneurship and labor). The indicators were developed in collaboration with academics, experts, local and regional administrations, major groups and stakeholders. Some 100 indicators survived from previous editions, 60 new ones were introduced and around 40 indicators are still being developed.

The 4th and current edition of the City Monitor was published in December 2011. The results are based on a survey which was sent to some 40,000 people, a representative part of the population of the 13 center cities. The net response rate was 46% (Bral et al. 2011). Rather than being an instrument for measuring policy impacts or results, the objective of the City Monitor is to provide input for strategic planning and policy programs, and it will therefore be further updated and supplemented with additional indicators. The City Monitor can be seen as an evaluation tool at a very general level, as an evaluation of the collective effects on the urban society of all the actions and efforts by the different public and private actors. In the long term this instrument should become even more central to evidence-based policy making. Academic experts who were closely involved in developing the City Monitor have stated that it should only be regarded as one element in complex urban decision-making processes. The design methodology of the City Monitor also addresses typical tensions brought about by such wide-ranging terms as 'urban sustainability'. Because of the complex and normative character of such a concept, an intensive co-design approach was chosen for defining such concepts for the City Monitor. This co-design process involved hundreds of stakeholders, potential users and 'experts' such as local civil servants, local politicians and representatives of local organizations. Because of their experiential knowledge those key actors are very well able to stipulate what urban sustainability should look like in the fields of mobility, housing, culture, economy, care, education, etc. The design process included dozens of stakeholders and expert meetings (workshops or focus groups) to develop the vision matrix, to select relevant indicators and to fill in standard forms for each indicator. In this way the City Monitor acknowledges complexity and this approach can be used to complement more traditional monitoring approaches (Block et al. 2011b).

In order to measure progress regarding the strategic goal that "Cities become important centers of creativity and entrepreneurship", the Policy Note on Urban Development 2009–2014 refers to five indicators of the City Monitor in order to discover whether the economic attractiveness of cities is improving. The indicators used are:

1. Increase of the share of employees in knowledge intensive and creative industries;
2. Net growth of the number of urban enterprises;
3. Improvement of the survival rate of start-ups;
4. Increase of economic growth power
5. Increase of the share of foreign investments in cities

However the 2011 edition of the City Monitor no longer contains all five indicators, conclusions can be quite difficult to draw, and it seems more appropriate to look at all 40 indicators in order to get a more balanced impression of the Flemish cities entrepreneurial and labor situation.

4 Multi-Level Governance and Stakeholder Involvement

The current urban policy contains some additional operational goals. The core of these objectives concerns co-operation: between cities, between a city and its suburban areas (city-regional cooperation) and with other stakeholders and policy sectors. Examples of this include the Flemish environment administration which provides for a voluntary cooperation agreement between the regional government and the municipalities and provinces (co-operation covenant, LNE 2008). This includes opportunities for additional funding of local level efforts. Cross-border or transnational cooperation between the municipal, regional and federal levels of different nations are also taking place for example in the form of an EGTC (European Grouping on Territorial Co-operation as agreed within the European Union). There are two EGTCs currently operational that include Flemish and French authorities.

In Flemish urban policy each policy level has a distinctive role. The development of regional urban policy was not a top-down approach but a rather inter-active one. The current policy is a result of a continuous process that involves all stakeholders and partners. This includes the local, municipal, as well as the Flemish Association of Municipalities and Cities (VVSG) level, which is the official representative organization of the local authorities. Furthermore Flemish policymaking builds upon a strong tradition of consultation, which is evidenced by the existence of a number of Strategic Advisory Councils.

Regarding the impacts of policies on the local level, the Strategic Advisory Council on Public Governance (VLABEST) is the official consultative institution. It has to be mentioned that each decision of the Flemish Government has to indicate decisions that might impact the local government level.

Next to policy development the regional administration, the Flemish Agency for the Interior, is responsible for the program management of the urban policies. The "Urban Team" of this Agency works in close cooperation with the local level.[9] The urban team is supported by the Knowledge Centre for the Flemish Cities, an initiative by the 13 center cities and the Flemish Association of Municipalities and Cities (VVSG). Before the regional elections in mid 2009, the Knowledge Centre

[9] The tasks of the Urban team are: (1) preparation, implementation, support and evaluation of the regional urban policy; (2) management of the Cities Fund; (3) supporting the urban renewal projects; (4) development and management of the City Monitor; (5) co-ordination of the follow-up of the City contracts; (6) communication campaigns ("At home in the city") and the award competition.

for the Flemish Cities and the Flemish Association of Municipalities and Cities (VVSG) published a Memorandum for innovative and attractive cities as an engine for Flemish prosperity.

The development of participatory approaches and neighborhood management have been explored and elaborated upon in the past years (De Rynck and Dezeure 2009; De Rynck 2008).

5 Conclusion: Some Achievements and Challenges

Flemish regional urban policy definitely embodies a positive and constructive approach, but reflects only what a good government should do: supporting and facilitating urban renewal and innovation as a joint or shared exercise. The regional policy has enhanced the ability and even obliged actors to co-operate. As the recent Visitation Synthesis Report indicates, the regional and local authorities need to enhance their partnerships. Flemish urban policy has also stimulated local participatory processes. As a result the development of Flemish urban policy during the past decade not only illustrates innovation but has also stimulated innovation in local good governance and local (economic) development. The latter has gradually incorporated "greening" objectives as required by both sector and integrated plans.

Regional urban policy is however still not a completely integrated regional policy (covering all policy domains) and better coordination of policy development and implementation within the Flemish administration is an ongoing challenge. Other regional sector policies remain important and create planning and reporting burdens for city and municipal authorities. This will no longer be the case given the new legislation adopted in 2011, which introduces single planning and reporting at the local level. Furthermore the regional level does not always accept the outcomes of local participative processes, so there is a need for more structural consultation (Van Dooren 2010; Van Humbeeck et al. 2012). Evaluation criteria regarding sustainability lack integration in content and approach, and more monitoring would be useful. Local, innovative approaches could be more widespread and there is a need for more capacity building. Given the content of the Urban Policy Brief 2011–2012, none of these challenges are being overlooked.

References

A green and dynamic urban region. Available via: http://vlaandereninactie.be/wp-content/uploads/2011/04/Groen-en-dynamisch-stedengewest-overzichtstabel-monitoring-ViA-regeerakkoord-September-2011.pdf. Last accessed on 21 Aug 2012

Beuselinck E, Verhoest K (2008) Coördinatie in een gefragmenteerd overheidslandschap: internationale trends en lessen voor Vlaamse en lokale besturen. Burger, bestuur & beleid 5(1):76–92

Block T, Steyers K, Oosterlynck S, Reynaert H, De Rynck F (2011a) When strategic plans fail to lead. A complexity acknowledging perspective on decision-making in urban development projects. The case of Kortrijk (Belgium). Eur Plan Stud 19:981

Block T, Van Assche J, Goeminne G (2011b) Unravelling urban sustainability: how the Flemish City Monitor acknowledges complexities. Ecol Inform. Special issue: Unravelling complexity and supporting sustainability

Boudry L, Cabus P, Corijn E, De Rynck F, Kesteloot C, Loeckx A (2005) The century of the city. City republics and grid cities. White Paper, Urban Policy Project. Ministry of the Flemish Community, Brussels. Available via: http://www.thuisindestad.be/Publicaties.html Last accessed on 21 Aug 2012

Bral L, Jacques A, Schelfaut H, Stuyck K, Vanderhasselt A (2011) Stadmonitor 2011. Een monitor voor leefbare en duurzame Vlaamse steden, Agentschap voor Binnenlands Bestuur/Studiedienst Vlaamse Regering, Brussel. Available via: http://www.thuisindestad.be/stadsmonitor/overzicht.aspx. Last accessed on 21 Aug 2012

De Rynck F (ed) (2008) De stad in de wijk. Gebiedsgericht werken in Vlaamse steden en gemeenten. Kenniscentrum Vlaamse Steden/Politeia, Brussels

De Rynck F, Dezeure K (2009) Burgerparticipatie in Vlaamse steden. Naar een innoverend participatiebeleid. Rapport Werkgroep Participatie, Brussels

Flemish Parliament (2011) Policy brief 2011–2012 by the Flemish Minister for Urban Policy 2011. Available via: http://docs.vlaamsparlement.be/docs/stukken/2011-2012/g1323-1.pdf. Last accessed on 21 Aug 2012

Hammer S, Kamal-Chaoui L, Robert A, Plouin M (2011) Cities and green growth: a conceptual framework. OECD regional development working papers 2011/08. OECD Publishing, fig. 2 at p 18. Available via: http://dx.doi.org/10.1787/5kg0tflmzx34-en. Last accessed on 21 Aug 2012

LNE (2008) Samenwerkingsovereenkomst Gemeenten 2008-2013, Cooperation covenant with local authorities. adopted by the Flemish Government in December 2007 and being implemented by the Environment Administration as from 1st January 2008. Available via: http://www.lne.be/doelgroepen/lokale-overheden/so_2008-2013/handleiding-gemeenten. Last accessed on 21 Aug 2012

Loeckx A, De Meulder B, Patteeuw V (2009) Stadsvernieuwingsprojecten in Vlaanderen. SUN, Amsterdam

Loopmans M, Uitermark J, De Maesschalck F (2003) Against all odds: poor people jumping scales and the development of an urban policy in Flanders, Belgium. Belgeo 2(3):243–258

Monitoring results of ViA/Pact 2020. Available via: http://ikdoe.vlaandereninactie.be/wp-content/uploads/2011/02/Samenvatting-eenmeting-2011.pdf. Last accessed on 21 Aug 2012

Paredis E (2011) Transition management as a form of policy innovation. A case study of Plan C, a process in sustainable materials management in Flanders. Steunpunt Duurzame Ontwikkeling, Working paper no. 26, October 2011. Available via: http://www.steunpuntdo.be/papers/Working%20paper%2026_Paredis.pdf. Last accessed on 21 Aug 2012

Price L, Heyligen J (2012) Het planlastendecreet. Een herijking van interbestuurlijke relaties in Vlaanderen. Vlaams Tijdschrift voor Overheidsmanagement 2012(2):9–18

Van Dooren W (2010) Na Oosterweel: participatie onder hoogspanning. Sampol 2010(3):29–37

Van Den Bossche F (2011) Policy Brief 2011–2012 of the Flemish Minister for Urban Policy. Available via: http://docs.vlaamsparlement.be/docs/stukken/2011-2012/g1323-1.pdf. Last accessed on 21 Aug 2012

Van Humbeeck P, Popelier P, Van Aeken K, Meuwese A (2012) Hoe beter consulteren. Vlaams Tijdschrift voor Overheidsmanagement 2012(2):19–31

Vermeulen S (2009) Needed: an intelligent and integrated vision for Brussels' urban planning. Paper, 4th international conference of the international forum on urbanism (IFoU), Delft, 26–28 Nov 2009. The New Urban Question – Urbanism beyond Neo-Liberalism. Available via: http://www.cosmopolis.be/cms/det_index_en.php?det=704. Last accessed on 21 Aug 2012

Vermoesen E (2011) Regulatory management in Flanders. Presentation at smart regulation and multilevel governance seminar, Brussels, November 2011. Available via: http://www.governance-flanders.be/seminar. Last accessed on 21 Aug 2012

VSDO (2011) Vlaamse Strategie Duurzame Ontwikkeling 2 (2011) Vision 2050. Flemish sustainable development strategy, Government of Flanders Sustainable Development Team, Brussels. Text available via: http://sta.dar-001.ys.be/sites/default/files/VSDO2_0.pdf. Last accessed on 21 Aug 2012

ViA (2006) Vlaanderen in Actie/Flanders in Action (2011) Action for 2020. The future plan for Flanders in progress, Brussels, Aug 2011. Available via: http://vlaandereninactie.be/?lang=en. Last accessed on 21 Aug 2012

Greening the Local Economy Through Municipal Sustainable Procurement Policies: Implementation Challenges and Successes in Western Canada

Andrew Kemp and Amelia Clarke

Abstract This chapter increases our understanding of procurement policies as a means of greening the local economy. The content provides an overview of the literature on current sustainable procurement practices being undertaken in Europe and North America. In addition, it details the sustainable procurement policy implementation experiences of municipalities in Western Canada, and offers practical lessons.

Keywords Purchasing policy • Sustainable procurement • Ethical procurement • Local government • Green economy

1 Introduction

Sustainable procurement is the purchasing of goods and/or services that are consistent with the values of sustainable development. The 2002 World Summit on Sustainable Development's call-to-action promotes "public procurement policies that encourage development and diffusion of environmentally sound goods and services"

A. Kemp
Green Municipal Fund, Federation of Canadian Municipalities, Ottawa, ON, Canada

University of Waterloo, Waterloo, ON, Canada
e-mail: andrewkemp@hotmail.com

A. Clarke (✉)
Master of Environment and Business (MEB) Program, School of Environment, Enterprise and Development (SEED), University of Waterloo, Waterloo, ON, Canada
e-mail: Amelia.Clarke@uwaterloo.ca

R. Simpson and M. Zimmermann (eds.), *The Economy of Green Cities: A World Compendium on the Green Urban Economy*, Local Sustainability 3, DOI 10.1007/978-94-007-1969-9_36,

(United Nations 2004). Both private and public organizations have implemented sustainable procurement policies in recent years (Carter and Rogers 2008; Coggburn and Rahm 2005; Saha and Paterson 2008), but to date the focus of analysis has mainly been on private sector procurement and supply chain strategy. "A knowledge gap exists in our understanding of government procurement strategies" (Purchase et al. 2009). Many of the policies are newly adopted and only a small amount of research on Canadian best practices for procurement practitioners has been completed, evidenced by the first Canadian local best practice procurement guide only being released in December 2010 (Reeve Consulting 2011).

The focus of this chapter is on sustainable public procurement. The size and amount of procured goods and services from the public sector should not be underestimated. "Public sector expenditure is substantial: government organizations across the world tend to spend between 8 and 25% of GDP on goods and services" (Walker and Brammer 2009). It is estimated that combined, all levels of Canadian government spend $100 billion (Canadian dollars) on goods and services (Boutin 2009). As a large procurer of goods and services, governments can potentially leverage their influence on the private sector by requiring higher social or environmental standards and by creating a market for sustainable products and services (Preuss 2007), hence encouraging a greener economy. These public sector institutions are responsible for providing a wide range of services with a direct implication on local sustainability and its social, environmental and economic spheres.

At the local level this is equally true; efforts to make municipal procurement along sustainability criteria can stimulate the local green economy by driving market demand and acting as a leader and catalyst for the development of new green products and services. This kind of influence on sustainability is widened as government procurers purchase products in a global marketplace. With the larger global supply chains, global sustainability efforts can also be enhanced.

This study focuses on the sustainable procurement efforts being undertaken by municipal lower-tier (e.g. City of Calgary) and upper-tier (e.g. Metro Vancouver Regional District) governments in Canada. Local governments in Canada have had a difficult time over the last two decades in meeting the amount of services the populace expects with the lack of fiscal resources at hand (Benner 2010). A Federation of Canadian Municipalities policy report illustrates this: "Over the past 20 years, as successive federal and provincial governments have off-loaded responsibilities to municipal governments and cut transfer payments, it is estimated that the municipal infrastructure deficit has grown to more than $123 billion" (Federation of Canadian Municipalities 2011). With the challenge of having to deliver more services in a sustainable manner with a decreased budget, local governments have turned to sustainable procurement policies to help address the challenge.

Given the current knowledge gap on municipal sustainable procurement, there is little known about the implementation of municipal sustainable procurement policies. This study therefore focuses on the implementation phase of these policies and the 'how-to' of overcoming challenges in implementing a new policy. Specifically, the addressed research questions are:

1. What are the motivations for local and regional governments to implement an eco-procurement policy?
2. What are the challenges to eco-procurement policy implementation, and which methods have been employed to overcome these challenges?

2 Background

Historically the role of the procurer has been to purchase goods and services based on price, quality, availability and functionality (Boutin 2009; Purchase et al. 2009). The historical role of the procurer is directly related to how the private sector has defined the role for procurement in their organization, which is one of profit maximization and efficiency (Murray 2001; Purchase et al. 2009). This procurement role ensures that private wealth is created for the organization's stakeholders (Purchase et al. 2009).

In recent years the trend of procuring only for profit maximization has changed due to increased pressure from consumers, social groups, and government (Markley and Davis 2007; Seuring and Müller 2008). As environmental degradation concerns grew through the 1990s, purchasers of goods increasingly acknowledged that environmental costs relating to procuring goods are important factors in the procurement process (Min and Galle 1997). This was the beginning of a market change, where those procuring goods increasingly began to incorporate strategies for decreasing environmental damage through two procurement procedures: source reduction and waste elimination (Min and Galle 1997). Depending on the product or service being supplied these procurement policies can stipulate specific details or rules which the supplying company must comply with or otherwise regulate their organization in order to meet. These policies may for instance declare that their procurement only occurs within the legislation of a particular convention or code (e.g. International Labour Organization's labour standards).

At the outset of this market change, many procurement policies contained 'green' procurement rules defined as "the procurement of products and services that have a lesser or reduced effect on human health and the environment when compared with competing products or services that serve the same purpose" (Chamberland 2008). More recently, sustainable procurement has emerged with greater detail and to a greater extent. It adopts a triple bottom line analysis to ensure that the procurement rules incorporate a much more holistic approach (Steurer 2007). Private organizations have employed a variety of methods to advance the social and environmental aspects of their organizations' procurement, such as management systems (e.g. ISO 14001,

SA 8000), supplier reporting and evaluations structures, and the integration of standards into larger corporate policy (Seuring and Müller 2008).

In contrast to private organizations, the public sector has a greater responsibility to ensure that the wider public interest is served (Purchase et al. 2009). In addition, large spending budgets allow the government to drive and influence the market for greater environmental and social quality by demanding higher standards. This can lead to the creation of new markets and show government as a leader by setting an example for the local citizenry and businesses (Boutin 2009; Purchase et al. 2009).

This has spurred on the growth of sustainable purchasing, which has been eloquently defined by one author as looking at "what products are made of and packaged in (and how services are delivered), where they have come from, who has made them, how they will be ultimately disposed of, and whether the purchase needs to be made at all" (Boutin 2009). In essence, such policies consider the entire life cycle of the product from extraction of the materials to manufacturing, distribution, use and disposal (Barber 2007).

In the public sector sustainable procurement policies can be very formal and include larger corporate social responsibility, or be of a smaller scale and incorporated into planning or operational documents. These policies can involve both products and services purchased by the government including: waste management services, energy/electricity, apparel, custodial services, food and catering services, green buildings, chemicals, transportation services and maintenance, and office products (paper, electronics) (Barber 2007).

3 Methodology

The study was conducted in the summer of 2011. Municipal governments that serve larger populations have significant budgets to procure goods. Therefore this study considered municipal and regional governments that serve a population above 50,000. To further frame the context of the study, the two Canadian provinces of Alberta and British Columbia (B.C.) were selected for this research. In this part of Canada several municipalities have acted as early adopters of sustainable procurement policies. These criteria resulted in ten qualifying municipal and regional governments, of which five chose to participate.

Data was collected from seven interviewees through semi-structured telephone interviews of about 30 min in length (Flick 2006). The interviews were audio recorded and hand-written notes were taken. The audio recordings were transcribed and codes were inductively applied (Flick 2006). The grounded theory process was considered complete when no new codes were emerging from the analysis (Glaser 1978). These codes were further compared with the literature to build the theory about motivating factors and barriers for the implementation of municipal sustainable procurement policies (Flick 2006).

4 Results and Discussion

The governments that were interviewed vary in geographic location and size and represent both lower-tier and upper-tier governments. The table below provides more detail including the overarching sustainability policies that are in place in these communities (Table 1).

The table below shows the studied literature on motivations for implementing sustainable procurement in comparison with the results of the interviews. Four of the themes found in the literature were also found in the selected cities, while one of the literature categories was not found. Two new themes were identified, thus making a contribution from this study to the literature (Table 2).

Organizations that have overarching corporate or government policies related to sustainable development have greater success implementing sustainable procurement policies. For example Japan has successfully incorporated green purchasing into macro-level policy. This has translated into successful purchasing policy implementation at the lower government levels, because of the larger amount of support that has been provided by the national government (eco-labeling, purchasing guidelines, product lists, economic instruments, a mandatory reporting system, lifecycle analysis information and a public awareness program) (Thomson and Jackson 2007). Throughout the interviews the cities that had historical and organizational commitment to sustainability have more robust sustainable procurement plans. Calgary and Metro Vancouver, the two largest interviewed communities, had the most formalized policies and these were spurred on by the development of longer-term sustainability plans, of which the sustainable procurement policies were one sub-set of organizational policies.

A second motivating factor, which is also discussed as a barrier later in this chapter, is leadership and management support. The literature and the interviewees

Table 1 Participating governments and their related procurement policies

City	Population[a]	Name of sustainable procurement policy	Other sustainability policies
Kamloops, British Columbia	80,376	Environmental\ethics statements	Sustainable Kamloops plan
Richmond, British Columbia	174,461	Environmental purchasing guide	Charting our path
Metro Vancouver, British Columbia	2,116,581	Sustainable procurement policy	Metro Vancouver sustainability framework
Strathcona County, Alberta	82,511	Purchasing policy	Municipal development plan
Calgary, Alberta	988,193	Sustainable ethical and environmental procurement policy	Imagine Calgary, the 2020 sustainability direction

[a]Based on 2006 Canadian census data

Table 2 Motivations identified in the literature and by interviewees

Motivations – literature	References	Motivations – interviewees	Cities
Organizational commitment to sustainability (over-arching corporate social responsibility policy or system in place)	Clarke (2006), Leire and Mont (2010), TerraChoice Environmental Marketing Inc (2009), and Thomson and Jackson (2007)	Organizational commitment to sustainability (over-arching corporate social responsibility policy or system in place)	Calgary, Kamloops, Richmond, Strathcona County, Metro Vancouver
Strong leadership and management support	Boutin (2009), Bruel et al. (2009), Murray (2001), Saha and Paterson (2008), and TerraChoice Environmental Marketing Inc (2009)	Strong leadership and management support (bottom-up staff, external leader and community pressures)	Calgary, Richmond, Metro Vancouver, Kamloops, Strathcona County
Financial savings	Boutin (2009) and TerraChoice Environmental Marketing Inc (2009)	Financial savings (market development of low cost environmental products)	Calgary
Local economic development	Murray (2001) and Preuss (2007)		
Increasing environmental and social health	Boutin (2009), Morgan (2008), Murray (2001), Preuss (2007), TerraChoice Environmental Marketing Inc (2009), and Thomson and Jackson (2007)	Increasing environmental and social health (change to legal standards)	Kamloops, Richmond, Metro Vancouver, Calgary
		New product to market	Metro Vancouver, Calgary
		Strengthening relationship with suppliers	Richmond, Metro Vancouver, Calgary

acknowledged that having management support drives policy implementation; whereas if the local leadership does not support or understand the concept of sustainable procurement then this is a barrier. Several authors as well as each of the interviewed cities identified this as a motivating factor, but the precise motivating factors or leaders identified were not always similar. Richmond, British Columbia (a suburb of greater Vancouver) identified the recent winter Olympics as a driver for public and staff support towards the implementation of sustainable procurement. Kamloops and Metro Vancouver both acknowledged that the public itself and the internal staff were motivators to drive sustainable thinking and change in their communities.

Two authors and the City of Calgary identified cost savings or cost competitiveness as a motivating factor for policy implementation. Shannon Guterson, Environmental Health and Safety Team Lead from the City of Calgary states,

> People were not as endearing to the idea of the SEEP [Sustainable Ethical and Environmental Procurement Policy] when asked to cut their budgets, because when there's a preconceived notion that if it's environmentally preferred or ethically preferred, then it's got to be more expensive. Don't get me wrong. There are certainly products like that. But our role … is that we've got to ensure that we justify any type of price premium. And to date, I don't think we've accepted a single contract where we see something in the 10 percent range for price premium …

Local economic development is a motivating factor, as employing local businesses and suppliers to provide goods can increase local employment and recycling of spending within the local economy. Although identified in the literature (Murray 2001; Preuss 2007), none of those interviewed indicated that this was a policy driver.

The City of Calgary and Metro Vancouver did acknowledge that sustainable procurement policy is an evolutionary process and that as new products have come to the market, the motivation to increase the number of products purchased with a sustainable procurement lens has inherently increased as well. Therefore the new products coming to market are motivators for policy implementation. In addition, the governments of Richmond, Metro Vancouver, and Calgary all identified that the sustainable procurement policy provided the opportunity to work with their suppliers to enhance their relationship through policy and product education. This has a duel positive effect of increasing suppliers' knowledge of meeting the sustainability criterion in the bidding process and increasing supplier and consumer knowledge of new product development. This all helps to green the local economy. Finally, the environment and health improvements that result from greening the local economy are motivating factors unto themselves.

The table below lists the reviewed literature on barriers for implementing sustainable procurement in comparison with the results from the interviews. Two of the themes from the literature were found in the case cities, while one was not found. Two new themes were also identified. This study thus makes a new contribution to the literature (Table 3).

One barrier several authors noted is the lack of municipal funds in the budget for sustainable procurement. The interviewed communities all noted the perception of (or actual) higher costs when purchasing the goods as a barrier to policy

Table 3 Barriers identified in the literature and by interviewees

Barriers – literature	References	Barriers – interviewee	Cities
Lack of resources in the municipal budget	Boutin (2009), Coggburn and Rahm (2005), European Commission (2010), Retzlaff (2009), TerraChoice Environmental Marketing Inc (2009), and Thomson and Jackson (2007), and Walker et al. (2009)	Lack of resources in the municipal budget (perception of increased cost)	Calgary, Kamloops, Strathcona County, Richmond
Lack of political and administrative support	Boutin (2009), European Commission (2010), Leire and Mont (2010), Michelsen and de Boer (2009), Morgan (2008), Preuss (2007), Saha and Paterson (2008), Walker and Preuss (2008)		
Lack of staff knowledge regarding sustainable procurement	Boutin (2009), Leire & Mont (2010), Murray (2001), Saha & Paterson (2008), Thomson & Jackson (2007), Walker & Brammer (2009)	Lack of staff knowledge regarding sustainable procurement ('green-washing', ever-evolving social and environmental standards)	Calgary, Kamloops, Strathcona County, Richmond, Metro Vancouver
		Lack of manufacturer or supplier knowledge	Calgary, Metro Vancouver
		Lack of public support for the perceived higher cost option	Strathcona County

implementation. The perception is strong that the sustainable purchased product will always be more expensive. There are however several examples where the costs are only slightly higher or the same, as is noted in the quote above from the City of Calgary. There is a contradiction that exists when the perception is that it is always higher, but those that are familiar with the real costs are aware of the potential for cost savings. In addition, several interviewees also included the fact that there is lack of budget for staff time to complete or implement the policy itself. There is therefore a barrier created by the higher cost of implementing a new sustainable procurement policy over traditional purchasing processes due to increased staff time.

The lack of knowledge of the sustainable procurement process is another well-identified barrier to policy implementation. The procurement staff and their management often do not have the necessary skills or knowledge to appropriately identify the sustainable product or service. As one respondent in Leire and Mont's article has noted, "Formulating what information is needed is very difficult, because it is a 'moving target' with continuously developing new products and continuous expansion of our knowledge about effects and impacts of various materials" (Leire and Mont 2010). This lack of skill set is not just an issue for the procurers but the suppliers as well, as they have to create the products to meet the most recently imposed standards. This is further elucidated by Purchasing and Risk Manager for Metro Vancouver Tracy Husoy, who stated, "Lack of knowledge may be one of the biggest initial barriers. When we introduced this [10% weighting for social and environmental factors as part of the bid process] to our suppliers ... I had a few respondents, during post award de-briefings say, 'You're kidding. You really meant that'. So maybe there were a few suppliers that did not take us seriously."

In Calgary and Metro Vancouver, the need for suppliers that have a higher knowledge level of sustainable products and the process of sustainable procurement policy implementation is acknowledged as a barrier. These cities have well-developed policies that are beyond the low-hanging fruits of initial policy implementation. They have expressed the need to educate suppliers on alternative procurement processes and products to ensure policy implementation of the more ambitious goals. In addition, Strathcona County indicates that the public may not be committed to sustainability and, therefore, may be a barrier to policy implementation, which was not identified in any of the literature.

Another barrier identified in the literature which has been identified as a motivation both in the literature and by several of the interviewees is the lack of political will and administrative support from senior levels to implement and maintain these policies. This is due to a variety of issues including the lack of education senior leaders have on the concept of sustainable procurement and how it can change the business-as-usual for their government (Leire and Mont 2010). None of the case cities identified this as a barrier.

5 Conclusion

The analysis of the cities and literature indicate that there are several best practice lessons for municipalities and suppliers:

- The policy must be a 'living document' that continues to evolve, as standards and specifications change.
- A set of standardized sustainability-related supplier information is useful as an educational tool to overcome the staff knowledge gap (Preuss 2007).
- Staff and senior leadership education is a key driver to ensuring effective uptake and implementation. Educating senior staff and politicians on the advantages of sustainable purchasing is important. Advantages include: greening local economic development through local sourcing initiatives, environmental improvements in the supply chain, political advocacy, and better product quality through supplier development (increased standards equate to better quality of product) (Murray 2001).
- Briefing the suppliers, by educating them on the sustainable procurement policy standards and the scoring systems, will enhance the quality of bids received. Pre-bid meetings allow the suppliers to ask any questions at the same time and staff can highlight the importance of the sustainable procurement policy in terms of the bid scoring.
- Extended producer responsibility should be a future focus of procurement officers. Working with suppliers to increase cradle-to-grave or cradle-to-cradle production processes will help establish the parameters of what is part of the sustainable procurement policy.

In conclusion, this chapter has provided an extensive overview of the literature on sustainable procurement policies, and the motivations and barriers to their implementation in local communities. This study has identified two motivations (new product to market, and strengthening relationship with suppliers) and two barriers (lack of manufacturer or supplier knowledge of sustainability concepts, and lack of public support for the perceived higher cost option) that have not been identified in the literature, further contributing to the subject matter on sustainable procurement policies.

References

Barber J (2007) Mapping the movement to achieve sustainable production and consumption in North America. J Clean Prod 15(6):499–512

Benner A (2010) Filling the gaps in social safety net. Welland Tribune. Available via: http://www.wellandtribune.ca.proxy.lib.uwaterloo.ca/ArticleDisplay.aspx?e=2507457&archive=true, CR003618. Accessed 10 Apr 2011

Boutin C (2009) Public procurement: the power to create change. Municipal World 119(11):9–11, p 11

Bruel O, Menuet O, Thaler P (2009) Sustainable procurement: a crucial lever to end the crisis? (White Paper). HEC; EcoVadis, Paris. Available via: http://www.buysmartbc.com.proxy.lib.uwaterloo.ca/UserFiles/File/HECEcovadis2009_EN.pdf

Carter CR, Rogers DS (2008) A framework of sustainable supply chain management: moving toward new theory. Int J Phys Distrib Logist Manage 38(5):360–387. doi:10.1108/09600030810882816

Chamberland D (2008) Green procurement requires more transparency. Municipal World 118(12):25, p 25

Clarke A (2006) Regional sustainable development strategies: variations in formulation and content in nine Canadian case studies and the implications for eco-procurement. Commission for Environmental Cooperation, Montreal

Coggburn JD, Rahm D (2005) Environmentally preferable purchasing: who is doing what in the United States? J Public Procure 5(1):23–53

European Commission (2010) Buying social – a guide to taking account of social considerations in public procurement. Publications Office of the European Union, Luxembourg. doi:10.2767/18977

Federation of Canadian Municipalities (2011) Policy statement on municipal infrastructure and transportation policy. Retrieved 15 Apr 2011 from http://fcm.ca.proxy.lib.uwaterloo.ca/Documents/corporate-resources/policy-statements/2011_Municipal_Infrastructure_and_Transportation_Policy%20Statement_EN.pdf. Quotation=p. 1

Flick U (2006) An introduction to qualitative research, 3rd edn. SAGE Publications, London

Glaser BG (1978) Chapter 1. In: Theoretical sensitivity, 1st edn. The Sociology Press, Mill Valley, pp 1–17

Leire C, Mont O (2010) The implementation of socially responsible purchasing. Corp Soc Responsib Environ Manage 17(1):27–39. doi:10.1002/csr.198. Quotation=p. 32

Markley MJ, Davis L (2007) Exploring future competitive advantage through sustainable supply chains. Int J Phys Distrib Logist Manage 37(9):763–774. doi:10.1108/09600030710840859

Michelsen O, de Boer L (2009) Green procurement in Norway: a survey of practices at the municipal and county level. J Environ Manage 91(1):160–167. doi:10.1016/j.jenvman.2009.08.001

Min H, Galle WP (1997) Green purchasing strategies: trends and implications. Int J Purch Mater Manage 33(3):10–17

Morgan K (2008) Greening the realm: sustainable food chains and the public plate. Reg Stud 42(9):1237–1250

Murray JG (2001) Improving purchasing's contribution – the purchasing strategy of buying council. Int J Public Sect Manage 14(5):391–410. doi:10.1108/EUM0000000005567

Preuss L (2007) Buying into our future: sustainability initiatives in local government procurement. Bus Strat Environ 16(5):354–365. doi:10.1002/bse.578

Purchase S, Goh T, Dooley K (2009) Supplier perceived value: differences between business-to-business and business-to-government relationships. J Purch Supply Manage 15(1):3–11. doi:10.1016/j.pursup. 2008.11.003. Quotation = p. 3

Reeve Consulting (2011) New report reveals trends & best practices in Canadian municipal sustainable purchasing and ethical sourcing. Available via: http://blog.reeveconsulting.com.proxy.lib.uwaterloo.ca/2011/05/11/new-report-reveals-trends-best-practices-in-canadian-municipal-sustainable-purchasing-and-ethical-sourcing/. Cited 20 June 2011

Retzlaff R (2009) The use of LEED in planning and development regulation. J Plann Educ Res 29(1):67–77. doi:10.1177/0739456X09340578

Saha D, Paterson R (2008) Local government efforts to promote the "Three Es" of sustainable development. J Plann Educ Res 28(1):21–37. doi:10.1177/0739456X08321803

Seuring S, Müller M (2008) From a literature review to a conceptual framework for sustainable supply chain management. J Clean Prod 16(15):1699–1710. doi:10.1016/j.jclepro.2008.04.020

Steurer R (2007) From government strategies to strategic public management: an exploratory outlook on the pursuit of cross-sectoral policy integration. Eur Environ 17(3):201–214. doi:10.1002/eet.452

TerraChoice Environmental Marketing Inc (2009) EcoMarkets summary report. Author, Ottawa. Available via: http://www.buysmartbc.com.proxy.lib.uwaterloo.ca/UserFiles/File/EcoMarkets_Summary_Report_2009.pdf

Thomson J, Jackson T (2007) Sustainable procurement in practice: lessons from local government. J Environ Plann Manage 50(3):421–444

United Nations (2004) Johannesburg plan of implementation – changing unsustainable patterns of consumption and production. Available via: http://www.un.org.proxy.lib.uwaterloo.ca/esa/sustdev/documents/WSSD_POI_PD/English/POIChapter3.htm. Cited 23 Nov 2010

Walker H, Preuss L (2008) Fostering sustainability through sourcing from small businesses: Public sector perspectives. Journal of Cleaner Production, 16(15):1600–1609. doi:10.1016/j.jclepro.2008.04.014

Walker H, Brammer S (2009) Sustainable procurement in the United Kingdom public sector. Supply Chain Manage Int J 14(2):128–137. doi:10.1108/13598540910941993, Quotation = p. 128

Walker H, Gough S, Bakker E, Knight L, McBain D (2009) Greening operations management. J Manage Educ 33(3):348–371. doi:10.1177/1052562908323190

City of Raleigh Office of Sustainability: Green Building Training Program

Lynn Graham, Emily Steele, Anna Leonard, Jen Baker, and Julian Prosser

Abstract The City of Raleigh's commitment to sustainability is the cornerstone of its vision for the future. In December 2009 Raleigh's City Council reaffirmed this vision in its mission statement, declaring that Raleigh is a "City of Innovation focusing on environmental, cultural and economic sustainability," adding, "We work with our universities, colleges, citizens and regional partners to promote emerging technologies, create new job opportunities and cultivate local businesses and entrepreneurs."

As part of this commitment, the City of Raleigh launched its innovative Green Building Training Series, with 12 different courses in varying topics related to the built environment. Ensuring Raleigh has a "green collar" workforce trained in cutting-edge technologies enables the city and its residents to build both greener and smarter. To date, the program has provided free training to more than 300 individuals.

These free courses provided Raleigh residents and businesses with the tools needed to compete in the burgeoning green market. Attendees' stories are personal; many individuals took advantage to retool their skills in a challenging economic environment. Courses such as Solar Thermal Installation, Green Plumbing, Leadership in Energy and Environmental Design® (LEED®) and Building Performance Institute's Building Analyst provide a competitive edge in today's workforce. Each class in the series also trained City of Raleigh employees, upgrading internal knowledge to more easily meet the new LEED Silver standard mandate for all new municipal City facilities.

L. Graham • E. Steele • A. Leonard • J. Baker (✉)
Office of Sustainability, P. B. 590, 222 West Hargett Street, Raleigh, NC 27602-0590, USA
e-mail: Lynn.Graham@raleighnc.gov; jen.baker@raleighnc.gov

J. Prosser
Office of the City Manager, P. B. 590, 222 West Hargett Street, Raleigh, NC 27602-0590, USA
e-mail: citymanager@raleighnc.gov

R. Simpson and M. Zimmermann (eds.), *The Economy of Green Cities: A World Compendium on the Green Urban Economy*, Local Sustainability 3,
DOI 10.1007/978-94-007-1969-9_37,

Keywords Training • Workforce • Jobs • Skills • Innovation

1 Introduction: Necessity of a Trained Green Workforce in Raleigh, North Carolina

Located in the region known as "The Triangle," Raleigh, North Carolina has benefited from long-term investments from major educational institutions and the Research Triangle Park, an epicenter for technological innovation. While these investments resulted in tremendous growth – Raleigh's population increased more than 46%, from 276,093 residents in 2000 to 403,892 in 2010 – they have also created a great need for responsible growth management and sustainable building practices. Green building certifications, such as Leadership in Energy and Environmental Design® (LEED®), have emerged to recognize environmentally responsible and healthy building and construction practices.

Sustainable building practices also suffuse economic strength and social integrity into a community and its workforce. A 2011 study released by McGraw-Hill Construction projected that 45% of all design and construction jobs will be green by 2014 (Portero 2011). To meet this need, a large segment of the construction and real estate workforce will need to be trained in green building practices. The Building Performance Institute (BPI), for example, trains construction professionals in residential energy efficiency building audits. These audits provide consumers with knowledgeable assessments of a home's energy usage, enabling them to make informed decisions regarding updates and projects that will allow their home to safely and effectively perform with optimal energy efficiency while deciding their preferred payback period of investment. Such practices save residential homeowners from costly heating and cooling energy bills while sustaining a workforce with evolving skills and knowledge (Fig. 1).

2 Case Study: City of Raleigh's Green Building Training Series

The City of Raleigh's Office of Sustainability was created in 2008 to develop a sustainability plan that coordinates all initiatives throughout the City of Raleigh's numerous departments, becoming a resource to the Raleigh community. After securing a $3.82 million Energy Efficiency and Conservation Block Grant (EECBG) in December 2009 from the U.S. Department of Energy (DOE) (as part of the federal stimulus package), the City created its Green Building Training Series. The series provided, at no cost, training and certification targeting the construction industry hardest hit by the economic downturn. By December 2011, more than 300 individuals

Fig. 1 According to the Urban Land Institute magazine, Raleigh took fifth place among metropolitan areas posting the highest total economic growth. From 2007 to 2010, Raleigh realized 10.5% growth in its gross domestic product. *Source*: City of Raleigh

were trained through the series in such classes as LEED, BPI Analyst/HERS, Green Plumbing, Solar Thermal Installation, and Solar Photovoltaic, among others.

A "train the trainer" component, incorporating educational partners into the series, was crucial to the success of the Green Building Training Program. The City partnered with two local higher education institutions, Wake Technical Community College and North Carolina State University, offering several available seats to staff members of both organizations. This component helped to institutionalize innovative curricula, such as Solar Thermal Installation and Green Home Appraisal Awareness. Several seats in these classes were also reserved for City of Raleigh employees, each an expert in his or her respective field. Following their participation, an internal training component was added to train other City employees in green practices, energy efficiency and renewable energy knowledge learned from the series. This "train the trainer" approach complemented the Green Building Training Series and allowed more employees to incorporate these skills into their work for the community they serve, making them better stewards to the public.

Partnerships were also formed with both St. Augustine's College and the City's own Community Development department. Both provided empty residential properties for several BPI courses, a necessary aspect to provide hands-on training for BPI participants. Several private citizens of Raleigh made their homes available for auditing as well. This opportunity provided homeowners a free energy audit to assess the efficiency of their homes while providing the BPI trainers with a crucial tool to practice their skills (Figs. 2 and 3).

Fig. 2 Students perform hands-on installation of rooftop solar panels at Wake Technical Community College. *Source*: City of Raleigh

Fig. 3 BPI participants identify air leakage and solutions during Building Analyst training. *Source*: City of Raleigh

3 Individual Narratives

In all, participant feedback from the series was eminently positive. In an anonymous survey response from an August 2011 BPI Analyst Course, a participant stated the course had helped to enhance his or her business, adding, "Additional knowledge is critical in expanding new opportunities." Because the Green Building classes accept indiscriminately on a first-come, first-served basis to all who live, work, or do business within the city of Raleigh, the program has attracted participants from diverse professional backgrounds. Many participants used the courses to develop and improve their business and professional skills. While some participated in the classes to broaden their green building knowledge, others took them as an opportunity to change careers entirely. Three Green Building Training participants – Robert Funck, Monica Barrett, and Vincent Taylor – shared their stories of how the program positively impacted both their personal lives and their careers.

Robert Funck: As a social worker during the downturn in the economy, Durham native Robert Funck found himself pursuing a different career path in the building and energy auditing industry. Recalling a few previous years spent in construction, Funck began to investigate a career in sustainable construction and decided to become a certified Building Performance Institute (BPI) Energy Auditor. After initially considering other cost-prohibitive training courses, in 2009 a friend recommended Raleigh's Green Building Training Series. Since then, Funck successfully completed the training and certifications for both BPI Building Analyst and Building Envelope.

"The skills and certifications from these classes really helped to get things jump-started for my career," said Funck, who obtained a position as an energy analyst at Southern Energy Management, a North Carolina sustainable energy company, following his Green Building Training.

More than simply a tool for job creation, Funck sees the skills from these classes as useful to everyone, adding, "They increase awareness. Energy auditing and weatherization are some of the most cost-effective improvements for homes or buildings. Knowledge can really change how you approach a home and your impact on the environment."

Monica Barrett: Friends dubbed Monica Barrett the "Green Girl" since she opened her own green consulting firm, Fini'me Green, in 1999. "It's a 360 green consulting firm, covering every aspect of construction, from furniture to energy usage," according to Barrett.

An active participant in the Green Building Training Series, Barrett completed courses in solar installation, BPI Energy Auditing, and Green Real Estate Agent Training, among others. The Series has impacted her business since she has "been able to get into the details of green building." As an owner of a start-up company, Barrett would not have been able to afford this training on her own. Deeming them "priceless," Barrett saw these opportunities as a way to maintain a competitive edge in a rapidly advancing field. "You can never stop learning," she said, "Innovation comes in every area; you'll get left behind if you stop."

Vincent Taylor: As a professional at Capstone Civil Group, the largest African American civil engineering firm in North Carolina, Vincent Taylor sought training in the solar installation field. After discovering the Solar Thermal course offered in June 2011 as part of the Green Building Training Series, Taylor was able to use the course as a job creation opportunity.

Following the Solar Thermal course, Taylor partnered with the Raleigh Business and Technology Center for a solar installation, placing 80 workers on the project. He is currently in negotiations for projects in both Durham and Charlotte to put 50 more people to work; both will require skills learned in the Green Building Training Series. "I saw the class as a great way to network; I've been able to work with several of my classmates on solar projects since," Taylor stated, adding, "Having the 'green factor' gives you the extra edge needed in this economy. It's not just about work – it's about an evolving world."

4 Conclusion: Next Steps

Green Building participants such as Taylor, Barrett, and Funck have channeled knowledge imparted from the series into their professions, with great success. The "train the trainer" component of the Green Building Training Program has also given the series a sense of permanence in the Raleigh community. By training staff from local educational institutions in classes such as BPI's Building Analyst and Solar Installation, green training practices are now embedded into the organizations educating the Raleigh community. Notably, the seed money made available from the series allowed North Carolina State University to further develop its own sustainably focused curriculum, including a course on Real Estate Energy Awareness. The series will continue as an internal program to train and educate City of Raleigh employees – sharpening their skills and allowing them to more efficiently use community resources (Fig. 4).

Other initiatives originating from the City's Office of Sustainability provide continued support of the green urban economy through public-private partnerships; best management practices in water, energy and waste management; and public sustainability education. Partnerships with innovative, locally based companies – such as Cree and Advanced Energy – integrate smart grid technology, which encompasses all aspects of a community's energy consumption, giving the Raleigh community an economically competitive edge as well as a healthier, more environmentally sound future. The Office continues to be recognized nationally for such partnerships and achievements in green innovation. As a recipient of the national 2011 Siemens Sustainable Community Award, Raleigh was commended for its Green Building Training program and the development of a "green economy."

Ultimately, energy awareness and green building and technology knowledge will make Raleigh community members more viable candidates to fill the growing demand of careers in the sustainable construction and real estate industries. Green building and conservation projects are the building blocks for increased economic vitality, energy security, and environmental quality.

Fig. 4 Raleigh's leadership in LEDs is dramatically illustrated by the iconic Cree Shimmer Wall adorning the west face of the LEED Silver-certified Raleigh Convention Center. *Source*: City of Raleigh

Acknowledgments The Green Building Training Program was made possible by funding provided by the Energy Efficiency and Conservation Block Grant through the American Recovery and Reinvestment Act.

References

City of Raleigh (2011) The 2030 comprehensive plan for the City of Raleigh
Portero A (2011) Green jobs make up one-third of construction workforce: McGraw-Hill
United States Census Bureau (2011) 2010 cCensus
Weisbecker L (2011) Raleigh lauded as most 'sustainable.' Triangle Bus J

Changing to a Sustainable Socio-Economic System: A Challenge for Kyoto as a Historic City

Makoto Ueda

Abstract In 2010, the City of Kyoto revised its regulations on climate change mitigation to move towards a more holistic sustainable socio-economic development goal. Kyoto showed an improved framework to realize multiplier effects of all policies and programs grounded on the view of the environment. In the new regulation, Kyoto sets the ambitious goal to reduce the city's total GHG emissions by 25% by 2020 and 40% by 2030 compared to 1990 levels. Although economic growth is anticipated to increase energy demand, Kyoto believes it can achieve this goal with the current level of technology through various measures. That is why we – citizens, businesses, environmental conservation groups and the local government – will pull together to achieve the goals, while spreading the watchword "DO YOU KYOTO?" synonymous with "Do you do something good for the environment?"

Keywords Historic city • Public involvement • Economic instruments

1 Introduction

The third Session of the Conference of the Parties to the United Nations Framework Convention on Climate Change (COP3) was held in Kyoto with the subsequent Kyoto Protocol adopted in 1997. Since then, the word "Kyoto" has become a verb or a pronoun for "doing something good for the environment" and it has spread throughout the world. While this is a great honor, it also puts Kyoto under great pressure in their fight against global warming.

M. Ueda (✉)
Climate Change Prevention Office, Environment Policy Bureau, Kyoto, Japan
e-mail: ge@city.kyoto.jp

R. Simpson and M. Zimmermann (eds.), *The Economy of Green Cities: A World Compendium on the Green Urban Economy*, Local Sustainability 3,
DOI 10.1007/978-94-007-1969-9_38,

Subsequently, Kyoto's slogan "DO YOU KYOTO?" equates to "Do you do something good for the environment?" By featuring the phrase "DO YOU KYOTO?" the Kyoto City seeks to join forces with its citizens, businesses, and nonprofit organizations (NPOs) to address global warming.

Measures taken by Kyoto City can inspire neighboring cities and beyond, along with influencing national projects and the entire country. Moreover, measures taken by cities to address environmental problems are essential for tackling global warming. Thus, focusing on city level measures is crucial.

As a representative of Kyoto, the host city of the UN Framework Convention on Climate Change (UNFCCC) COP3 in 1997, this chapter will report on Kyoto's policy framework, enacted ordinances, plans, and practical measures towards building a sustainable socio-economic system in Kyoto.

2 Kyoto

Kyoto has a 1,200 year history. Several strata of cultural and historic properties of different ages still exist and are maintained in the city today. Kyoto has prospered while seeking to maintain harmony with the surrounding mountain ranges and the local wood culture. Kyoto's citizens value the wisdom of their forefathers and have sought to nurture the legacy of the local natural beauty to maintain it for future generations.

At present the population of Kyoto is 1.47 million and it is the sixth largest city in Japan. It has an area of 828 km^2 of which three-fourth are forests. The downtown area is mildly sloped from the north to the south. Kamo River, which runs through the center of downtown, features fish that only survive in clean water (e.g. sweetfish). Kyoto holds 20% of national treasures, temples and shrines, including 14 World Cultural Heritage sites. The natural environment has been praised as the "Purple Mountains and Clear Water", and *Kyo-machiya* town houses symbolize the traditional local living culture. These assets attract annually 50 million tourists.

Kyoto has additional attractions. For example, it is known as an innovative manufacturing city, where traditional industrial skills and advanced technologies have been fused to create various innovations. It is also a university city with 37 universities and junior colleges. The student college population accounts for 10% of the total population.

3 Target Setting

Kyoto City adopted the "Ordinance on Countermeasures on Global Warming" in Japan in 2004. In 2010, the ordinance was amended following a decision by the city to become a low carbon society and to reduce greenhouse gases by more than 80% in the long term. The amended ordinance advocates the target of "building a sustainable socio-economic system" and calls for including environmental aspects in every policy and measure implemented in Kyoto.

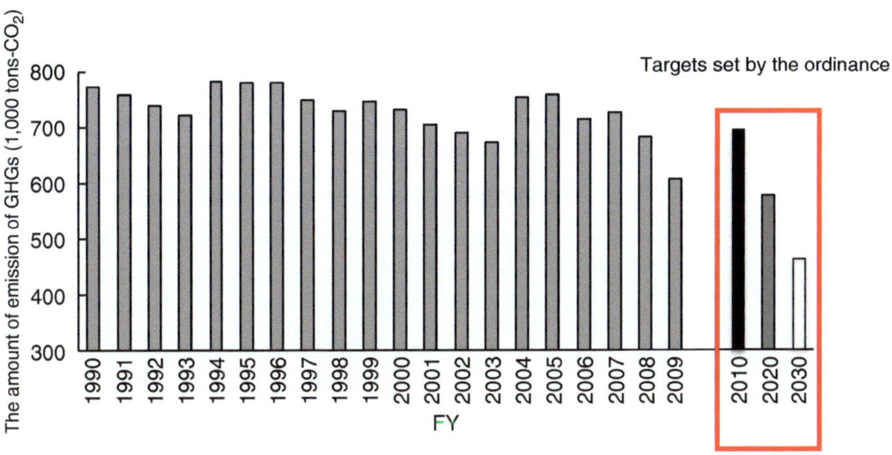

Fig. 1 Change in the amount of emission of greenhouse gases (GHGs) in Kyoto

The city identified the need to establish a system or framework that allows creating synergy effects. In addition, numerical targets were set. For example, the greenhouse gas (GHG) reduction of 25% by 2020 and 40% by 2030 compared to 1990 levels. This is one of the highest targets in the world. To establish a low carbon society the ordinance includes the obligation that large enterprises use eco-friendly cars, adopt Environmental Management Systems (EMS), use local timber for large building constructions, and use renewable energy.

Figure 1 shows the change of greenhouse gas emissions of Kyoto since 1990. By 2009, Kyoto achieved a 21.3% reduction of GHG emissions compared to 1990 levels. This was the result of cooperation among and enthusiasm of citizens and enterprises. Currently, Kyoto has set an even more ambitious target.

Figure 2 shows the reduction potential by 2030 for various sectors and measures. The figure provides a scenario for reducing greenhouse gases by 40% target by 2030.

The potential of the service sector has the largest reduction potential with 34.2%. It is followed by transport and households. Measures for reduction include the installation of high efficiency electrical equipment by for example using automobiles of the next generation, and sourcing energy from renewables. Kyoto is working on a strategy to reach the new targets by understanding the problems and potential of each sector first.

4 Practical Measures

4.1 Popularizing the Solar Photovoltaic System

The project "Promoting the Solar Photovoltaic System" is the core project for the promotion of renewable energy. In 2003, Kyoto City introduced this subsidy program to enable citizens to cover a portion of the costs of a solar photovoltaic

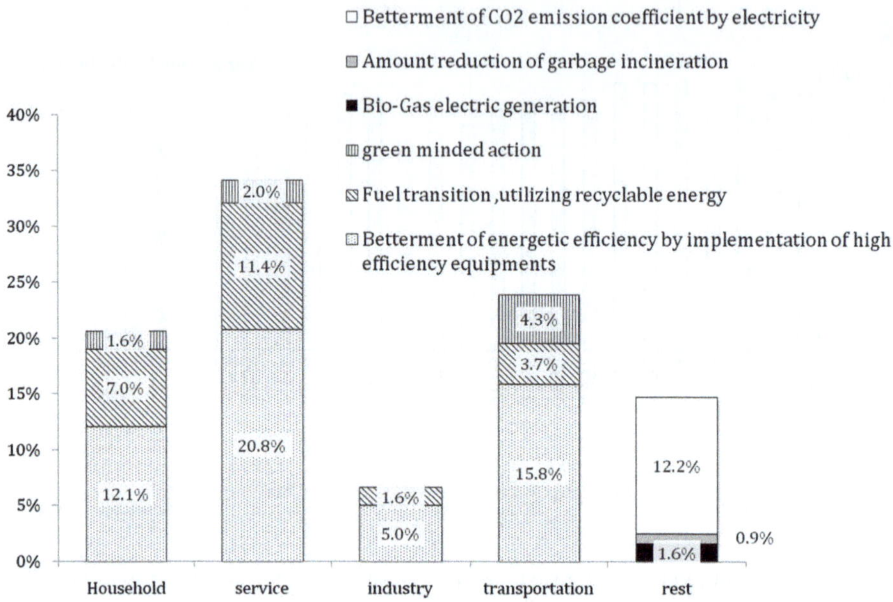

Fig. 2 Departmental and modes of reduction countermeasure's reduction potential (2030)

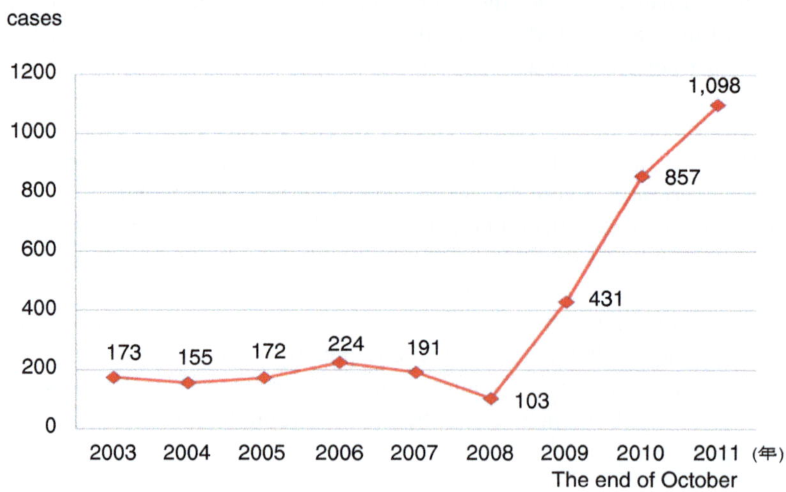

Fig. 3 Changes taken place in the number of subsidy cases for the house solar photovoltaic system

system installation. As of November 2011, the city has paid Yen 40,000 (US$520) per 1 kW. By the end of 2010, the total number of cases paid out was 857 worth Yen 213 million (US$2.76 million). Comparing this figure with 2009 figures, it indicates that the number of cases, the amount, as well as the generation output, have doubled (see Fig. 3). This project resulted in a reduction of 3,700 t of annual CO_2.

Fig. 4 A photovoltaic panel roof installation

The benefits of this subsidy project include:

1. The subsidy is given directly to citizens, so that the program quickly spreads. It also contributed to the revitalization of the local economy.
2. Those who have installed a solar photovoltaic system practice energy saving more often than before.

However, before a solar photovoltaic system is installed, the city verifies if the installation meets the landscape regulations to ensure that the installation adheres to historic and environmental landscape requirements. This is an indispensable element for the sustainable and harmonic development of a historic city (see Fig. 4).

4.2 Local Credit System "DO YOU KYOTO? Credit"

The second important initiative is the local credit system. The "DO YOU KYOTO? Credit" system started in 2011. The aim of this credit is to promote greenhouse gas reduction initiatives among citizen groups, events, shopping-mall associations, and medium and small-sized enterprises. In this system, the amount of CO_2 emission reductions, which are released through the activities of these groups, is recognized by the city as a credit. This reduction credit is the used within the city and by large businesses to offset their carbon consumption. According to the reduction amount of greenhouse gases, a reward/credit is given by the city (see Fig. 5).

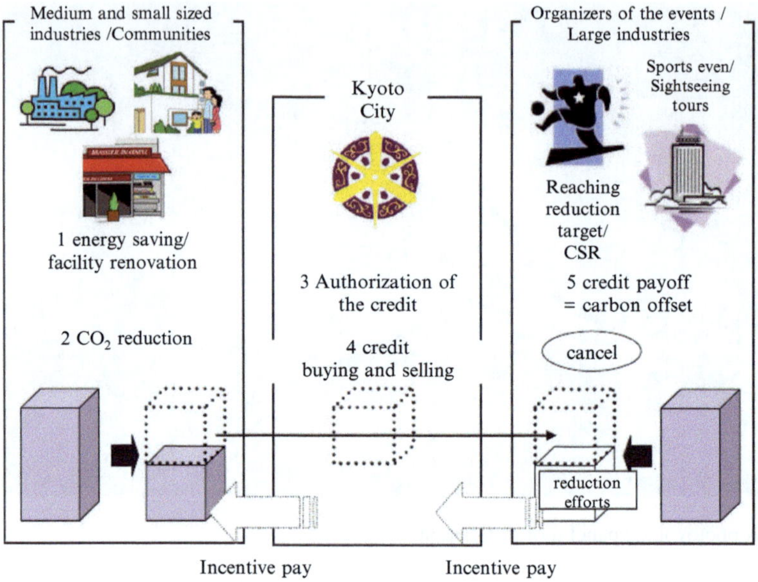

Fig. 5 Scheme of the system of DO YOU KYOTO? Credit

The city then sells these credits to organizers of different events and large enterprises, so that they can offset their carbon emission with the purchase of these credits. Kyoto City has launched the "Local Production for Local Consumption" project so that credits created in Kyoto can be used in Kyoto, and big enterprises financially support local communities, and medium and small enterprises.

The advantage of this system is that the reduced amount is valued as an economic asset, and contributes to the development of a sustainable local economy. Moreover, efforts to reduce CO_2 emissions by local communities can strengthen the bond and increase trust among the community. The budget for 2011 was Yen 14.2 million (US$0.18 million).

4.3 Bio-Diesel Fuel Production Project

A third project is the production of bio-diesel fuel from cooking oil. In 1997 the city began to produce bio-diesel fuel utilizing used cooking oil from homes. The city used the produced fuel to run garbage collection vehicles and city buses. At present, used cooking oil is collected at 1,600 points amounting to 190,000 l annually as of 2010 (see Fig. 6, not Fig. 8).

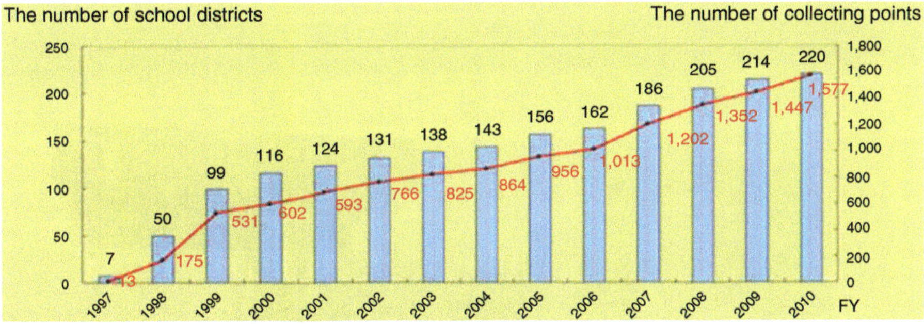

Fig. 6 Changes in the number of used cooking oil collecting points

Fig. 7 A used cooking oil collection point

In Kyoto, 150 garbage collection vehicles use 100% bio-diesel fuel. Moreover, 93 city buses run on a mixture of 20% bio-diesel fuel and 80% light diesel. With this, 4,000 t of CO_2 emissions are saved annually.

The important point of this project is not "the production technique" but the re-use of cooking oil through a collection system based upon the autonomous support by citizens. The involvement of communities in such projects is likely to continue (Figs. 7, and 8).

Fig. 8 A Kyoto City facility for used cooking oil fuel established in 2004, producing 5,000 l per day

4.4 *Electric Vehicle Initiative*

The fourth project is the use of electric vehicles. The aim of this project is to promote the use of electric vehicles (EV) and plug-in hybrid vehicles (PHV) to reduce greenhouse gas emissions. The city uses three principle measures to promote the use of EV in addition to encouraging citizens to use fewer automobiles, more public transportation, and enjoy more walking.

1. Installation of charging stations for EV vehicles.
2. Program to support citizens and EV enterprises to promote EV.
3. The city uses EV.

The budget of this project was Yen 109 million (US$1.4 million) in 2009, Yen 49 million (US$0.64 million) in 2010, and Yen 35 million (US$0.45 million) in 2011.

At present there are 46 charging stations for EV vehicles including 43 200 V stations and three rapid charge stations. The city launched the program by providing subsidies to taxi and rental car companies and giving financial support to medium and small-sized enterprises, as well as providing tax exemptions for light automobiles. In this program, the city purchased five electric automobiles, which the city shares with citizens and enterprises. With this car sharing program 8.9 t of CO_2 emission were cut in 2010. Promoting cars of the next-generation, like EV, is indispensable for establishing a sustainable social and economic system, but should not be the sole focus of transport policies (Fig. 9).

Fig. 9 EV car sharing (renting to citizens and enterprises)

4.5 Child Education: Children's Eco-Life Challenge

Global warming endangers the survival on earth. Children education is very important to promote and build a sustainable social and economic attitude and system. In 2005 Kyoto City started a global warming curriculum in elementary schools with a 4-h-class. The pupils learn in the classes about the relationship between global warming, their lifestyles, and encourages them to practice eco-friendly lifestyles at their home during summer or winter vacation. Such a class is not included in the national curriculum, but is an initiative developed by the city and nonprofit organizations (NPOs). At first, Kyoto tried this program in one school and later the number of schools increased. In 2010 all municipal elementary schools (173 schools) adopted this curriculum. The budget for this project was Yen 19 million (US$0.25 million) in 2010, and Yen 21 million (US$0.27 million) in 2011.

The features of this project are:

- Inclusion in the elementary school curriculum.
- Some members of NPOs serve as teachers on environmental issues.
- Cooperation with volunteering local residents.

With this project 11,000 pupils take the same environmental education classes per year. Although it is difficult to measure the effects of this project in terms of greenhouse gas reductions, its economic effects are immeasurable when considering that these children will play an active role in various fields of the future. In addition,

Kyoto believes that creating a cooperative society, where children education is supported by NPOs and the wider community supports an eco-friendly sustainable society of the future.

5 Conclusion

This chapter reported on selected efforts of Kyoto City to change the city towards a more sustainable socio-economic system through a framework of ordinances and plans, and five practical measures. These measures are not intended to bring about environmental improvements by investing enormous amounts of financial capital. They are neither there to drastically build a green economic system in the city. They are intended to change the citizens' sense of value in their daily lives. They are encouraged to change their lifestyles through economic instruments such as subsidies, rewards and loans.

It is needless to say that actions by individuals play a crucial role in changing the socio-economic system. The sense of cohesion among citizens, called "*Kizuna*" in Japanese, and the norms and values arising from it to solving various problems of cities becomes then especially effective. Japan has suffered an astronomical loss of property through the East Japan Earthquake in March 2011. In return, we have learned many important facts. One of them is to review the economic system, which gives priority to efficiency and speed to use this for changing citizens' lifestyle. From now on Japan needs to be more concerned about the Earth and its global environment, value its limited resources, love communities and their people, and develop a sustainable economy and society.

In order to build an eco-friendly socio-economic system, the role of cities is of great importance, because they are closest to citizens, enterprises and NPOs. Kyoto City talks, thinks, and acts together with the people in various fields. Such collaboration is of great value in building a sustainable socio-economic system. We hope that "DO YOU KYOTO?" will become a slogan to link cities and countries.

Index

R. Simpson and M. Zimmermann (eds.), *The Economy of Green Cities: A World Compendium on the Green Urban Economy*, Local Sustainability 3,
DOI 10.1007/978-94-007-1969-9,

MIX
Papier aus verantwortungsvollen Quellen
Paper from responsible sources
FSC® C105338

If you have any concerns about our products,
you can contact us on
ProductSafety@springernature.com

In case Publisher is established outside the EU,
the EU authorized representative is:
Springer Nature Customer Service Center GmbH
Europaplatz 3, 69115 Heidelberg, Germany

Printed by Libri Plureos GmbH
in Hamburg, Germany